Rise of the Machines: Transforming Industries with Service Robotics

From Automation to AI: The Revolution that's Reshaping Our World

Trizula Digital Solutions' AI Team

Our aspiration must be to reform, upgrade and enlarge our education system - and to make it relevant to 21st century realities of the digital economy, genomics, robotics and automation.

-Ram Nath Kovind
Former President Of India

Authors

Ravishekar Gupta Thallapally
CEO & Founder,
Trizula Digital Solutions Pvt Ltd

Bhuvaneshwari | Monika | Akhila| Sai Sri TDS

AI TEAM

Contents

About Us

At Trizula Digital Solutions Pvt Ltd, a forward-thinking startup established in 2018, we have been at the forefront of providing IT services in the realms of data sciences and ERP to multinational corporations. In our relentless pursuit of innovation, we embarked on a transformative journey into the realm of Artificial Intelligence and found AITech.Studio.

The AITech.Studio portal stands as the ultimate resource for a profound and all-encompassing exploration of the ever-evolving universe of artificial intelligence. We offer a wealth of AI-related content, ranging from product insights, breaking news, tool analyses, career support, and expert perspectives.

The dedicated team at TDS is thrilled to introduce this comprehensive book dedicated to the world of Service Robotics. Our extensive research and a robust exploration of this dynamic field have fueled our commitment to provide you with invaluable insights into the exciting realm of Service Robotics careers.

Our journey involved rigorous research and an in-depth investigation, harnessing the capabilities of popular AI tools to collect, assess, and deliver information comprehensively and insightfully. We pored over extensive datasets, tapped into state-of-the-art natural language processing models, and harnessed machine learning algorithms to ensure the content of this book remains both current and reliable.

Our team is an amalgamation of AI enthusiasts, seasoned researchers, and industry experts who are ardent about sharing their knowledge and expertise with you.

This book serves as an indispensable roadmap, invaluable not only for those seeking a career in Service Robotics but also for CXOs, IT leaders, robotics project managers, service robot recruiters, and product developers in this domain. Whether you aim to enhance your existing role in Service Robotics or embark on a novel career journey, this book imparts essential guidance and insights for success in the realm of Service Robotics.

Our book, **"Rise of the Machines: Transforming Industries with Service Robotics"** is a testament to our contribution to AI technologies. This book serves as a comprehensive guide, designed to unravel the intricacies of service robotics, making it accessible to everyone, from beginners to seasoned professionals.

We delve into the fascinating world of AI-driven service robotics, covering everything from the fundamentals to the latest trends, applications, and emerging possibilities. Whether you're a curious mind, a student, a professional, or a tech enthusiast, or AI expert or IT leader responsible for Service Robotics adoption or there implementation this book will equip you with insights, knowledge, and a deeper appreciation for the service robotics landscape.

Join us on this thrilling journey as we uncover the dynamic fusion of service robotics and AI, shaping the future and leaving an indelible mark on the way we interact with technology. Together, we'll explore the incredible potential of service robotics, making the complex simple and the future tangible.

Preface

In the not-so-distant past, the idea of robots serving in our daily lives was confined to the realms of science fiction. Today, service robots have emerged from the pages of novels and the screens of Hollywood to become an integral part of our reality. These remarkable machines are no longer figments of our imagination; they are tangible, versatile, and transforming the world around us.

As we stand on the precipice of a new era, it's evident that the rapid advancements in robotics and artificial intelligence are reshaping industries, redefining human-robot interactions, and offering innovative solutions to age-old challenges. This book, **"Rise of the Machines: Transforming Industries with Service Robotics,"** delves into the heart of this technological revolution, providing you with a comprehensive and insightful journey into the world of service robotics.

With a profound focus on both the theoretical foundations and practical applications, this book aims to serve as your guiding light through the labyrinth of service robotics. Whether you are an eager student, a curious researcher, an industry professional, or simply an enthusiast eager to understand the profound changes taking place in our world, this book has something valuable to offer.

Understanding Service Robotics: The Essence

We begin by laying a solid foundation, starting with an exploration of the very essence of service robotics. From defining the term "robotics" itself to tracing the intriguing history of this field, we embark on a journey that helps you comprehend the roots of this transformative technology. We delve into the various types of robots, uncovering their diverse forms and functions, and explore the myriad applications that make them indispensable in our modern world.

The Evolving Landscape: Service Robotics Unveiled

Our exploration doesn't stop at the basics. We guide you through the intricate terrain of service robotics, shedding light on the various components and software that breathe life into these machines. From sensors that act as their senses to control systems that govern their actions, we decipher the intricate anatomy of service robots. The software that fuels these robotic wonders, from operating systems to machine learning algorithms, is demystified to showcase how they adapt, learn, and interact with the world.

From Concept to Operation: Design, Development, and Deployment

Designing, building, and deploying service robots is no small feat. We take you behind the scenes, revealing the meticulous processes involved in bringing these robotic entities to life. You'll journey through the stages of problem identification, conceptualization, and detailed design. Prototyping, simulation, and testing become your companions as we traverse the critical paths to perfection.

Manufacturing and assembly processes come into sharp focus, highlighting the importance of material selection, procurement strategies, and quality control. Our discussion extends to the operation and maintenance of service robots, where you'll learn about their operational components, safety measures, and data management.

The crucial aspect of deployment strategies is uncovered, with insights into needs assessment, integration planning, and user acceptance. We delve into the ethical considerations that permeate the world of service robotics, ensuring that technology serves humanity while respecting its values.

Career Guide: Navigating the Robotics Frontier

For those inspired to embark on a career in service robotics, we offer a comprehensive guide. You'll discover the educational pathways, relevant courses, and essential skills needed to thrive in this field. Practical experiences, certifications, and hands-on training are discussed in detail to help you build a successful career in the dynamic world of service robotics.

Real-Time Applications and Revolutionary Outcomes

The heart of this book lies in the exploration of real-time applications. We uncover the pivotal role of service robots in healthcare, customer service, agriculture, disaster response, space exploration, military operations, education, entertainment, and the domestic sphere. Through compelling case studies and success stories, you'll witness the tangible impact of these machines on our lives.

Exploring the Market: Popular Products and Companies

The service robotics market is thriving, and we guide you through its growth trends, popular products, and key companies shaping its landscape. From household service robots that simplify daily chores to healthcare robots that assist in hospitals, we unveil the impressive lineup of robots and the companies behind them.

Peering into the Future: Emerging Trends and Challenges

As the world of service robotics hurtles toward an exciting future, we provide you with a glimpse of the emerging trends that will define this journey. Human-robot interaction, swarm robotics, ethical considerations, and the integration of 5G and connectivity are just a few of the captivating themes we explore. We conclude with a call to action, encouraging you to be a part of this transformative era.

A Collaborative Endeavor

Writing a book of this magnitude and significance is a collaborative endeavor, made possible by the dedication, expertise, and passion of many individuals. Researchers, engineers, innovators, and educators from around the world have contributed their knowledge and insights to create a comprehensive resource that encapsulates the current state and future potential of service robotics.

Your Journey Begins Here

As you turn the pages of this book, you embark on a journey through the exciting world of service robotics. It's a world where innovation knows no bounds, where technology serves humanity, and where the future unfolds before our eyes. Join us in this exploration, and together, let's discover the incredible possibilities that service robotics hold for our ever-evolving world.

Welcome to "**Rise of the Machines: Transforming Industries with Service Robotics.**" The future is here, and it's robotic.

Who Should Read This Book?

This comprehensive book on Service Robotics is suitable for a diverse range of readers. It serves as an invaluable resource for students, researchers, and professionals in the fields of robotics and engineering. Additionally, it caters to robotics enthusiasts, educators, and individuals engaged in AI and machine learning.

Furthermore, the book offers profound insights that extend to service robotics practitioners, entrepreneurs, policymakers, and anyone interested in exploring the ethical and legal aspects of this technology. It is a comprehensive reference that spans various interests and expertise within the realm of robotics and artificial intelligence.

Why Should We Read This Book?

This book is a valuable resource for anyone interested in service robotics due to its comprehensive coverage, educational value, practical guidance, market insights, exploration of emerging trends, ethical and legal considerations, and diverse applications. It caters to students, researchers, professionals, and enthusiasts, offering knowledge and insights into the dynamic field of service robotics.

Furthermore, the book provides case studies and impact assessment, highlighting its emphasis on interdisciplinary collaboration. With its wealth of knowledge and insights, it serves as a dynamic guide to the world of service robotics.

Acknowledgment

We would like to express our sincere gratitude to all those who contributed to the creation of this book on service robotics. The journey of compiling this work would not have been possible without the support, expertise, and dedication of numerous individuals in our organization.

We extend our heartfelt thanks to the researchers, engineers, and innovators in the field of robotics who have tirelessly pursued advancements in service robotics. Your groundbreaking work has not only inspired this book but has also propelled the field forward, shaping its future.

We would like to acknowledge the valuable insights, feedback, and contributions provided by our colleagues and peers who reviewed and offered their expertise to enhance the quality of this book. Your input has been instrumental in ensuring the accuracy and relevance of the content.

A special thanks to our families and friends for their unwavering encouragement and understanding during the often demanding process of writing and researching this book. Your support has been a constant source of motivation.

I would like to extend my heartfelt gratitude to my co-authors of this book and the unwavering support of the TDS AI team in bringing this project to fruition.

We appreciate the service robotics community's engagement with this book. We aim for the knowledge shared here to support the field's ongoing growth and progress.

Thank you all for your invaluable contributions and support.

RaviShekar Gupta Thallapally

CEO & Founder

Trizula Digital Solutions Pvt Ltd

www.aitech.studio

Chapter 1: An Introduction to Robotics and Service Robots

1.1. Introduction to Robotics

Robotics represents a diverse and multifaceted domain situated at the crossroads of engineering, computer science, and artificial intelligence (AI). Its core objective is the conception, fabrication, programming, and management of autonomous entities commonly referred to as robots. These robots, whether tangible physical entities or software-based constructs, are devised to carry out tasks with varying degrees of autonomy, from full independence to semi-autonomy, or even under the remote guidance of human operators. Over the years, the field of robotics has undergone substantial evolution, revealing a broad spectrum of applications encompassing manufacturing, healthcare, exploration, and entertainment, among others.

Introduction to Robotics

History Of Robotics

The word "robot" was first coined by the Czech playwright Karel Čapek in his 1920 play "R.U.R." (Rossum's Universal Robots). In the play, robots are artificial beings created to serve humans. While Čapek's play was a work of fiction, it helped to popularize the idea of robots and laid the foundation for the field of robotics.

The first practical robots were developed in the early 20th century. These robots were typically used in manufacturing to perform repetitive tasks such as welding and assembly. In the 1960s, the development of computers led to the development of more sophisticated robots that could be programmed to perform a wider range of tasks.

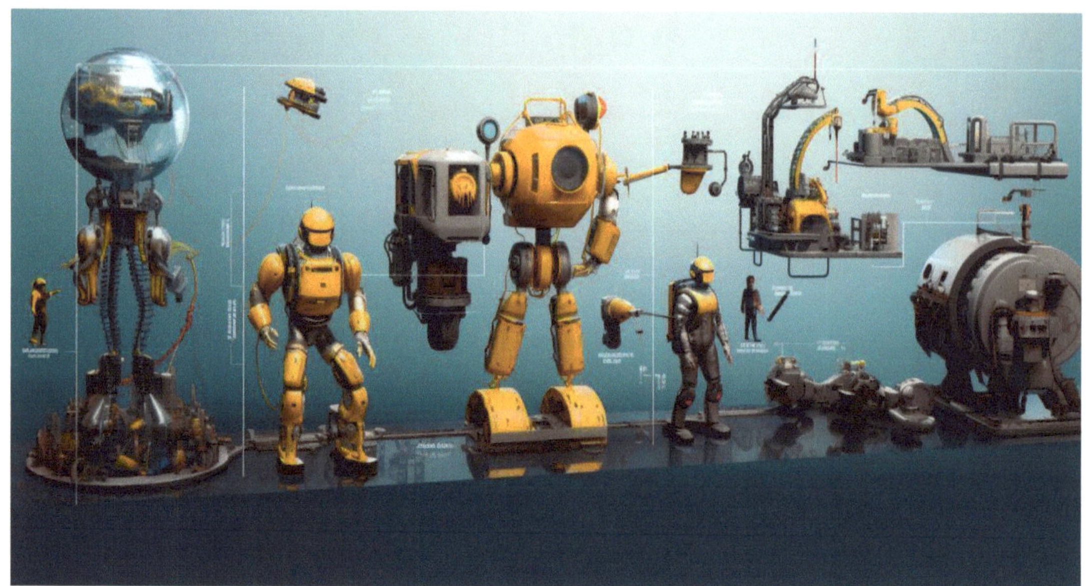

The history of robotics is a fascinating journey through human ingenuity, creativity, and technological progress. It spans centuries and encompasses a wide range of innovations and inventions. Here is a brief overview of the key milestones in the history of robotics:

Ancient Dreams And Automata (Antiquity To The Middle Ages)

The concept of artificial beings that could mimic human or animal movements dates back to ancient civilizations. In ancient Egypt, around 2,500 years ago, water clocks were built with automated figurines that moved as time passed. Similarly, in ancient Greece, inventors like Archytas of Tarentum and Hero of Alexandria created mechanical devices, often referred to as "automata," including birds that could sing and move. These early automatons were powered by mechanisms such as water, steam, and simple gears.

Context Of Ancient Indian Scriptures

In the context of ancient Indian scriptures like the Mahabharata, Ramayana, Shiva Purana, the mention of advanced technologies and weaponry is often depicted in a symbolic manner, rather than as a direct parallel to modern concepts of robotics and technology.

The Sudarshan Chakra, associated with Lord Krishna in the Mahabharata, represents a celestial and divine weapon. It is described as a powerful, spinning disc that symbolizes the cosmic order and the divine's ability to maintain justice in the universe.

Similarly, the Brahmastra, mentioned in both the Mahabharata and Ramayana, is an idealized and potent weapon capable of immense destruction. It is associated with Lord Brahma and is considered a symbol of the ultimate destructive force.

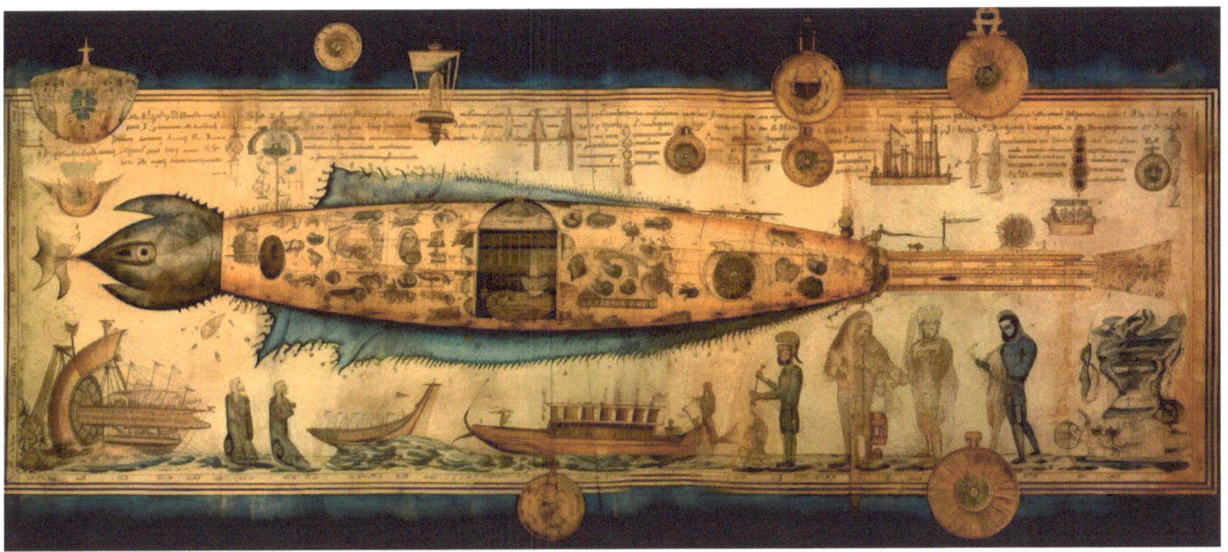

Context of ancient Indian scriptures

These descriptions are rich in symbolism and conveying deeper spiritual and moral messages as the existence of such advanced civilizations were rather than describing literal technological advancements. They emphasize the extraordinary powers of divine beings and the consequences of wielding such immense forces.

In ancient Hindu texts like the Puranas, we find intriguing tales of powerful deities and demons. One such story revolves around Goddess Kali and the Demon King Mahishasura. Mahishasura possessed an extraordinary ability – if a drop of his blood fell on the earth, it would spawn a clone of himself. However, Kali, with her divine prowess, triumphed over Mahishasura without letting his blood touch the ground.

Now, fast forward to the 21st century, where we are witnessing remarkable advancements in AI technology and medical science. While we may not be dealing with demons, we are certainly on the path to creating **bioengineered features** that might resemble the awe-inspiring powers we read about in the ancient texts.

AI is enabling us to push the boundaries of what we once thought was possible. Through AI and biotechnology, we are exploring ways to enhance human capabilities. This includes developing bioengineered features that could empower individuals with incredible skills and attributes, much like the deities and demons of old.

The future could hold exciting possibilities, where AI-driven enhancements are integrated with our biology to create superhuman abilities, just as the gods and demons possessed extraordinary powers in ancient times. These advancements might one day blur the line between vedic age and r, as we delve deeper into the realms of AI and biotechnology.
From ancient tales of goddesses slaying demons to the future of AI and medical advancements in 2050 – could we soon see bio-enhancements with extraordinary powers? Just as Kali defeated Mahishasura without spilling his blood on Earth, our advanced AI and medical science might

lead to astonishing abilities. What if the legends of the past become the reality of the future?

In contrast, the introduction of the term "robot" by Karel Čapek in the 1920s and the subsequent development of practical robots in the 20th century are part of modern technological history. While they both involve artificial beings created to serve humans, they belong to different cultural and historical contexts.

In summary, while there may be intriguing parallels between ancient Indian scriptures and modern technological advancements, it's important to recognize that the scriptures primarily serve as symbolic narratives with deeper philosophical and moral meanings, rather than providing literal descriptions of robotics and technology as we understand them today.

The Renaissance And Mechanical Automata (15th To 18th Century)

During the Renaissance, mechanical automata became more sophisticated. Leonardo da Vinci designed a humanoid automaton known as "Leonardo's robot knight," although it was never built during his lifetime. In the 18th century, Swiss watchmakers created intricate clockwork automata that could perform various tasks, such as drawing pictures and playing musical instruments. These automata were often showcased as marvels of engineering and craftsmanship.

Industrial Revolution And Early Robotics (Late 18th To 19th Century)

The Industrial Revolution marked a significant shift in the development of robotics. In 1770, Swiss-born inventor Pierre Jaquet-Droz created "The Writer," a mechanical doll that could write custom text. This was a precursor to programmable machines. Charles Babbage, considered the "father of the computer," designed the Analytical Engine in the 1830s, which had some elements of modern computing and could be seen as a conceptual ancestor to robots.

20th Century: Birth Of Modern Robotics

1920s-1930s: Czech writer Karel Čapek introduced the term "robot" in his 1920 play "R.U.R. (Rossum's Universal Robots)." The play explored the idea of artificial, human-like workers.

1930s-1940s: Isaac Asimov, a prolific science fiction writer, introduced the concept of robotics as we understand it today. He formulated the "Three Laws of Robotics," which set ethical guidelines for robot behavior.

1950s-1960s: George Devol and Joseph Engelberger developed the Unimate, the world's first industrial robot. It was used for tasks like loading and unloading materials in manufacturing.

1970s-1980s: The development of microprocessors and computer control systems led to significant advancements in robotics. Robots became more versatile and started to be used in various industries, including automotive manufacturing and space exploration.

21st Century: Robotics In Everyday Life

Early 2000s: The emergence of Roomba, an autonomous vacuum cleaner, brought robotics into homes and marked the beginning of consumer robotics.

2010s-Present: Robotics has continued to advance rapidly, with robots playing roles in healthcare (surgical robots), logistics (warehouse automation), and agriculture (autonomous tractors). The development of humanoid robots like Boston Dynamics' Atlas and SoftBank's Pepper has pushed the boundaries of what robots can do. The history of robotics is an ongoing story of innovation, with robots increasingly becoming integrated into various aspects of our lives. From manufacturing and healthcare to space exploration and entertainment, robots continue to evolve, offering new opportunities and challenges for society. As technology continues to advance, it's likely that the next chapters in the history of robotics will be even more exciting and transformative.

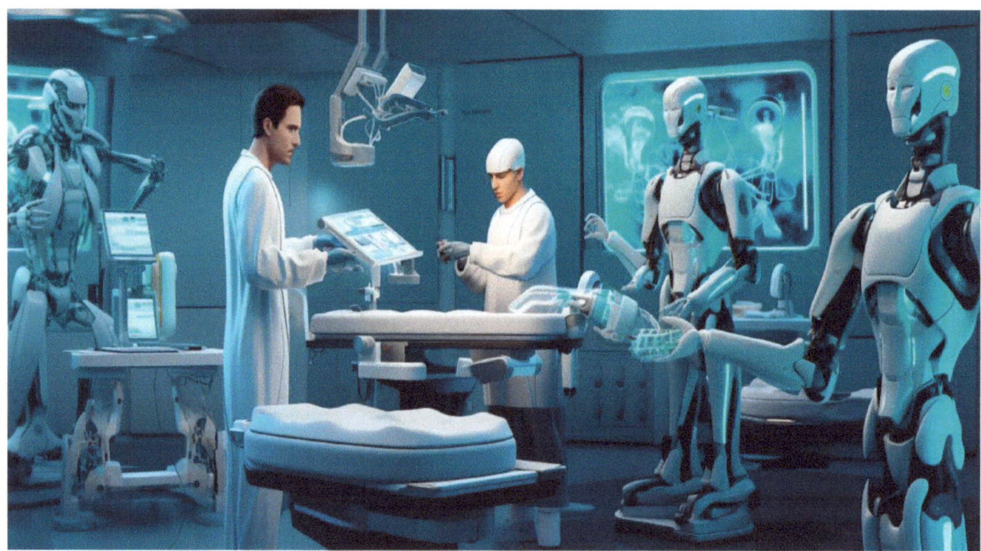

21st Century: Robotics in Everyday Life

Different Types Of Robots

In the heart of the 21st century, the world of robotics has evolved into a mesmerizing tapestry of innovation and automation. It's a realm where science fiction meets reality, where machines designed to mimic and augment human capabilities have become an integral part of our daily lives. In this chapter, we embark on a journey into the diverse and captivating universe of robots.

There are many different types of robots, each meticulously crafted for a specific purpose, whether it's to enhance our productivity, provide essential services, explore uncharted territories, or even replicate the intricacies of the human form. As we delve into the realm of robotics, we will uncover the fascinating array of these mechanical marvels, including industrial giants, friendly assistants, collaborative comrades, nimble wanderers, life-saving surgeons, spacefaring pioneers, and eerily human-like companions. So, fasten your seatbelts as we explore the captivating world of robots and the myriad roles they play in our modern society.

There Are Many Different Types Of Robots, Each Designed For A Specific Purpose. Some Common Types Of Robots Include:

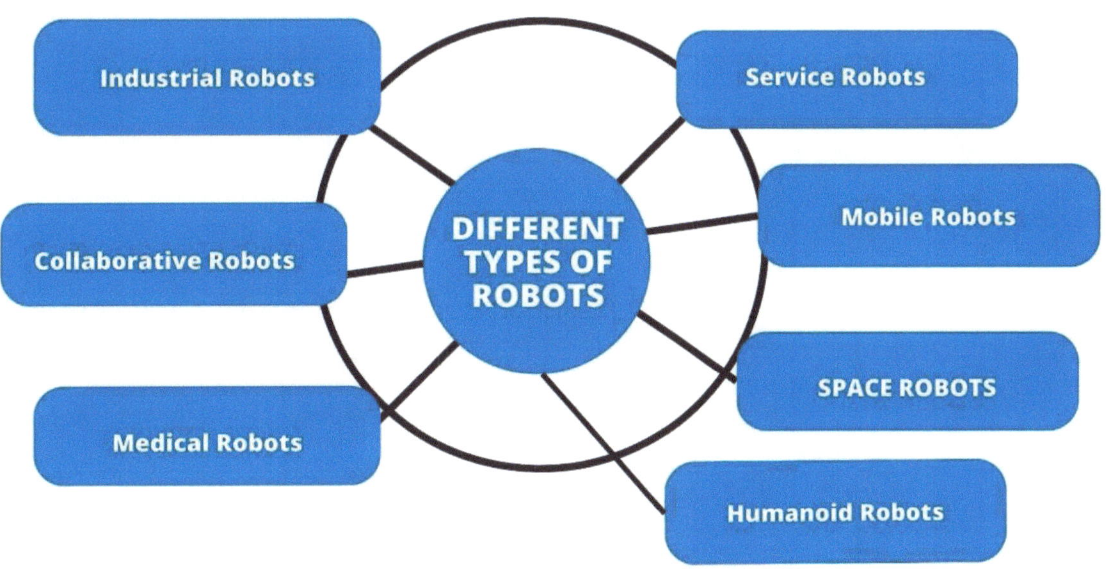

Industrial Robots:

In the sprawling factories and manufacturing facilities that underpin our modern world, there exists a silent army of precision and power: industrial robots. These mechanical giants, often towering over their human counterparts, are the unsung heroes of mass production. They possess a unique blend of strength and precision, making them indispensable in a wide range of industrial applications. From welding intricate metal components to assembling complex electronics, and even wielding paintbrushes with artful finesse, industrial robots are the artisans of the assembly line.

Their ability to work tirelessly and with unwavering accuracy has revolutionized the manufacturing industry, enabling companies to produce goods with unparalleled consistency and quality. What sets them apart is their programmability – their capacity to execute tasks with a level of precision that exceeds human capabilities. To know more about the world of industrial robots, we need to understand the mechanics behind their operations, the pivotal role they play in shaping industries, and the astonishing feats they accomplish on the factory floor.

Industrial Robots

Service Robots:

Amid the whirlwind of modern life, a silent revolution is unfolding, one powered by mechanical marvels known as service robots. These versatile companions have seamlessly woven themselves into the fabric of our daily existence, finding their place in our homes, offices, and bustling public spaces. Their mission? To simplify our routines, enhance our convenience, and elevate our efficiency. Imagine the gentle hum of a robotic vacuum cleaner effortlessly gliding across your floors, meticulously erasing the crumbs of yesterday's snack, or the autonomous lawnmower diligently sculpting the unruly edges of your lawn, granting you more leisure time to enjoy the outdoors. And then there are the friendly delivery robots, buzzing along sidewalks, bearing our packages with the precision of a postal worker. These are just a few of the many faces of service robots that have become an integral part of our modern landscapes.

What sets service robots apart is their compact and lightweight design, enabling them to navigate diverse environments with grace and agility. Unlike their hulking industrial counterparts, these robots are engineered to be intuitive and user-friendly in their interactions with humans. They are the bridge between technology and everyday life, a testament to how seamlessly innovation can integrate into our existence.

These mechanical marvels represent the forefront of a technological revolution aimed at enhancing our daily lives. In our fast-paced, interconnected world, the demands on our time and energy seem to grow unceasingly. In response, service robots have emerged as tireless allies, ever-ready to tackle a wide range of tasks and challenges. They embody the essence of convenience and efficiency, freeing us from the mundane so that we can focus on what truly matters.

The realm of service robots is as diverse as it is dynamic. Beyond the common household helpers like vacuum cleaners and lawnmowers, they span a wide spectrum of applications. In the bustling hospitality industry, service robots take on the roles of concierge, waitstaff, and even luggage carriers, ensuring guests enjoy a seamless and memorable experience. Healthcare facilities employ robotic aids to transport medications and assist with patient care, while retail businesses rely on them to manage inventory and streamline logistics. Even in agriculture, autonomous machines navigate fields, tending to crops and increasing yields.

The intelligence and adaptability of service robots continue to evolve at a rapid pace. They incorporate advanced sensors, machine learning, and artificial intelligence to better understand and respond to their surroundings. Many are equipped with natural language processing, allowing them to engage in meaningful conversations and provide valuable information. These machines are not merely tools but companions, working tirelessly to ease the burdens of our daily routines and enrich our lives in profound ways.

Service robots are not confined to a single industry or setting. They are the embodiment of innovation, continually pushing the boundaries of what is possible. As we delve deeper into this book, we will uncover the fascinating intricacies that power these robots, explore their evolving roles in various sectors, and glimpse into the promising future where service robots will continue to redefine convenience, efficiency, and the very nature of human-robot collaboration.

Service robots

Collaborative Robots:

In the ever-evolving landscape of automation, a remarkable category of robots has emerged as true partners to humans in the workplace – collaborative robots, often lovingly referred to as "cobots." These intelligent and adaptable machines are meticulously crafted to share our workspace, seamlessly blending the realms of technology and human expertise. Unlike their larger and more powerful industrial counterparts, cobots are designed with a focus on safety and user-friendliness, allowing them to operate alongside humans without posing any harm. Their agility and ease of programming make them indispensable in assembly line applications, where they assist with delicate tasks like picking and placing parts, all the while maintaining a harmonious coexistence with their human colleagues.

As we journey deeper into the realm of collaborative robots, we will unravel the remarkable synergy they bring to the workplace. Their ability to take on repetitive and physically demanding tasks liberates human workers to focus on more intricate and creative endeavors, thereby reshaping the way we think about productivity and collaboration. Collaborative robots are not just tools; they are companions, working hand in hand with us to forge a brighter future where technology augments our capabilities without diminishing our humanity.

Collaborative Robots

Mobile Robots:

In the ever-evolving landscape of robotics, there exists a category of machines that embody the spirit of autonomy and movement—mobile robots. These technological marvels possess the remarkable ability to traverse their surroundings independently, a feat that has opened the door to countless applications across various industries.

One of the most notable examples of mobile robots is the emergence of self-driving cars, a disruptive force poised to reshape the future of transportation. Equipped with a plethora of sensors, cameras, and sophisticated software, these vehicles can navigate busy streets, anticipate traffic patterns, and make split- second decisions to ensure the safety of passengers and pedestrians alike. Self-driving cars represent a leap forward in not only reducing accidents but also optimizing traffic flow and reducing the environmental footprint of transportation.

Beyond the realm of personal mobility, mobile robots have also found their niche in the world of logistics and delivery. Picture a sidewalk scene with a friendly, compact delivery robot weaving through the crowd, transporting packages with unwavering precision. These little robots are equipped with advanced sensors and sophisticated algorithms that enable them to navigate bustling urban environments, avoid obstacles, and deliver goods with efficiency, making the last mile of delivery faster and more convenient for consumers.

The common thread among mobile robots is their ability to perceive and interact with their environment. These robots rely on a combination of sensors such as LIDAR, cameras, and ultrasonic sensors, coupled with powerful software algorithms, to create detailed maps of their surroundings, plan optimal paths, and make real-time adjustments to avoid obstacles. This fusion of sensing and intelligence empowers mobile robots to operate safely and effectively in complex and dynamic environments.

As we delve further into the world of mobile robots, we will explore the cutting-edge technologies that propel self-driving cars and delivery robots forward, investigate the exciting possibilities they offer for industries and urban planning, and contemplate the profound impact they have on the way we move, work, and live in our increasingly interconnected world. Mobile robots are not just machines; they are the embodiment of progress, transforming our vision of mobility and automation.

Mobile Robots

Humanoid Robots:

The quest to replicate the complexity and grace of the human form has given rise to a fascinating category of robots known as humanoid robots. These machines are designed with a striking resemblance to humans, both in appearance and movement, and they stand as a testament to our boundless curiosity and desire to push the boundaries of robotics.

Humanoid robots have long been a cornerstone of research and development in the field of robotics. They serve as invaluable tools for scientists and engineers, offering a platform to explore the intricate interplay of mechanics, electronics, and artificial intelligence required to mimic human actions and gestures. This research extends into various domains, from biomechanics and neuroscience to human-robot interaction and social robotics, all with the ultimate goal of understanding and replicating human capabilities.

Beyond the confines of laboratories and R&D centers, humanoid robots are beginning to step into the limelight of commercial applications. In the realm of entertainment, they make appearances as captivating performers in theaters and theme parks, enchanting audiences with their lifelike movements and expressions. Some forward-thinking businesses are also employing humanoid robots in customer service roles, creating engaging and interactive experiences for patrons in hotels, shopping malls, and even restaurants.

The journey of humanoid robots is a testament to our relentless pursuit of innovation and our desire to create machines that can seamlessly integrate into our human-centric world. As we delve deeper into this chapter, we will unravel the intricate mechanisms and technologies that breathe life into these mechanical marvels, explore their current and potential roles in our society, and contemplate the ethical and philosophical questions they raise about the intersection of man and machine. Humanoid robots are more than just robots. They are our reflection in the world of technology, pushing the boundaries of what it means to be human.

Humanoid Robots

Medical Robots:

In the ever-advancing realm of healthcare, a revolution quietly unfolds, one guided by the steady hands of medical robots. These extraordinary machines have carved out a crucial niche in the world of medicine, performing tasks ranging from surgery and rehabilitation to patient care, all with precision and unwavering accuracy.

One of the most significant contributions of medical robots is their role in surgery. Surgeons have long relied on these advanced tools to assist them in performing complex procedures with unparalleled precision. Robotic surgical systems offer the ability to magnify movements and provide a stable, tremor-free hand, allowing surgeons to operate with incredible accuracy. They also allow for minimally invasive procedures, reducing patient trauma, pain, and recovery times. This fusion of human expertise and robotic precision ensures not only better outcomes but also greater safety for patients.

Medical robots extend their influence beyond the operating room, with applications in rehabilitation therapy. These robots assist patients in regaining mobility and strength after injury or surgery.

They provide consistent and personalized therapy regimens, tracking progress with meticulous detail. Such robotic aids empower individuals on their journey toward recovery.

In patient care, robots are increasingly used for tasks like medication delivery and monitoring vital signs, offering a helping hand to healthcare professionals and ensuring timely and accurate care for patients. These machines enhance the efficiency of healthcare facilities, allowing human staff to focus on more complex and compassionate aspects of patient care.

The world of medical robots embodies the union of cutting-edge technology and human empathy. As we delve into this chapter, we will uncover the intricate mechanisms that power these robots, explore their evolving roles in healthcare, and appreciate the profound impact they have on improving the quality of medical procedures, patient outcomes, and the overall patient experience. Medical robots are not just machines; they are instruments of healing, blending the best of human compassion with the precision of technology to provide a brighter future for healthcare.

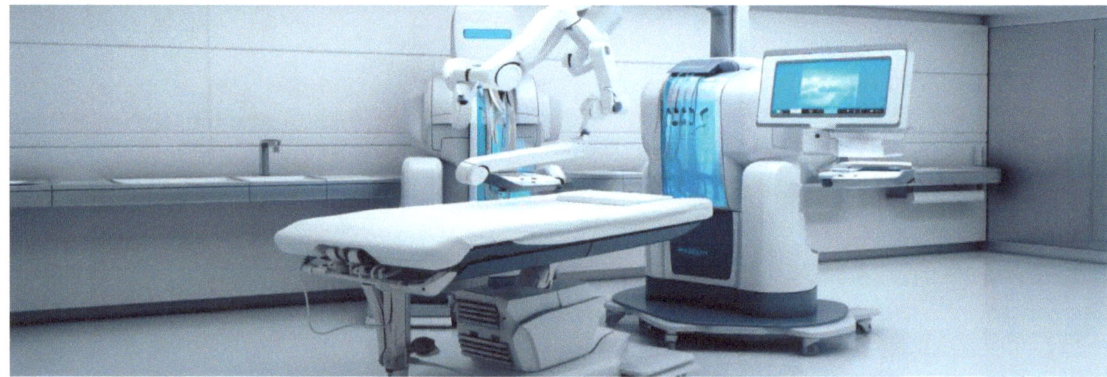

Medical robots

Space Robots:

In the boundless expanse of the cosmos, where humans cannot tread, a unique breed of explorers takes center stage—space robots. These remarkable machines are the vanguards of interstellar exploration, designed to traverse the unforgiving terrains of planets and moons, undertake satellite maintenance and repair, and tackle a myriad of complex tasks in the harsh vacuum of space.

One of the most awe-inspiring roles of space robots is planetary exploration. These mechanical astronauts journey to distant celestial bodies, where extreme conditions, such as scorching heat, bitter cold, and treacherous terrain, would be insurmountable challenges for humans. Equipped with an array of sensors, cameras, and scientific instruments, they unravel the mysteries of other worlds, sending back invaluable data and images that expand our understanding of the cosmos.

Satellite maintenance and repair is another vital task undertaken by space robots. They extend the lifespan of orbiting satellites and spacecraft by conducting intricate repairs or servicing tasks.

Often operated remotely from mission control centers on Earth, these robots exhibit remarkable dexterity in performing delicate operations in the microgravity environment of space.

The design of space robots is a testament to human ingenuity. They are engineered to withstand extreme temperatures, radiation, and vacuum conditions, often operating in environments where the margin for error is virtually nonexistent. These machines are equipped with a plethora of specialized sensors and tools that enable them to manipulate objects, collect samples, and execute complex tasks with the utmost precision.

The cutting-edge technologies that empower them, delving into their fascinating missions, and reflecting on the profound impact they have on our understanding of the universe. Space robots are not just explorers; they are pioneers pushing the boundaries of human knowledge, venturing into the cosmic unknown and returning with the treasures of discovery.

space robots

Application of Robots

Robots have transcended the boundaries of their traditional roles, reaching far beyond the sectors of industry, healthcare, exploration, and beyond the appearance and capabilities of human-like machines. Their utility extends to a vast array of applications that touch nearly every aspect of our lives, shaping the future in ways we are only beginning to grasp.

In agriculture, autonomous drones equipped with advanced imaging technologies help farmers monitor crop health and optimize yields, revolutionizing the way we cultivate food. In disaster relief efforts, specialized robots are deployed to navigate hazardous environments, search for survivors, and assist with recovery operations, offering hope in the darkest of times. In the field of education, robots serve as engaging tools for interactive learning, capturing the imagination of students and inspiring the next generation of innovators.

Beyond Earth, robots venture into the depths of the ocean to explore its mysteries and conduct crucial research. Submersibles equipped with cameras and sensors provide scientists with invaluable data, deepening our understanding of the oceans' ecosystems and their impact on our planet.

Moreover, robots have made their mark in the realm of art and creativity, with robotic arms assisting artists in crafting intricate sculptures and paintings, blurring the lines between technology and artistic expression. In logistics and warehousing, automated robots tirelessly shuffle goods, speeding up order fulfillment and ensuring efficient supply chains.

These versatile machines have found their place in countless industries and applications, redefining the limits of what's possible. As we journey further into the world of robotics, we'll continue to explore the innovative and unexpected ways in which robots are reshaping our world, pushing boundaries, and propelling us into a future where their presence is increasingly woven into the fabric of our daily lives.

Application of Robots

A World of Robot Applications:

Manufacturing and Assembly: Robots are commonly used in manufacturing industries for tasks like welding, painting, assembling, and quality control. They increase production speed and consistency while reducing errors.

Logistics and Warehousing: Autonomous robots assist in material handling, inventory management, and order fulfillment in warehouses and distribution centers. They optimize storage and retrieval processes.

Agriculture: Robots are employed in agriculture for planting, harvesting, weeding, and monitoring crops. They can work autonomously and help increase crop yields and reduce labor costs.

Space Exploration: Robotic spacecraft and planetary rovers are sent to explore outer space and other celestial bodies. They gather data, conduct experiments, and search for signs of life.

Mining: Robots are used in mining operations to explore and extract resources from deep underground or in challenging conditions. They enhance safety and efficiency.

Construction: Construction robots can perform tasks such as bricklaying, concrete pouring, and site inspection. They improve construction speed and precision.

Defense and Military: Robots are used for tasks like reconnaissance, bomb disposal, and demining. Unmanned aerial vehicles (UAVs) and ground robots enhance military capabilities.

Entertainment: Robots are used in the entertainment industry as performers, animatronics, and interactive exhibits in theme parks, museums, and entertainment venues.

Food Industry: Robots are used in food processing and packaging, including tasks such as sorting, cutting, and quality inspection.

Oil and Gas: Robots are employed in the oil and gas industry for tasks like pipeline inspection, maintenance, and underwater exploration.

Research and Development: Robots are used in research laboratories and industries to conduct experiments, test prototypes, and develop new technologies.

What Are Service Robots?
Service Robots: Enhancing Lives, Redefining Convenience

In an era characterized by the relentless pace of modern life, service robots stand as the vanguard of a technological revolution aimed at enhancing our daily existence. As demands on our time and energy continue to grow, the world of robotics has responded with ingenious solutions that not only alleviate our burdens but also open doors to a new era of convenience and efficiency. Beyond their role as mere mechanical marvels, service robots have emerged as tireless allies, ever-ready to tackle a wide range of tasks and challenges, allowing us to reclaim our most precious resource: time.

The realm of service robots is as diverse as it is dynamic, extending far beyond the realms of household chores such as vacuuming and lawn mowing. These remarkable machines span a wide spectrum of applications, seamlessly integrating into industries as varied as hospitality, healthcare, retail, and agriculture. They have become the concierges, waitstaff, and luggage carriers of the bustling hospitality industry, ensuring that guests experience seamless and memorable stays. In the healthcare sector, robotic aids transport medications and assist with patient care, augmenting the capabilities of medical professionals and enhancing patient outcomes. Meanwhile, in retail, service robots have streamlined inventory management and logistics, optimizing supply chains to meet the demands of the modern consumer. Even in agriculture, autonomous machines navigate fields, tending to crops and boosting yields to feed our ever- growing population.

The intelligence and adaptability of service robots continue to evolve at a rapid pace, driven by advancements in sensors, machine learning, and artificial intelligence. Many of these robots are now equipped with natural language processing, enabling them to engage in meaningful conversations and provide valuable information. Service robots are not merely tools; they have evolved into indispensable companions, working tirelessly to ease the burdens of our daily routines and enrich our lives in profound ways.

However, the impact of service robots extends well beyond their respective sectors. They represent the embodiment of innovation, continually pushing the boundaries of what is possible and challenging preconceived notions of what robots can achieve. As we embark on this exploration of service robots, we will delve deeper into their fascinating intricacies, unravel the technological marvels that power them, and uncover their evolving roles in various industries. Moreover, we will glimpse into the promising future where service robots are poised to redefine not only convenience and efficiency but also the very nature of human-robot collaboration.

Service Robot Types: Diverse Applications and Roles

Service robots have evolved significantly over the years, and their diverse types have emerged as a result of technological advancements that have expanded their capabilities and widened their applications. The journey of service robots traces back to the convergence of several groundbreaking technologies, including advancements in sensors, artificial intelligence, machine learning, and natural language processing. These breakthroughs have empowered robots to become more than just machines; they have become intelligent, adaptable companions in various aspects of our lives.

As these technological strides continue to push the boundaries of what service robots can achieve, their applications have grown exponentially. They are no longer confined to specific niches but have found their way into a wide array of sectors, transforming industries and enriching our daily experiences. From personal service robots that simplify our daily routines to healthcare robots that assist medical professionals in life- saving tasks, from agricultural robots that optimize farming practices to hotel and hospitality robots that enhance guest experiences, the spectrum of service robot types reflects the remarkable journey of innovation that has shaped the modern world.

We will also introduce and delve into various types of service robots, each designed to cater to specific needs and challenges. From enhancing our educational and research endeavors to ensuring the security of our surroundings, from revolutionizing retail to contributing to environmental monitoring and even adding flair to our dining experiences, these service robots exemplify the profound impact of technology on our ever-evolving relationship with robotics.

There are many different types of service robots, each designed for a specific purpose.

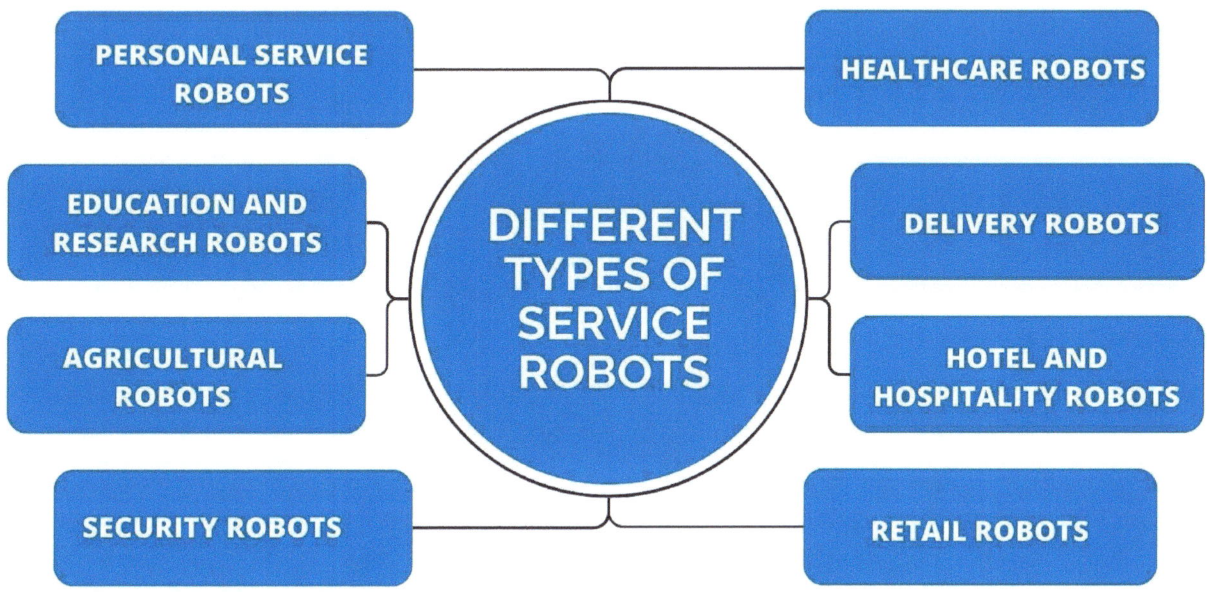

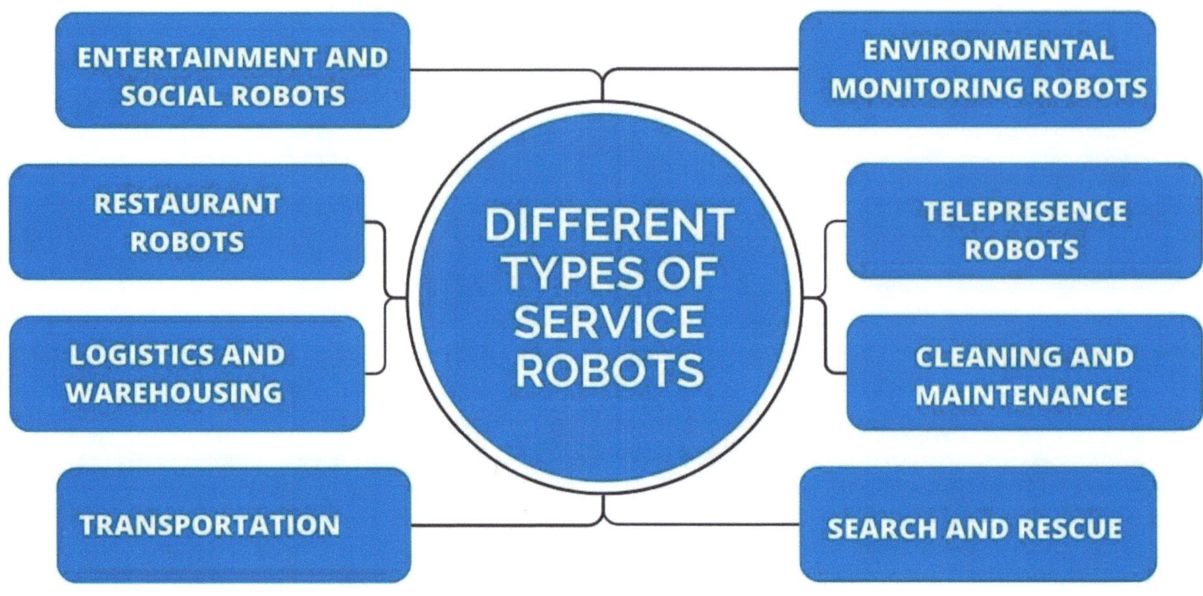

Service Robot Types

1.2 Applications of Service Robots

Personal Service Robots: These robots are designed to assist individuals in their daily lives.

Home Cleaning Robots: Such as robotic vacuum cleaners (e.g., Roomba) and floor mopping robots (e.g., Braava).
Robotic Lawn Mowers: Autonomous machines that mow lawns without human intervention. **Personal Assistants:** Robots like Pepper and Jibo are designed for companionship, entertainment, and providing information.

Healthcare Robots: Healthcare robots assist in medical tasks like surgery, patient care, and rehabilitation, enhancing efficiency and precision in healthcare settings.

Surgical Robots: Assist surgeons in performing precise and minimally invasive surgeries (e.g., da Vinci Surgical System).

Rehabilitation Robots: Aid in physical therapy and rehabilitation exercises for patients recovering from injuries or surgeries.

Telemedicine Robots: Enable remote medical consultations and examinations, allowing doctors to interact with patients through a robot equipped with cameras and sensors

Robotic Prosthetics: Assist individuals with limb loss in regaining mobility and functionality.

Education and Research Robots: Education and Research Robots assist in teaching and scientific investigations, enhancing learning and conducting experiments in educational and research settings.

Educational Robots: Used in classrooms to teach robotics and programming concepts to students (e.g., LEGO Mindstorms).

Research Robots: Customizable platforms for researchers to conduct experiments and studies in various fields.

Delivery Robots: Delivery robots are autonomous machines designed to transport goods and packages, typically within defined areas, providing convenient and efficient deliveries.

Autonomous Delivery Robots: Used by companies for last-mile delivery of packages, groceries, and food orders in urban environments.

Hospital Delivery Robots: Transport medical supplies, medications, and documents within hospitals.

Agricultural Robots: Agricultural robots are machines used in farming for tasks like planting, harvesting, and monitoring crops, improving efficiency and productivity.

Autonomous Tractors: Used for plowing, planting, and harvesting crops with precision. **Agricultural Drones:** Monitor crop health, collect data, and perform tasks like crop spraying. **Harvesting Robots:** Automate the picking and harvesting of fruits and vegetables.

Hotel and Hospitality Robots: Hotel and hospitality robots assist with guest services, room delivery, and concierge tasks to enhance the guest experience in hotels.

Room Service Robots: Deliver food, towels, and other items to hotel guests.

Concierge Robots: Provide information, check-ins, and recommendations to hotel guests.

Restaurant Robots: Prepare and serve food or assist in kitchen operations.

Hotel Service Robots: Deliver amenities and room service orders to hotel guests.

Security Robots: Security robots are automated devices designed for surveillance, monitoring, and safeguarding, often used in locations like malls, airports, and industrial sites.

Patrol Robots: Patrol areas, monitor security, and provide surveillance in commercial buildings, malls, and public spaces.

Bomb Disposal Robots: Designed to handle and safely dispose of explosive devices.

Retail Robots: Retail robots are automated machines used in stores for tasks like inventory management, customer assistance, and shelf restocking to enhance retail operations.

Inventory Management Robots: Scan and track inventory levels in stores and warehouses. **Customer Service Robots:** Assist shoppers with information, directions, and product recommendations.
Checkout Robots: Assist with self-checkout and payment processes.

Entertainment and Social Robots:
Entertainment and social robots are designed for recreational and interactive purposes, engaging with humans for fun, companionship, and amusement.

Robotic Companions: Designed to interact with people, provide companionship and entertain users.
Animatronics: Used in theme parks and entertainment venues to create lifelike characters and exhibits.

Environmental Monitoring Robots:
Environmental monitoring robots are automated machines used to collect data and assess conditions in natural ecosystems, aiding conservation and research.

Underwater Robots (ROVs and AUVs): Explore and collect data in underwater environments. **Environmental Monitoring Drones:** Collect environmental data, study wildlife, and monitor ecosystems.

Restaurant Robots:
Cooking and Food Preparation Robots: Assist chefs and kitchen staff in preparing dishes.

Waitstaff Robots: Deliver orders to tables and provide customer service in restaurants.

Telepresence Robots:
Telepresence robots are remotely controlled machines with cameras and screens, enabling users to interact and navigate in distant locations.

Remote Presence Robots: Enable users to be virtually present in a distant location through a robot equipped with cameras and screens.

These types of service robots showcase the versatility of robotic technology in various industries and everyday life. As advancements in robotics continue, new applications and specialized service robots are continually emerging to address a wide range of tasks and needs.

Logistics and Warehousing:
Logistics and Warehousing robots automate tasks like goods transportation, picking, and packing in warehouses and distribution centers, improving efficiency.
Autonomous Guided Vehicles (AGVs): Transport goods within warehouses and factories.
Parcel Sorting Robots: Sort and organize packages for shipping and distribution.
Material Handling Robots: Load and unload items in distribution centers.

Cleaning and Maintenance:
Cleaning and Maintenance robots are autonomous machines designed to clean, repair, or upkeep various spaces and equipment, reducing human intervention.

Robotic Vacuum Cleaners: Automatically clean floors and surfaces in homes and offices.
Window Cleaning Robots: Clean windows and glass surfaces on tall buildings.
Industrial Cleaning Robots: Clean and maintain industrial facilities.

Transportation:
Transportation robots are autonomous machines designed to move goods or people from one place to another, streamlining logistics and mobility.

Autonomous Vehicles: Self-driving cars and delivery vehicles revolutionize transportation.
Passenger Shuttles: Provide automated transportation services in controlled environments.

Search and Rescue:
Search and Rescue robots are designed to locate and assist people in disaster- stricken areas, enhancing safety and saving lives.

Search and Rescue Robots: Assist in locating and rescuing individuals in disaster-stricken areas.

Food Industry:
Food industry robots automate tasks like food processing, packaging, and quality control, enhancing efficiency, consistency, and safety in food production.

Cooking Robots: Assist chefs in food preparation and cooking processes.
Waitstaff Robots: Deliver orders to tables and provide customer service in restaurants.

These applications demonstrate the versatility and potential of service robots to enhance productivity, safety, and quality of life in a wide range of industries and everyday situations. As technology continues to advance, service robots are likely to find even more diverse applications in the future.

Delivery Robotics

Education and Research Robots

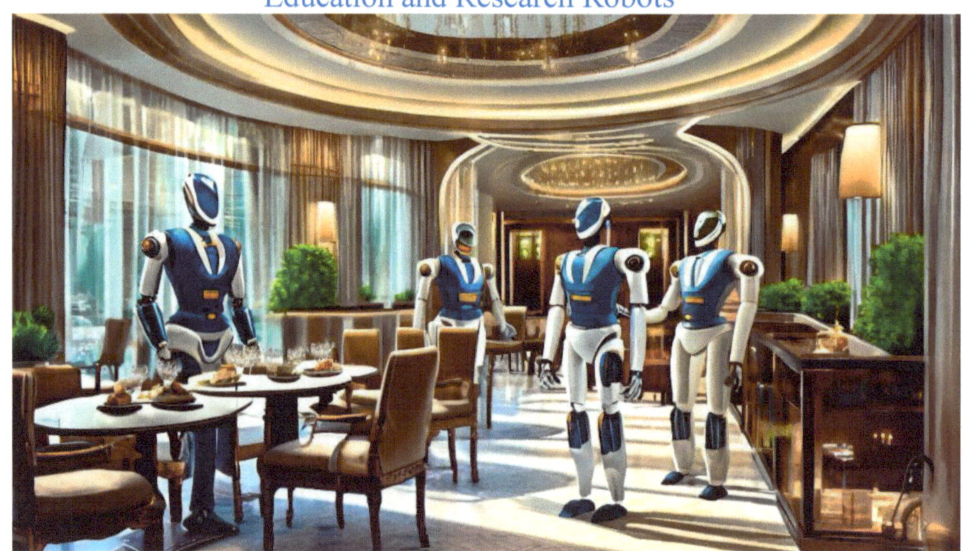

Hotel and Hospitality Robots

1.3. The Future of Service Robots

The future of service robots is promising. Advancements in AI, machine learning, and robotics will lead to more capable and adaptable service robots. This holds a promising landscape of technological advancements poised to transform the way we live, work, and interact with the world around us. As we peer into the horizon of robotics, several compelling trends emerge, each bearing the potential to redefine the role of service robots in our daily lives.

First and foremost, we anticipate a profound surge in the intelligence of service robots. These mechanical marvels are poised to evolve into highly sophisticated beings, equipped with the cognitive abilities required to perform intricate tasks and navigate complex environments. This surge in intelligence is set to broaden the spectrum of applications, enabling service robots to venture into new territories, where their capabilities were once deemed out of reach.

Furthermore, the cost of service robots is expected to undergo a steady decline. This reduction in cost will render them more accessible, not only to businesses but also to individuals, ushering in an era where service robots become commonplace tools in our daily routines. The democratization of this technology promises to empower a wider audience with the advantages that service robots bring to various domains.

The Future of Service Robots

Lastly, the trajectory of increased affordability and enhanced capabilities is poised to drive greater adoption across industries and settings. Service robots will no longer be confined to specific niches but will infiltrate diverse sectors, catalyzing transformative changes in healthcare, education, agriculture, hospitality, and more. In this exploration of the future of service robots, we will delve into these key trends, unraveling the technological marvels that underpin them and forecasting the ways in which service robots are set to revolutionize our world.

They will play a crucial role in healthcare, assisting with surgeries, caregiving, and monitoring patients. In logistics, they'll optimize warehouses and enhance last-mile delivery. In homes, they'll offer increased automation and companionship for the elderly. Service robots will also contribute to agriculture by improving crop management and harvesting. Moreover, their applications in education, entertainment, and environmental conservation will expand. Overall, the future holds a diverse array of service robots that will transform industries and enhance our daily lives.

Future Trends in Service Robot Technology

Increased intelligence: Service robots will become more intelligent and capable of performing complex tasks. This will allow them to be used in a wider range of applications.

Reduced costs: The cost of service robots is expected to decrease over time. This will make them more affordable for businesses and individuals.

Increased adoption: As service robots become more affordable and capable, we can expect to see them adopted more widely in a variety of industries and settings.

Anticipating Future Applications of Service Robots:

Explore Potential Use Cases

Elderly Care: Service robots could assist the elderly by providing companionship, helping with daily tasks, and monitoring their health.

Autonomous Taxis: Self-driving cars equipped with service robots could provide transportation services, such as helping passengers with luggage and navigation.

Personal Shoppers: Robots could shop for groceries and other goods on behalf of customers, saving time and effort.

Disaster Response: Robots capable of navigating disaster-stricken areas could assist in search and rescue missions, providing aid to survivors.

Telemedicine Assistants: Robots could aid doctors during remote medical consultations, performing tasks like taking vitals and displaying patient records.

Agricultural Pollination: With declining bee populations, robots could be used for pollination in agriculture, ensuring crop yields.

Language Translation: Service robots equipped with advanced language processing could serve as real-time translators, bridging language barriers.

Educational Tutors: Robots could provide personalized tutoring and assist students in their learning, adapting to individual needs.

Construction and Demolition: Robots could handle dangerous tasks in construction and demolition, improving safety for workers.

Virtual Reality Guides: In tourism and museums, robots could provide interactive guided tours in virtual reality.

Mental Health Support: Robots with AI-driven emotional intelligence could provide mental health support and therapy.

Personal Fitness Coaches: Robots could guide individuals through exercise routines, promoting health and fitness.

Environmental Conservation: Robots could be used for reforestation, monitoring endangered species, and cleaning up polluted areas.

Space Mining: Robots might play a role in mining resources on celestial bodies, like asteroids and the Moon, for use on Earth.

Retail Customer Service: Robots could assist shoppers in stores, answering questions and providing recommendations.

Waste Management: Robots could efficiently sort and process recyclables and reduce landfill waste.

Public Transportation Services: Autonomous buses and trains with onboard robots could assist passengers and ensure safety.

Archaeological Excavations: Robots could aid archaeologists in delicate excavations, preserving historical artifacts.

Sports Coaching: Robots could provide real-time coaching feedback and analysis to athletes. **Emergency Response:** Swarming robots could quickly assess and respond to natural disasters, providing crucial information to emergency teams.

These examples demonstrate the potential for service robots to transform various aspects of our lives and industries in the future, improving efficiency, safety, and convenience.

Elderly Care of Service Robots

Educational Tutors of Service Robots

Environmental Conservation of Service Robots

Chapter 2: Service Robotics: Components and Software of Robotics

2.1. Components of Service Robots

The components of service robots refer to the essential parts that constitute these versatile machines. They include sensors and actuators, enabling perception and interaction with the environment. Power systems provide the necessary energy for functioning, while manipulators and effectors facilitate precise movements and actions. Control systems, comprised of algorithms and software, govern the overall behavior and decision-making processes. These interlinked components synergize to create a functional service robot capable of performing designated tasks efficiently and accurately.

Components of Service Robots

Service Robots Typically Consist Of Several Key Components:

Sensors: Sensors, such as cameras, ultrasonic sensors, and lidar, play a crucial role in perceiving the robot's surroundings. Cameras capture visual data, while ultrasonic sensors detect distances from objects, and lidar helps create 3D maps. These inputs enable the robot to navigate, avoid obstacles, and interact with the environment effectively.

Actuators: Actuators, including motors and servos, convert electrical signals into mechanical motion. They allow the robot to move its limbs, manipulate objects, and perform specific tasks. Actuators are pivotal in giving robots the ability to interact with their surroundings and execute actions with precision.

Control System: The control system, comprising central processing units (CPUs) and microcontrollers, acts as the robot's brain. It processes data from sensors, making sense of the environment and executing commands accordingly. The control system is responsible for coordinating movements, actions, and responses based on the gathered information.

Power Supply: Power supply, typically in the form of batteries or energy sources, is essential for the robot's mobility and functionality. It provides the necessary electrical power to run motors, actuators, and the control system, allowing the robot to perform its designated tasks effectively.

End-Effector: The end-effector refers to the manipulation tools or arms equipped with grippers, specialized tools, or instruments necessary for specific tasks. It is the part of the robot that directly interacts with objects or the environment, enabling actions such as cleaning, surgery, or other precise operations.

Communication Interface: The communication interface facilitates interaction between the robot and humans or other devices. This can include voice commands, gestures, or screens, enabling intuitive and efficient communication between the robot and its operators or users.

Software: Software, encompassing algorithms and programming code, governs the behavior and actions of the robot. This includes navigation algorithms for movement, decision-making processes based on sensor data, and instructions for task execution. Software is fundamental in defining how the robot operates and responds to various inputs.

Safety Features: Safety features are vital components designed to ensure the robot's safe operation. These include sensors that detect potential hazards, emergency stop mechanisms that halt operations in critical situations, and collision avoidance systems that prevent accidents, particularly in dynamic or crowded environments.

Each of these components contributes to the overall functionality and capabilities of service robots, allowing them to perform a wide range of tasks efficiently and safely.

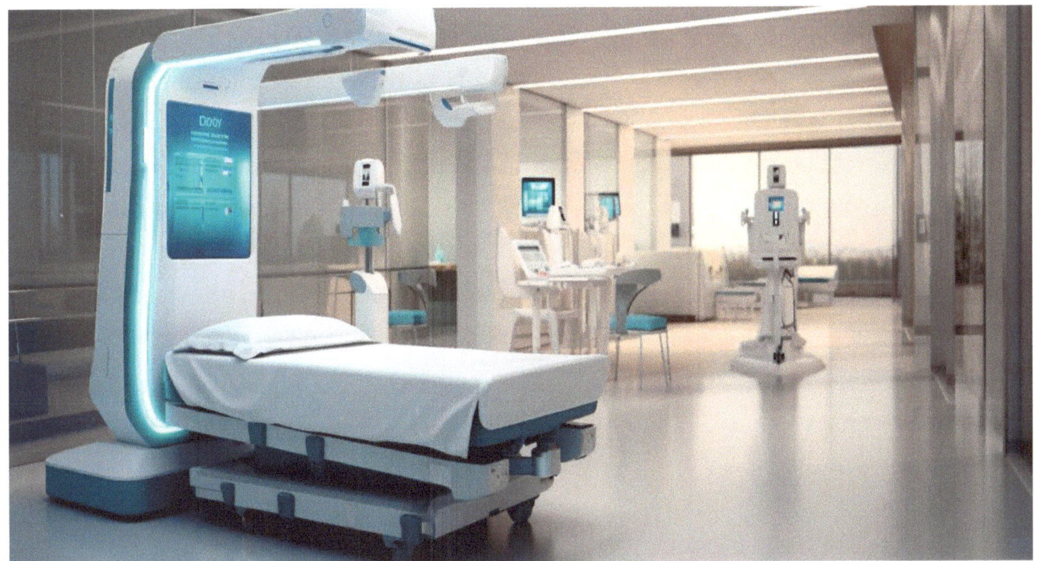

Service robots typically consist of several key components

Sensors: The Robot's Senses

At the heart of every service robot lies an array of sensors that serve as its sensory organs. These sensors provide the robot with the ability to perceive and interact with its environment. Common types of sensors include cameras for visual input, lidar for distance measurement, ultrasonic sensors for obstacle detection, microphones for audio input, infrared sensors for heat and motion detection, and touch sensors for physical interaction. Sensors serve as the eyes and ears of service robots, allowing them to perceive and interact with their environment effectively.

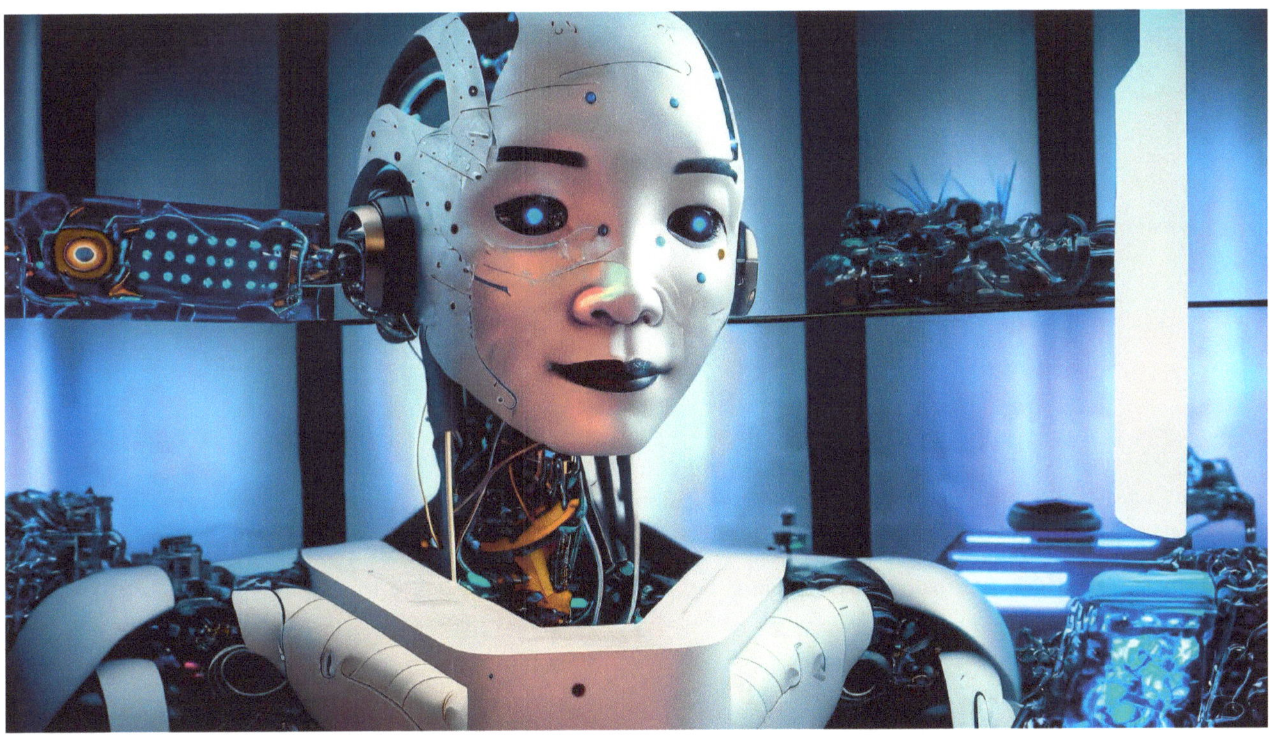

Sensors: The Robot's Senses

Different Types Of Sensors Are Employed, Including

Camera: Cameras in service robots function as the eyes, capturing visual data in the form of images and videos. They are crucial for tasks that require visual perception, navigation, object recognition, and various applications in which understanding the environment is vital. **Lidar:**Lidar, which stands for Light Detection and Ranging, employs lasers to measure distances and create detailed 3D maps of the surroundings. It aids in obstacle detection, mapping environments for navigation, and enabling robots to understand and navigate through complex terrains.

Ultrasonic sensors: Ultrasonic sensors emit sound waves and measure the time it takes for them to bounce back, determining distances to objects. This data is crucial for proximity sensing, obstacle detection, and collision avoidance, enabling the robot to move safely within its environment.

Infrared sensors: Infrared sensors detect infrared radiation, often associated with heat. They are used to detect heat sources, motion, or changes in temperature. These sensors play a crucial role in applications like tracking and identifying living beings or determining temperature variations in the robot's vicinity.

Microphones: Microphones in service robots serve as the ears, capturing audio data from the environment. This data can be used for various purposes, including voice recognition, understanding commands, ambient sound analysis, and providing auditory feedback or responses.

Touch sensors: Touch sensors are designed to detect physical contact with objects, humans, or surfaces. They are essential for tasks requiring interaction with the environment, object manipulation, or providing a response based on touch or pressure, enhancing the robot's ability to engage with its surroundings.

These components collectively empower the service robot to perceive its environment, make informed decisions, navigate through spaces, and interact with objects and individuals, enabling a wide range of applications in different domains.

Actuators: Bringing Robots to Life

Actuators are responsible for translating the robot's decisions into physical actions. These components enable the robot to move, manipulate objects, and perform tasks. Among the most common actuators are motors, which power mobility systems like wheels, tracks, or legs, and also drive the robotic arms and grippers used for manipulation. Servo motors offer precision in joint movements, while pneumatic actuators and linear actuators serve specific purposes in extending or lifting components.

Actuators Types:

Motors: Motors are essential actuators that convert electrical energy into mechanical motion. Various types include DC motors for precise control, AC motors for constant speed, and stepper motors for incremental rotation. Servo motors offer accuracy, while brushless motors provide longevity.

Electric Motors: Electric motors are commonly used in robotics for mobility and manipulation. They convert electrical energy into mechanical motion. Depending on the application, various types of electric motors can be used, including brushed DC motors, brushless DC motors, and stepper motors. These motors offer good control and can be used in various configurations such as wheels, tracks, or arms.

Hydraulic Motors: Hydraulic motors use pressurized fluid to create motion. They are often chosen for heavy-duty applications where high torque is required, such as in construction or industrial robotics. Hydraulic systems are known for their strength and durability, although they can be complex and require maintenance.

Servo Motors: Servo motors are specialized electric motors used in robotic joints and limbs. They offer precise control over position, speed, and torque. This precision is essential for tasks that require accurate movement, such as pick-and-place operations and robotic arms.Servo motors are commonly used in combination with feedback systems like encoders to achieve high accuracy.

Pneumatic Actuators: Pneumatic actuators use compressed air to generate motion. They are often chosen for applications that require relatively simple and quick movements, such as in robotic arms used in manufacturing and assembly lines. Pneumatic actuators are valued for their speed and simplicity, but they may lack the fine control and precision of other types of actuators.

Linear Actuators: Linear actuators are devices that provide linear motion, meaning they move in a straight line. They are commonly used in robotics to lift, extend, or retract parts. Linear actuators can be electric, hydraulic, or pneumatic, depending on the application's requirements. They are often used in tasks like adjusting the height of a robot's platform or extending a gripper to grasp an object.

Each of these components has its own advantages and disadvantages, making them suitable for different robotic applications based on factors like precision, speed, power requirements, and environmental conditions. The choice of which component to use depends on the specific needs of the robot and the tasks it is designed to perform.

Control System: The Robot's Brain

The control system is the brain of the robot, orchestrating its actions and decision-making processes. It consists of microcontrollers or microprocessors that process data from sensors and run the robot's software. Advanced algorithms for navigation, perception, and decision-making are essential components of the control system, alongside sensor fusion techniques that integrate data from multiple sensors. Machine learning and artificial intelligence play vital roles in tasks such as object recognition, path planning, and natural language processing.

Control System: The Robot's Brain

Control System Consists:

Microcontrollers vs. Microprocessors:

Microcontrollers: These are specialized integrated circuits designed for embedded systems and robotics. They typically include a CPU, memory, input/output interfaces, and sometimes specialized hardware for tasks like analog-to-digital conversion or motor control. Microcontrollers are often used in robots for real-time control tasks, such as motor control, sensor interfacing, and low-level decision-making. They are efficient and consume less power compared to microprocessors.

Microprocessors: Microprocessors are general-purpose computing units, like those found in personal computers. They are capable of handling complex algorithms and running high- level software. In a robot, microprocessors are often used for tasks that require significant computation, such as image processing, path planning, and artificial intelligence (AI). Microprocessors provide the computational power needed to process data from sensors, run algorithms, and make intelligent decisions.

Software: Robot software typically includes a range of algorithms for different functions

Navigation Algorithms: These algorithms help the robot determine its position, plan paths, and avoid obstacles. Common navigation algorithms include SLAM (Simultaneous Localization and Mapping) and A* pathfinding.

Perception Algorithms: These are used to interpret data from sensors like cameras, LiDAR, and ultrasonic sensors. Object recognition, feature detection, and image processing algorithms fall under this category.

Decision-Making Algorithms: These algorithms enable the robot to make intelligent decisions based on its perception of the environment and predefined goals. They can include rule-based systems, finite state machines, or more advanced AI- based decision-making.

Sensor Fusion: Sensor fusion is the process of combining data from multiple sensors to create a more comprehensive and accurate understanding of the robot's environment. This is crucial for robust and reliable robot operation.

Machine Learning and AI: Machine learning and AI are increasingly used in robotics to perform various tasks:

Object Recognition: Convolutional Neural Networks (CNNs) are often used to recognize objects in images or point clouds.

Path Planning: Reinforcement learning or classical algorithms like A* are employed for finding safe and efficient paths.

Natural Language Processing (NLP): In human-robot interaction, NLP techniques are used to enable robots to understand and respond to natural language commands or queries.

These technologies allow robots to adapt to changing environments, learn from their experiences, and interact with humans more intuitively.

Power Source: Fueling the Machine

Robots need a source of energy to function. Depending on the robot's design and purpose, this can take the form of batteries for mobility and portability, power cables for tethered robots, or more specialized options such as fuel cells or combustion engines for industrial or heavy-duty robots.

Power Source: Fueling the Machine

Service robots require a source of energy to operate, which can be in the form of:

Batteries: Batteries are a common power source for portable and mobile robots. They provide a compact and self-contained energy solution, making them ideal for robots that need to operate without being tethered to a fixed power source.Portable robots, such as consumer-grade robotic vacuum cleaners, drones, and even some smaller industrial robots, often rely on rechargeable batteries. These batteries can be lithium-ion, nickel-metal hydride, or other chemistries depending on the specific requirements of the robot.The advantage of batteries is their mobility and flexibility. They enable robots to move freely and operate in various environments without being restricted by power cables.

Power Cables: Tethered robots are those that are physically connected to an external power source through power cables. This arrangement allows them to draw power continuously, which can be essential for certain applications.Tethered robots are commonly used in situations where uninterrupted operation is critical, such as in manufacturing, laboratory automation, or medical robotics. These robots are often stationary or have limited mobility within the range of their power cables.While tethered robots may lack the mobility of battery-powered ones, they have the advantage of never running out of power as long as they remain connected to their power source.

Fuel Cells or Combustion Engines: Some industrial or heavy-duty robots require more substantial and long-lasting power sources, such as fuel cells or combustion engines.

> **Fuel Cells:** Fuel cell-powered robots use electrochemical reactions to convert hydrogen or other fuels into electricity. These are often used in applications where extended runtime is needed, like underwater exploration robots or unmanned aerial vehicles (UAVs).

> **Combustion Engines:** Combustion engines, such as gasoline or diesel engines, are occasionally employed in large, heavy-duty robots like construction equipment or agricultural machinery. These engines generate mechanical power, which can be converted into electrical power for electrically-driven components.

2.2. Manipulation Components: The Hands of the Robot

When a robot's tasks involve interacting with objects, it often features specialized manipulation components. Robotic arms, equipped with multiple joints, are employed for tasks like grasping and moving objects. These arms can be further enhanced with end-effectors, specialized tools, or grippers designed for specific tasks. For robots designed to interact with objects, specialized manipulation components may be included, such as:

> **Robotic arms:**Robotic arms are essential components of various automation and industrial processes. They are designed to mimic the functionality of human arms but are equipped with multiple joints, making them versatile tools for a wide range of tasks. These multi-jointed limbs are commonly used in manufacturing, assembly lines, healthcare, and even space exploration, where precision and dexterity are crucial.

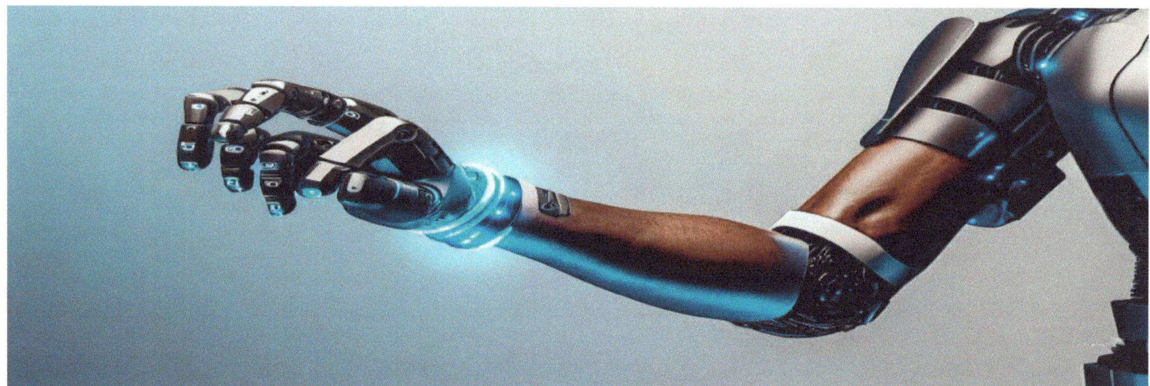

Robotic arms

> **End-effectors:** End-effectors are the functional extensions of robotic arms, designed to perform specific tasks or functions. These tools or grippers are attached to the end of the robotic arm and are crucial for achieving the desired outcome in various industries and applications.

End-effectors

Mobility System: Getting Around

Many service robots are mobile, necessitating a means of transportation. The choice of mobility system depends on the robot's intended environment and terrain. Options include wheels for ground-based robots, tracks for rough terrain traversal, legs for walking robots, and propellers or thrusters for aerial or underwater robots.

Mobility System: Getting Around

Some of the items include:

Wheels: Wheels are a common choice of locomotion for ground-based robots. They offer stability, efficiency, and precise control over movement. Wheeled robots are commonly used in indoor environments, factories, warehouses, and even on roads and sidewalks. They are well-suited for applications that require smooth and predictable motion, such as material handling, surveillance, and autonomous vehicles. Wheels come in various sizes and types, making them adaptable to different terrain and load requirements.

Tracks: Tracks are an ideal choice for robots that need to traverse rough or uneven terrain. Unlike wheels, which can struggle on uneven surfaces, tracks provide better stability and grip. This makes them suitable for off-road applications like agriculture, construction, and military operations. Track- based robots distribute their weight over a larger area, reducing the likelihood of getting stuck or losing traction. This makes them highly capable in environments with obstacles, mud, sand, or snow.

Legs: Legs are a preferred choice for walking robots, also known as legged robots or quadrupeds. These robots mimic the locomotion of animals and can navigate a wide range of terrains, from flat surfaces to complex and uneven environments. Legged robots are advantageous in scenarios where wheeled or tracked robots may struggle, such as climbing stairs, crossing rubble, or exploring rough outdoor terrain. Their ability to adapt to different surfaces and step over obstacles makes them valuable for search and rescue missions and planetary exploration.

Propellers or thrusters: Propellers or thrusters are essential for robots that operate in aerial or underwater environments. In aerial robotics, such as drones, quadcopters, or fixed-wing UAVs (Unmanned Aerial Vehicles), propellers generate lift and thrust to achieve controlled flight. These robots are used for tasks like aerial photography, surveillance, and environmental monitoring.

Communication Interface: Interaction with the World

Robots frequently require communication interfaces to interact with users or other systems. These may include displays for information presentation, speakers for audio output, and wireless communication protocols such as Wi-Fi and Bluetooth for remote control or data exchange. Service robots may have communication components to interact with users or other systems, including:

Displays: Screens for Showing Information or Facilitating Social Interaction

Displays encompass visual interfaces, commonly screens, that present information, graphics, or interactive elements. They are pivotal for showcasing data, enabling user interaction, and enhancing user experiences. Displays are utilized in devices like smartphones, computers, TVs, and public information systems, shaping modern communication and interaction methods.

Speakers: Audio Output and Alert Systems

Speakers serve as audio output devices, rendering sounds, speech, or alerts. They play a crucial role in conveying information audibly, adding depth to multimedia experiences, and ensuring accessibility. Speakers are integral in devices ranging from smartphones and laptops to home entertainment systems, amplifying the auditory dimension of our interactions with technology.

Wireless Communication: Wi-Fi, Bluetooth, and Beyond

Wireless communication involves protocols like Wi-Fi and Bluetooth that enable devices to connect and exchange data without physical connections. This technology facilitates seamless remote control, data sharing, and networking across a myriad of devices, contributing to the ever-connected world we live in. It's a cornerstone of the Internet of Things (IoT) and modern connectivity.

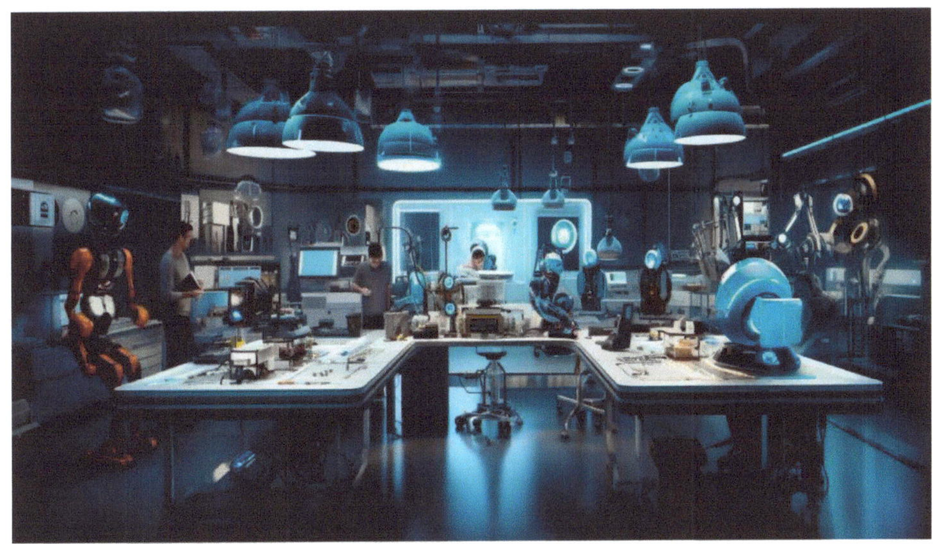

Wireless Communication: Wi-Fi, Bluetooth, and Beyond

Safety Systems: Ensuring Secure Operation

Safety Systems: Ensuring Secure Operation

Safety is paramount in the world of service robots. These machines often feature collision detection and avoidance systems, emergency stop buttons for manual shutdown in emergencies, and redundancy in critical components to enhance reliability.

To ensure safe operation, service robots often incorporate safety features, such as

Collision detection and avoidance: Collision detection refers to the process of identifying and determining when objects or entities in a given environment come into contact with each other. This is a critical aspect of various fields such as robotics, autonomous vehicles, video games, and industrial automation. The primary goal of collision detection is to prevent or respond to collisions effectively.

Emergency stop buttons: Emergency stop buttons are essential safety devices designed for manual shutdown during critical situations. They provide a quick and reliable means to halt machinery or processes to prevent accidents or respond to emergencies effectively.

Redundant systems: Redundant systems involve incorporating backup components to enhance reliability, ensuring that if a primary component fails, an alternative is available, thereby minimizing downtime and ensuring uninterrupted operation.

User Interface: Connecting with Humans

Some service robots incorporate user interfaces to facilitate interaction with humans. These interfaces can encompass touchscreen displays, voice recognition, or gesture recognition systems, enhancing the robot's usability and user experience.

User Interface: Connecting with Humans

Key uses of User Interface are:

Touchscreen Displays: These intuitive interfaces allow users to interact with electronic devices by physically touching the screen. They offer a visually engaging and responsive experience, making them popular in smartphones, tablets, and kiosks. Users can tap, swipe, and pinch to navigate, select options, and input data effortlessly.

Voice Recognition: Voice recognition technology enables hands-free interaction with devices and software. It listens to spoken commands and translates them into actions, making tasks like calling a contact, setting reminders, or controlling smart devices more convenient. This technology has seen significant advancements in recent years, enhancing accessibility and user convenience.

Gesture Recognition: Gesture recognition systems analyze hand and body movements to interpret user intentions. This technology is employed in gaming consoles, virtual reality setups, and interactive displays. Users can control applications, play games, and navigate interfaces through natural gestures, fostering immersive experiences.

Haptic Feedback: Haptic feedback provides tactile sensations to users, enhancing their engagement with digital interfaces. Devices equipped with haptic technology can simulate physical sensations, such as vibrations or texture feedback, to make interactions more realistic and informative. This technology is integral to virtual reality, gaming controllers, and touchscreen keyboards.

AR Interfaces: Augmented Reality (AR) interfaces blend digital information with the real world, typically viewed through smartphone apps or AR glasses.

They enhance user perception by overlaying graphics, text, or interactive elements onto the user's surroundings. AR interfaces find applications in navigation, education, gaming, and remote assistance, offering a new dimension of interaction.

Mobile Apps: Mobile applications, or mobile apps, provide a versatile means of controlling and monitoring various devices and systems through smartphones. Users can access features, receive updates, and remotely manage connected devices via these apps. They serve as a central hub for controlling smart homes, tracking health metrics, and managing productivity, simplifying daily tasks and enhancing convenience.

Navigation System: Finding the Way

For robots that need to move autonomously, navigation systems are indispensable. These systems employ mapping, localization, and path-planning algorithms to enable the robot to navigate its environment with precision and efficiency.

Navigation System: Finding the Way

Navigation System Applications are:

Mapping: Creating and Maintaining a Digital Map
Mapping involves the creation and continuous updating of a digital representation of the environment in which a robot operates. This digital map serves as a crucial reference, allowing the robot to navigate and interact with its surroundings accurately.

Localization: Determining the Robot's Position within the Map

Localization is the process of precisely determining the robot's position within the digital map. Utilizing various sensors and algorithms, the robot aligns its real-world location with the corresponding location on the digital map, enabling accurate movement and decision- making.

Path Planning: Generating Safe and Efficient Routes

Path planning involves the generation of optimal routes for the robot to reach its desired destinations while considering obstacles, terrain, and other constraints. Algorithms analyze the digital map to chart the safest and most efficient paths, enhancing the robot's navigational efficiency.

Obstacle Avoidance: Adjusting Path to Avoid Collisions

Obstacle avoidance is a critical aspect of navigation where the robot dynamically adjusts its path to avoid colliding with obstacles in its environment. Utilizing sensor data and real-time analysis, the robot identifies obstacles and alters its trajectory to ensure a safe and obstacle- free journey.

GPS: Providing Global Positioning in Outdoor Environments

GPS (Global Positioning System) is a satellite-based navigation system that enables accurate positioning and tracking of the robot in outdoor settings. By receiving signals from multiple satellites, the robot determines its precise global coordinates, facilitating effective navigation and location-based tasks.

In the world of service robots, these components combine to create a symphony of technology and engineering, allowing these machines to perform tasks that range from household chores to complex surgical procedures and autonomous delivery services. Understanding the intricacies of these components is essential to grasp the remarkable capabilities of modern service robots.

2.3. Software for Service Robots

In this part, we delve into the critical software components that drive the functionality and intelligence of service robots. These software elements enable service robots to perceive their environment, make informed decisions, and interact effectively with both their surroundings and human users. A comprehensive understanding of these software aspects is crucial for engineers, developers, and enthusiasts working with service robots.

Software for Service Robots

Operating System (OS)

The operating system plays a critical role in a robot's functionality, serving as the essential software framework that oversees and allocates hardware resources while executing various software modules. It acts as the core foundation, ensuring efficient operation and coordination of all components within the robot's system.

Linux-Based Options: Linux distributions like Ubuntu and Debian are favored in robotics for their robustness and open-source ethos. Their stability provides a solid foundation for running complex robot systems, while their open nature allows for customization and community-driven development.

ROS Integration: The Robot Operating System (ROS) is a versatile framework for robotic development. It boasts a rich ecosystem of tools and libraries, facilitating various robot tasks from perception to control. ROS enables developers to create modular and adaptable robot software, enhancing efficiency and collaboration in the field.

Custom Real-time OS: Custom real-time operating systems are occasionally designed for specific robotic needs. These bespoke solutions prioritize precise timing and minimal latency, ensuring that robots can meet demanding performance requirements, especially in applications where timing is critical, such as industrial automation or autonomous vehicles.

Perception Software

Perception software plays a critical role in a robot's functionality, as it processes sensor data to enable the robot to comprehend its surroundings. This software is instrumental in recognizing objects, interpreting sensory inputs, and comprehending the environment. Some illustrative instances of perception software include computer vision algorithms, lidar processing systems, and sensor fusion techniques, all of which collectively empower the robot to navigate, interact, and make informed decisions within its environment.

Perception Software

Computer Vision: Libraries such as OpenCV empower robots to analyze visual information, including tasks like identifying objects, analyzing images, and recognizing faces. These capabilities are fundamental for enabling robots to interact with and navigate their environment effectively.

Speech Recognition: Systems like CMU Sphinx and Google's Speech Recognition API equip robots with the ability to comprehend spoken language, facilitating human-robot communication. This technology allows robots to receive verbal commands, answer questions, and engage in natural dialogues with users.

Object Detection: Algorithms like YOLO (You Only Look Once) are essential for real-time object detection, enabling robots to swiftly identify and locate objects in their surroundings. This functionality is crucial for tasks such as obstacle avoidance, object manipulation, and autonomous navigation.

Navigation and Localization

Navigation and localization components are critical for enabling a robot to efficiently operate in its environment. These components employ various techniques, including simultaneous localization and mapping (SLAM), sensor fusion, and path planning algorithms. Through SLAM, robots can build maps of their surroundings while simultaneously determining their precise position. This information is then used to plan optimal paths, ensuring the robot can navigate effectively and avoid obstacles, making it a versatile and adaptable autonomous system.

Navigation and Localization

SLAM Techniques: Simultaneous Localization and Mapping (SLAM) algorithms, such as Google's Cartographer, empower robots to create maps of their surroundings while simultaneously pinpointing their own location within these maps. These techniques leverage sensor data and complex algorithms to iteratively refine the robot's understanding of its environment, allowing it to navigate and interact effectively in unknown or changing spaces.

Path Planning: Path planning algorithms like A* and D* are fundamental tools in robotics for finding optimal routes through environments filled with obstacles. These algorithms consider both static and dynamic obstacles, enabling robots to adapt their paths in real-time as the environment changes. A* and D* employ heuristic-based search strategies to efficiently compute paths while avoiding collisions.

Path Planning

The software employs advanced algorithms to compute optimal paths for robots, considering safety and efficiency. It factors in various obstacles and operational constraints, ensuring the robot can navigate its environment smoothly. This intelligent routing system not only enhances the robot's ability to avoid collisions but also maximizes its productivity by selecting the most efficient routes for completing tasks.

Path Planning of service robotics

Dijkstra's Algorithm: Dijkstra's Algorithm is a well-established method for determining the shortest path between two points within a graph. It is particularly suitable for global path planning, such as route optimization in transportation networks. By iteratively evaluating the cost of reaching each node, Dijkstra's Algorithm guarantees the shortest path and is commonly employed in applications like GPS navigation systems.

Rapidly Exploring Random Trees (RRT): Rapidly Exploring Random Trees (RRTs) are highly effective algorithms for real-time motion planning in dynamic environments, notably in the field of robotics. RRTs use a randomized approach to rapidly explore the configuration space and generate feasible paths. They excel in adapting to changing conditions, making them valuable for tasks like autonomous navigation and obstacle avoidance in dynamic environments.

Control Algorithms

Control software plays a crucial role in orchestrating a robot's movements and interactions, guaranteeing meticulous and synchronized actions. This software employs various techniques and algorithms to govern the robot's behavior, enabling it to navigate, manipulate objects, and respond to its environment with precision and efficiency. These methods are essential for ensuring the robot's effectiveness and reliability in performing its designated tasks.

Control Algorithms of service Robotics

PID Controllers: Proportional-Integral-Derivative Controllers

PID controllers are a widely used control mechanism in robotics for achieving and maintaining desired positions and velocities of robotic systems. The 'P' (Proportional) component responds to the current error, the 'I' (Integral) component considers past errors, and the 'D' (Derivative) component anticipates future errors. This combination allows for precise and efficient control, aiding in the stability and accuracy of robotic movements.

Motion Planning Libraries: Simplifying Complex Robotic Motion Planning

Motion planning is a critical aspect of robotics, ensuring safe and efficient movement of robots within their environment. Motion planning libraries like MoveIt! are valuable tools that simplify complex motion planning tasks. They provide algorithms and tools to help design paths and trajectories, considering factors such as obstacle avoidance and dynamics, making it easier to achieve smooth and collision-free robotic motions. These libraries significantly contribute to streamlining the development and deployment of robotic applications.

Machine Learning and AI

Machine learning and artificial intelligence (AI) methods enable robots to acquire knowledge from data, identify recurring patterns, and execute informed actions. These techniques encompass various approaches such as neural networks, reinforcement learning, natural language processing, and computer vision. They equip robots with the ability to adapt, solve complex problems, and perform tasks with increasing autonomy, making them valuable tools in a wide range of applications, from manufacturing to healthcare and beyond.

Machine Learning and AI

Reinforcement Learning: Reinforcement Learning (RL) in robotics involves a learning paradigm where robots learn to perform tasks by interacting with their environment through a series of actions. The robot receives feedback in the form of rewards for desired actions and penalties for unfavorable ones. Over time, the robot adapts and refines its actions to maximize cumulative rewards, effectively learning and improving its performance in various tasks.

Deep Learning: Deep Learning (DL) is a subset of artificial intelligence that employs neural networks, which are inspired by the human brain's structure, to process vast amounts of data and extract patterns. In robotics, DL plays a crucial role in enabling tasks such as image recognition, object tracking, and natural language understanding. Neural networks can learn and understand complex features, allowing robots to interpret images, track objects in real-time, and comprehend human language, enhancing their capabilities and functionality.

Natural Language Processing (NLP): Natural Language Processing (NLP) equips robots with the ability to understand, interpret, and generate human language, enabling seamless communication between humans and machines. NLP models enable robots to comprehend spoken or written language, extract meaning, and generate appropriate responses. This facilitates natural and intuitive interactions, allowing users to communicate with robots in a manner similar to conversing with another person, enhancing user experience and usability in various applications.

Human-Robot Interaction (HRI)

Human-Robot Interaction (HRI) software plays a pivotal role in bridging the gap between robots and humans. It empowers robots to comprehend and react to human gestures, verbal commands, and actions effectively. Key components of HRI software encompass speech recognition, natural language processing, computer vision, and gesture recognition. This sophisticated technology enhances the fluidity of communication between humans and robots, promoting seamless cooperation and collaboration in various domains, from healthcare to manufacturing.

Human-Robot Interaction (HRI)

Speech Synthesis (Text-to-Speech): Speech synthesis, often referred to as Text-to-Speech (TTS), empowers robots to generate human-like speech. This technology enables robots to communicate with users in a relatable and informative manner. By converting written text into spoken words, TTS enhances the naturalness of interactions, making it easier for humans to engage with robots for tasks ranging from information dissemination to providing companionship.

Speech Recognition (Automatic Speech Recognition): Automatic Speech Recognition (ASR) is the backbone of a robot's ability to understand spoken commands and engage in conversations with humans. ASR technology enables robots to convert spoken language into text, allowing them to process and respond to user requests accurately. This capability enhances the efficiency and convenience of human-robot interactions, making robots more responsive and user-friendly.

Gesture Recognition: Gesture recognition technology equips robots with the ability to interpret and respond to human gestures. This feature enhances non-verbal communication between robots and humans, making interactions more intuitive and effective. Robots can detect and interpret a wide range of gestures, allowing them to understand user intentions and provide appropriate responses. Gesture recognition enhances the overall user experience, making interactions with robots more seamless and engaging.

Behavior Control

Behavior-based control systems establish the framework for a robot's responses to diverse stimuli and scenarios, directing its actions. Typical models employed in behavior control encompass various approaches, such as subsumption architecture and finite-state machines. These models enable robots to exhibit adaptive behaviors, allowing them to effectively navigate and interact with their environment, responding dynamically to changing circumstances.

Behavior Control of Service Robots

Finite State Machines (FSMs): Finite State Machines (FSMs) are widely employed in robotics to define discrete behavioral states and transitions. They offer a simple yet effective way to model robot behaviors, allowing for clear state representations and well-defined transitions between states. FSMs are particularly useful for tasks where the robot's behavior can be broken down into distinct, non-overlapping states, making them a valuable tool in robotics control.

Behavior Trees: Behavior Trees are a structured method for representing intricate behaviors in robotics, featuring hierarchical control flows. These trees organize actions and conditions into a tree-like structure, making it easier to design and manage complex robot behaviors. Behavior Trees provide a clear and intuitive way to handle decision-making processes, with the ability to create flexible, adaptable, and robust robot control systems.

Hierarchical Task Networks (HTNs): Hierarchical Task Networks (HTNs) serve as a valuable framework for representing high-level planning and task execution in a hierarchical manner. They are particularly useful for robots that need to perform tasks with varying levels of complexity and abstraction. HTNs enable the decomposition of tasks into sub-tasks, allowing for efficient planning and execution while maintaining a structured, hierarchical approach to managing robot actions and objectives.

Task Planning

Task planning software is instrumental in enabling robots to efficiently coordinate and perform intricate sequences of actions, ultimately accomplishing predefined objectives. These frameworks, such as specific examples if needed, empower robots to navigate complex environments, make decisions, and execute tasks with precision, making them indispensable tools in various industries, from manufacturing and logistics to healthcare and space exploration.

Task Planning of service Robots

Planning Domain Definition Language (PDDL):PDDL is a standardized, formal language designed for specifying planning problems and domain-specific knowledge in artificial intelligence. It allows practitioners to precisely define the initial state, goals, actions, and constraints within a planning domain. This language plays a crucial role in representing problems for automatedplanners, facilitating the development and communication of planning tasks in a clear and unambiguous manner.

STRIPS (Stanford Research Institute Problem Solver):STRIPS is a widely used notation for representing planning domains and problems. It focuses on defining the preconditions and effects of actions within a domain. STRIPS representations employ a simplified syntax, making it easier for automated planners to process and generate solutions. This notation helps capture the essential elements of a planning problem, enabling efficient problem-solving algorithms to operate effectively.

Fast Downward: Fast Downward is a prominent open-source planning system known for its remarkable efficiency in solving complex planning problems. Developed by the artificial intelligence community, it employs state-of-the-art techniques such as heuristic search and search space pruning to find optimal or near-optimal solutions. Fast Downward speed and effectiveness make it a valuable tool for addressing a wide range of planning challenges across various domains.

Robot Middleware

Middleware serves as a crucial intermediary in the realm of robotics, facilitating seamless communication and coordination among diverse robotic components. This technology bridges the gap between hardware and software, enabling robots to interact effectively with sensors, actuators, and control systems. Examples of middleware in robotics encompass Robot Operating System (ROS), Robot Middleware (RT- Middleware), and Middleware for Robotics (MIRA), all of which play a pivotal role in orchestrating the complex interplay of robot components for enhanced functionality and versatility.

Robot Middleware

ROS (Robot Operating System):ROS is a highly popular middleware framework in the field of robotics, boasting a vast repository of pre-built packages and tools that streamline the development of robotic applications. It offers a comprehensive ecosystem for robot software development, encompassing perception, control, simulation, and communication. ROS facilitates interoperability among various hardware and software components, making it a go-to choice for researchers and developers in the robotics community.

DDS (Data Distribution Service):DDS middleware is a dependable solution for facilitating real- time data distribution and communication within robotic systems. It provides a standardized, publisher-subscriber communication model that ensures efficient and reliable data exchange among distributed components in real-time. DDS is particularly well-suited for applications that demand low-latency, high-throughput data sharing, making it an integral part of many robotics projects where timely data dissemination is crucial for decision-making and control.

OROCOS (Open Robot Control Software):OROCOS serves as a robust framework designed specifically for real-time control and robot software development. It offers a comprehensive toolkit for developing and managing control systems in robotics, enabling engineers to implement complex control algorithms with precision. OROCOS emphasizes real-time performance, safety, and modularity, making it an ideal choice for applications such as industrial automation, autonomous vehicles, and robotics research. Its open-source nature encourages collaboration and innovation within the robotics community.

Programming Languages

Programming languages are essential tools for developing robot software components, greatly influencing efficiency and performance. Some of the widely adopted languages include Python, which excels in ease of use and versatility, making it suitable for various robot applications. C++ offers high performance and control over hardware, ideal for real-time tasks. Additionally, ROS (Robot Operating System) provides a framework for robot development, supporting multiple languages like C++ and Python, fostering interoperability and rapid prototyping. These language choices impact the success of robot software projects.

Programming Languages

Python in Robotics: Python is renowned for its simplicity and versatility, making it a popular choice in robotics. Its ease of use facilitates rapid prototyping and seamless integration with AI and machine learning libraries. This enables developers to quickly experiment with various robotic applications, from computer vision to natural language processing, making Python a go-to language for robot software development.

C++ in Robotics:C++ stands out in robotics due to its exceptional performance and precise control over hardware resources. It is preferred for real-time and computationally intensive applications, where speed and resource management are critical. C++ allows developers to harness the full potential of hardware, making it a top choice for tasks like sensor data processing and motion control in robotics.

Java in Robotics: Java's platform independence is a key advantage for cross-platform robotic development. It offers portability, making it suitable for applications that need to run on diverse hardware and operating systems. Java's robust ecosystem and extensive libraries also make it a valuable choice for developing cross-platform robotic applications that require compatibility across a wide range of devices.

ROS-specific DSLs in Robotics: The Robot Operating System (ROS) provides Domain Specific Languages (DSLs) designed specifically for robotics. These DSLs simplify the development of robot-specific tasks by offering specialized tools and abstractions. ROS DSLs enable developers to create complex robotic behaviors and control systems with greater ease and efficiency, streamlining the development process and enhancing the interoperability of robotic components within the ROS ecosystem.

Safety Software

Safety-critical software is essential for ensuring the safe operation of robots, in compliance with stringent safety standards and regulations. This software incorporates vital features to guarantee robot safety, such as real-time monitoring, fault detection, and emergency shutdown mechanisms. These measures are crucial to prevent accidents and uphold the highest safety standards in various industries where robots are deployed, including manufacturing, healthcare, and autonomous vehicles.

Safety Software

Emergency Stop Mechanisms: Safety in robotics is paramount. Emergency stop mechanisms are integral to ensuring swift and reliable shutdown of robot operations when unforeseen dangers arise. These mechanisms are designed to minimize potential harm by instantly halting the robot's movements and functions. They serve as a crucial fail-safe, activated manually or automatically, to protect both operators and the surrounding environment in emergency situations.

Fault Detection Algorithms: To maintain safe operation, fault detection algorithms are deployed to continuously monitor the robot's systems for any abnormalities or faults.

These algorithms employ real-time data analysis to identify deviations from expected behavior. When an issue is detected, they trigger predefined responses, such as error messages, system adjustments, or shutdown protocols. This proactive approach ensures that potential hazards are swiftly addressed to prevent accidents and injuries.

Safety-Certified Components: Robotics systems often rely on various components like sensors and actuators, some of which come with safety certifications. These certifications ensure that these components meet industry standards and are designed to operate reliably and safely in various conditions. Safety-certified components play a pivotal role in building robust and secure robotic systems, contributing to overall safety and compliance with regulatory requirements.

Data Logging and Analytics

Data logging software plays a crucial role in collecting and storing extensive data generated during a robot's operation. This data is then analyzed using specialized analytics tools and libraries to extract valuable insights. These tools enable businesses and engineers to better understand the robot's performance, identify areas for improvement, and optimize its operations for increased efficiency and productivity.

Data Logging and Analytics

Data Logging Tools:

Data logging is a fundamental aspect of robotics and automation, enabling the collection and analysis of critical information during various operations. One powerful tool in this domain is ROS Bags, provided by the Robot Operating System (ROS). ROS Bags serve as an efficient mechanism to record and replay data, which proves to be indispensable for debugging and in-depth analysis.

ROS Bags : ROS (Robot Operating System) Bags are a unique feature within ROS, allowing developers and researchers to record streams of messages, including sensor data, topics, and other ROS-specific information during the robot's operation. These bags can be stored for later use and analysis.

Efficient Data Recording:ROS Bags efficiently capture data in a structured format, ensuring that various types of data can be logged simultaneously. This includes sensor readings, camera feeds, lidar scans, and more. The organized manner of storing this data streamlines the subsequent analysis.

Playback and Analysis: These bags can be replayed at any time, enabling a detailed review of the robot's behavior during a particular time frame. This feature is crucial for debugging and understanding how the robot's sensors and algorithms respond to different situations.

Debugging and Troubleshooting: When developing robotics applications, debugging is a common requirement. ROS Bags simplify this process by allowing developers to replay recorded data, thus aiding in identifying and rectifying issues efficiently.

Offline Analysis: Researchers often need to perform extensive analysis on robot data. ROS Bags facilitate this by providing an offline mode for analysis, enabling researchers to delve deep into the recorded data and extract meaningful insights without the need for the robot to be operational.

Research and Algorithm Validation: For academic and research purposes, ROS Bags play a crucial role in validating algorithms and conducting experiments. Researchers can utilize recorded data for algorithm testing and comparing the performance of different approaches.In conclusion, ROS Bags are a valuable tool in the field of robotics, providing an efficient way to record, store, and analyze data essential for development, debugging, and research. Their versatility and functionality significantly contribute to the advancement and improvement of robotic systems and applications.

Data Analysis Libraries:

Pandas: Pandas is a robust and widely utilized open-source Python library that provides efficient and easy-to-use data structures and data analysis tools. It excels in processing and manipulating structured data, making it highly suitable for handling robot-generated data. Pandas offers features like DataFrame, a two-dimensional labeled data structure, and a plethora of functions for data cleaning, transformation, aggregation, and more. Its versatility makes it a fundamental tool for data analysis in various domains, including robotics.

NumPy: NumPy, short for Numerical Python, is a fundamental Python library for scientific computing, particularly in the domain of numerical and mathematical operations. It provides support for arrays and matrices, along with a vast collection of mathematical functions to operate on these data structures. NumPy complements Pandas by enhancing the efficiency of numerical operations and enabling complex mathematical computations. The seamless integration of Pandas and NumPy is a common practice in data analysis, allowing for comprehensive and powerful analytical capabilities when processing robot-generated data.

By exploring these software components, we gain a deeper insight into the intricate and interconnected systems that make service robots versatile and capable of fulfilling diverse tasks in a variety of environments. The choice and effective integration of these software elements are pivotal in the development and operation of service robots across different applications.

Chapter-3- Design, Development, Testing, and Operation

3.1. Design Process

The development of service robots begins with a comprehensive design process. This involves defining the purpose and goals of the robot. For service robots, these goals typically revolve around assisting humans in various tasks.The design process includes conceptualization, where engineers and designers brainstorm ideas, create sketches, and decide on the robot's form and function.

After conceptualization, a detailed design phase follows. This includes creating 3D models, specifying materials, and selecting components like sensors, actuators, and processors.

User interface and human-robot interaction design are also crucial, as service robots often need to interact seamlessly with humans.

Design Process of Service Robotics

Service Robot Design Process Overview:

Problem Identification: The design process begins with the identification of a specific problem or task that a service robot is intended to solve. This crucial step involves understanding the needs and challenges of the users, which can range from household assistance to industrial automation.

Conceptualization: Once the problem is identified, the next step is to conceptualize a robotic solution. This stage involves brainstorming and ideation to outline the core features and functionalities of the robot. It's about envisioning how the robot will interact with its environment and users.

Requirement Analysis: After conceptualization, a thorough analysis of the requirements is conducted. This involves breaking down the conceptualized idea into specific technical, functional, and operational requirements. Understanding these requirements is essential for the subsequent design phases.

Detailed Design: With the requirements in hand, the detailed design phase commences. Engineers and designers work on creating a comprehensive blueprint for the robot, including its mechanical structure, electronics, sensors, and software architecture. This stage requires meticulous attention to detail.

Software Design: Service robots heavily rely on software to control their actions, interact with users, and process data from sensors. In this phase, the software architecture is designed, encompassing algorithms for perception, decision-making, and control, ensuring the robot can perform its intended tasks efficiently.

Prototyping: Building a physical prototype is a critical step in the design process. Prototyping allows for testing and validating the design in a real-world setting. It helps identify potential flaws and areas for improvement before moving on to production.

Each of these stages in the design process plays a pivotal role in shaping the capabilities and performance of a service robot. By following this structured approach, designers and engineers can develop robots that are not only technologically advanced but also capable of meeting the needs and expectations of users in various domains.

Steps in Designing Service Robots: A Breakdown Problem Identification:

The first step in designing a service robot is problem identification i.e. to define the specific problem or task that the service robot is intended to address.

- What is the need for the robot?
- What are the challenges and constraints that it will face?

For example, a service robot could be used to:

Deliver food in a restaurant, freeing up staff to focus on other tasks.

- Provide companionship to the elderly, helping to reduce loneliness and isolation.
- Assist with surgery in a hospital, improving precision and efficiency.

- Clean and disinfect surfaces, helping to prevent the spread of disease.
- Perform dangerous or repetitive tasks in industrial settings, improving safety and productivity.

Understand the context and requirements of the problem to be solved. Once the specific problem or task has been identified, it is important to understand the context and requirements of the problem to be solved. This includes understanding the environment in which the robot will operate, the users of the robot, and the specific tasks that the robot needs to be able to perform.

For example, a service robot that is used in a restaurant will need to be able to navigate around people and obstacles, and it will need to be able to interact with people in a friendly and helpful way. A service robot that is used in a hospital will need to be able to operate in a sterile environment, and it will need to be able to handle delicate medical instruments.By carefully defining the problem or task that the service robot is intended to address, and by understanding the context and requirements of the problem to be solved, the team can develop a better understanding of the robot's design requirements.

Example 1: Problem Statement: Design a Service Robot for Elderly Care

Objectives: Develop a robot that can assist the elderly with tasks such as medication reminders, monitoring vital signs, and providing companionship.

Context: The robot should operate in home environments and ensure the safety and well-being of the elderly.

Constraints: The robot should be cost-effective, compact, and lightweight to fit within the budget and space constraints of elderly individuals' homes.

Requirements: The robot must be capable of providing medication reminders, monitoring vital signs such as heart rate and blood pressure, and engaging in natural conversation with the elderly. It must also have obstacle avoidance capabilities to ensure safety within the home environment.

Service Robot for Elderly Care

Example 2: Problem Statement: Autonomous Delivery Robot for Restaurants

Objectives: Create a robot capable of autonomously delivering food orders within a restaurant setting.

Context: The robot will navigate crowded dining areas, avoid obstacles, and ensure food orders are delivered accurately and on time.

Constraints: The robot must have a limited size to move seamlessly through narrow spaces, and it should be cost-effective for restaurant owners.

Requirements: The robot should have advanced navigation capabilities, a secure compartment for food delivery, and a user-friendly interface for restaurant staff to program and manage delivery routes.

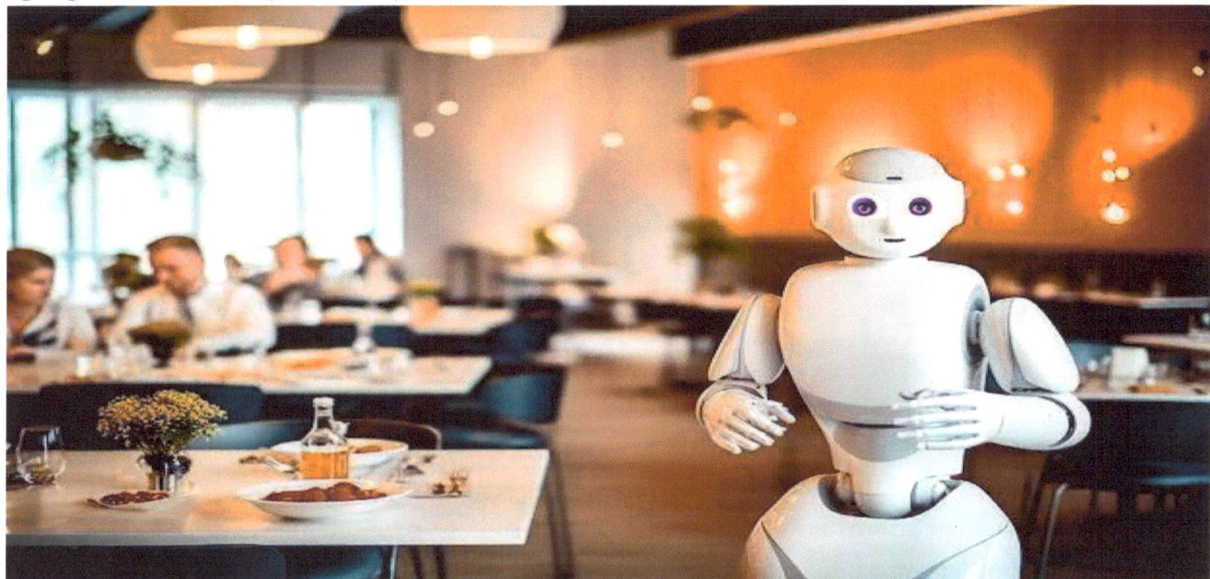

Autonomous Delivery Robot for Restaurants

Example 3:Problem Statement: Agricultural Robot for Crop Monitoring

Objectives: Develop a robot to monitor crop health, detect pests, and collect data on soil conditions.

Context: The robot will operate in large agricultural fields and greenhouses, helping farmers optimize crop yields.

Constraints: The robot should be able to cover large areas efficiently, operate for extended periods on a single charge, and remain within a reasonable budget.

Requirements: The robot must be equipped with sensors for crop health monitoring, pest detection, and soil analysis. It should also have GPS capabilities for precise mapping of the field and the ability to transmit data wirelessly to the farmer's control center.

Agricultural Robot for Crop Monitoring

Example 4: Problem Statement: Warehouse Inventory Management Robot

Objectives: Design a robot to automate inventory tracking and management within a warehouse. **Context**: The robot will need to navigate narrow aisles, scan barcodes, and update inventory databases in real-time.

Constraints: The robot should be compact, cost-effective, and capable of working alongside human warehouse staff.

Requirements: The robot should have barcode scanning capabilities, mapping and navigation systems for efficient movement in a warehouse, and the ability to integrate with the warehouse's inventory management software. It should also be designed for safe interaction with human workers.

Warehouse Inventory Management Robot

Example 5: Problem Statement: Underwater Exploration Robot

Objectives: Create a robot for underwater exploration, capable of conducting research and collecting samples in deep-sea environments.

Context: The robot will operate at significant depths, withstand high pressures, and capture images and data from the ocean floor.

Constraints: The robot's design should prioritize durability, waterproofing, and adherence to a strict weight limit for submersible vehicles.

Requirements: The robot must be equipped with high-resolution cameras, sonar systems, and sample collection tools. It should also have a robust pressure-resistant enclosure, efficient propulsion systems, and the ability to transmit data to a surface vessel or base station.

Underwater Exploration Robot

Example 6: Problem Statement: Security Patrol Robot for Large Facilities

Objectives: Develop a robot for security patrolling in large facilities like airports or warehouses. **Context:** The robot should autonomously patrol designated areas, detect intrusions, and provide real-time surveillance.

Constraints: The robot must be cost-effective, compact enough to navigate narrow corridors, and integrate with existing security systems.

Requirements: The robot should have advanced surveillance cameras, motion and intrusion detection sensors, and the ability to autonomously patrol predefined routes. It should also integrate with the facility's security network for real-time monitoring and alerting.

Security Patrol Robot for Large Facilities

Example 7: Problem Statement: Robot for Hazardous Waste Cleanup

Objectives: Design a robot capable of safely removing and disposing of hazardous materials in contaminated environments.

Context: The robot will operate in areas with chemical spills or radioactive materials, minimizing human exposure.

Constraints: The robot must be highly resistant to hazardous substances, adhere to strict safety standards, and be within a reasonable budget.

Requirements: The robot should have specialized tools for handling and disposing of hazardous materials, a sealed and ruggedized construction to prevent contamination, and the ability to remotely operate in hazardous zones while providing real-time feedback and data transmission to human operators.

Robot for Hazardous Waste Cleanup

Conceptualization

The second step in designing a service robot is to conceptualize it. This involves defining its purpose, brainstorming ideas, and creating initial concepts.

Define the Purpose

The first step is to identify the specific tasks or functions the service robot is intended to perform. What problem is it aiming to solve? For example, a service robot could be used to deliver food in a restaurant, provide companionship to the elderly, or assist with surgery in a hospital.Once the purpose of the robot is defined, it is important to develop a good understanding of the environment in which it will operate. What are the challenges and constraints that the robot will face? For example, a service robot that is used in a restaurant will need to be able to navigate around people and obstacles, and it will need to be able to interact with people in a friendly and helpful way.

Brainstorm Ideas

Once the purpose of the robot is defined, it is helpful to brainstorm ideas for its design. This can be done by gathering a team of engineers, designers, and domain experts. The goal is to generate as many ideas as possible, without judgment.

Some things to consider when brainstorming ideas include:

- What form should the robot take? (e.g., humanoid, mobile platform, etc.)
- How should the robot interact with its environment? (e.g., sensors, actuators, etc.)
- What materials should be used to build the robot?
- How will the robot be powered?
- How will the robot be programmed?

Create Initial Concepts

Once a number of ideas have been generated, the next step is to create initial concepts. This can be done by generating rough sketches or concept drawings. The goal is to visualize the robot's appearance and functionality.

The initial concepts should be shared with the team and other stakeholders for feedback. This feedback can be used to refine the concepts and develop a more detailed understanding of the robot's design.

The conceptualization phase is an important step in the development of a service robot. By carefully defining the purpose of the robot, brainstorming ideas, and creating initial concepts, the team can develop a solid foundation for the robot's design.

Some Examples Of Conceptualization For Designing A Robot:

Example-1: Design a Service Robot for Elderly Care

Purpose: The robot's purpose is to provide comprehensive support to elderly individuals. It will assist with medication management by dispensing medications at the appropriate times. It will also monitor vital signs such as heart rate and blood pressure, providing real-time data to caregivers or medical professionals. Additionally, the robot will offer companionship and engage in conversation to reduce feelings of loneliness and isolation among the elderly.

Brainstorm Ideas:

Idea 1: Implement machine learning algorithms to enable the robot to learn individual preferences and adapt its companionship and support based on the unique needs of each elderly person.

Idea 2: Incorporate a user-friendly mobile app that allows family members and caregivers to remotely check on the elderly individual and communicate with the robot.

Initial Concepts: The initial concept for this robot involves a compact, humanoid design with a gentle and approachable appearance. It will have a touchscreen interface for easy interaction and medication dispensing. Sensors will be integrated to monitor vital signs, and it will be programmed with natural language processing capabilities for conversation with the elderly.

Example-2: Agricultural Robot for Crop Monitoring

Purpose: The robot is designed to address the needs of modern agriculture by providing detailed crop monitoring and data collection services. It will regularly survey crops, assess their health, detect signs of disease or pest infestations, and collect data on soil conditions. This information will help farmers make data-driven decisions to optimize crop yields and resource allocation.

Brainstorm Ideas:

Idea 1: Implement AI-driven predictive analytics to provide farmers with insights on optimal planting times, irrigation schedules, and pesticide application.

Idea 2: Enable the robot to communicate with drones for aerial surveys to complement ground-based monitoring.

Initial Concepts: Initial concepts for this robot involve a rugged, wheeled or tracked design for stability and maneuverability in various field conditions. It will be equipped with cameras, multispectral sensors, and soil analysis tools. The robot will navigate autonomously or semi- autonomously through fields, transmitting data to a central farm management system.

Example-3: Warehouse Inventory Management Robot

Warehouse Inventory Management Robot

(Image Source: https://apexstoragesystems.co.uk/)

Purpose: The robot's primary purpose is to streamline inventory management in warehouse facilities. It will automate the process of tracking and locating products, reducing the likelihood of errors and improving efficiency. By regularly scanning barcodes and updating inventory databases in real-time, it will help warehouse staff easily manage stock levels and fulfill orders accurately.

Brainstorm Ideas:

Idea 1: Develop a system for the robot to autonomously restock inventory by coordinating with robotic forklifts or conveyor belts.

Idea 2: Use machine learning to predict demand and suggest inventory reordering based on historical data and current trends.

Initial Concepts: Initial concepts for this robot feature a compact and modular design for easy integration into existing warehouse layouts. It will be equipped with a barcode scanner and RFID reader for identifying products. The robot will have a user-friendly interface accessible via a mobile device, allowing warehouse staff to assign tasks and monitor inventory levels.

Example-4: Underwater Exploration Robot

Purpose: The robot's primary purpose is to facilitate scientific research and exploration in underwater environments. It will be capable of conducting deep-sea exploration missions, capturing high-resolution images, and collecting samples from the ocean floor. Researchers and scientists will use its data to study marine life, geology, and environmental conditions.

Brainstorm Ideas:

Idea 1: Equip the robot with modular sampling tools, such as a detachable remotely operated vehicle (ROV), to collect deep-sea samples with precision.

Idea 2: Implement swarm robotics concepts, allowing multiple underwater robots to collaborate on large-scale exploration missions.

Initial Concepts: Initial concepts for this robot envision a robust submersible design with advanced sensor packages, including sonar, cameras, and environmental sensors. It will be highly maneuverable with precise thrusters to navigate ocean depths. The robot will feature modular payload options, allowing researchers to customize it for specific missions.

Requirement Analysis

The third step in designing a service robot is requirement analysis. This involves defining the technical requirements for the robot, as well as understanding the needs and preferences of the end-users.

Define Technical Requirements

The technical requirements for a service robot will vary depending on its specific purpose. However, some common technical requirements include:

Size and weight: The robot should be sized and weighted appropriately for its intended environment and tasks.

Mobility: The robot should be able to move around its environment safely and efficiently. **Sensors:** The robot should be equipped with the necessary sensors to perceive its environment and interact with objects and people.

Actuators: The robot should have the necessary actuators to perform its intended tasks.

Computing power: The robot should have enough computing power to run its control software and perform its intended tasks.

User Needs Assessment

It is also important to understand the needs and preferences of the end-users of the service robot. This includes considering human-robot interaction aspects, such as the user interface and communication methods.

For example, a service robot that is used in a restaurant should have a user interface that is easy to use for both customers and staff. The robot should also be able to communicate with users in a way that is clear and concise. By carefully defining the technical requirements and understanding the needs and preferences of the end-users, the team can develop a better understanding of the robot's design requirements.

When conducting a requirement analysis for a service robot, consider these additional factors:
Safety: The robot should be designed with safety in mind. It should be able to operate safely around people and objects, and it should have safeguards in place to prevent accidents.
Reliability: The robot should be reliable and able to perform its tasks consistently. It should also be easy to maintain and repair.
Cost: The cost of developing and deploying the robot should be considered. The robot should be affordable to purchase and operate.

By carefully considering all of the requirements, the team can develop a service robot that meets the needs of both the users and the business.

Example-1: Design a Service Robot for Elderly Care Technical Requirements:

Size: Compact, should be able to move through doorways and narrow spaces in homes.
Weight: Lightweight for easy maneuverability and transport.
Mobility: Wheel-based or legged for smooth navigation on different surfaces.
Sensors: Proximity sensors for obstacle avoidance, vital sign monitoring sensors, and voice recognition.
Actuators: Robotic arms for assisting with tasks, like picking up objects.
Computing Power: Sufficient processing power for natural language understanding and vital sign analysis.

User Needs and Preferences:

- Easy-to-use interface for elderly users.
- Clear and audible communication methods, including speech and visual displays.
- Customizable features to accommodate individual preferences, such as medication schedules.

Example-2: Agricultural Robot for Crop Monitoring Technical Requirements:

Size: Variable sizes, depending on the application (e.g., small for greenhouses, larger for fields).
Weight: Lightweight to avoid soil compaction.
Mobility: Wheel or track-based for stability on uneven terrain. **Sensors**: Crop health sensors, soil condition sensors, GPS for mapping. **Actuators:** Precision spraying or harvesting mechanisms.
Computing Power: Data processing for real-time monitoring and decision-making.

User Needs and Preferences:

- Easy-to-use control interface for farmers.
- Remote monitoring and control capabilities.

- Data visualization tools for tracking crop health and conditions.

Example-3: Warehouse Inventory Management Robot Technical Requirements:

Size: Compact to navigate narrow warehouse aisles.
Weight: Lightweight to avoid damaging goods.
Mobility: Wheel-based with omnidirectional movement.
Sensors: Barcode scanners, RFID readers, inventory tracking cameras.
Actuators: Picking arms or conveyor belts for moving items.
Computing Power: Integration with warehouse management systems.

User Needs and Preferences:

- Integration with existing warehouse software.
- User-friendly interface for warehouse personnel.
- Efficient coordination with human workers to minimize disruptions.

Example-4: Underwater Exploration Robot Technical Requirements:

Size: The size of the underwater exploration vehicle will vary based on the depth of the mission and
specific requirements. Different depths may necessitate different sizes to accommodate the equipment, propulsion systems, and payload capacity required for the mission.

Weight: The vehicle's weight must be carefully managed to achieve neutral buoyancy or be adjustable for deep-sea operations. Maintaining neutral buoyancy is crucial for efficient navigation and maneuverability at different depths.

Mobility: The vehicle should be equipped with propellers or fins to facilitate precise underwater navigation. These mobility features are essential for maneuvering through challenging underwater environments and reaching targeted locations for data collection and sample retrieval.

Sensors: Incorporating a variety of sensors is vital for comprehensive data collection. These sensors may include sonar systems for mapping the underwater topography, cameras for visual data capture, and scientific instruments for collecting environmental data such as temperature, salinity, and pressure.

Actuators: Manipulator arms are indispensable for an underwater exploration vehicle, allowing it to collect samples, interact with the environment, and carry out tasks that require dexterity and precision. These manipulator arms are crucial for various scientific and research purposes.

Computing Power: High-performance computing capabilities are essential for real-time data processing, analysis, and decision-making during missions. The vehicle must be equipped with robust computing systems to handle the vast amount of data collected and perform sophisticated analyses.

User Needs and Preferences

Remote Control and Monitoring: Researchers require the ability to remotely control and monitor the vehicle during underwater operations. Remote control allows for real-time adjustments, ensuring that the vehicle responds to changing conditions or objectives as the mission progresses.

User-Friendly Interface:A user-friendly interface for mission planning and data visualization is crucial to enhance the efficiency and effectiveness of the mission. Researchers should be able to easily plan, execute, and analyze missions through an intuitive and accessible interface.

Reliable Communication Systems: Reliable communication systems are vital for maintaining a constant link between the vehicle and the researchers. This ensures seamless data transmission, control, and safety during deep-sea operations where communication can be challenging due to the underwater environment.

Detailed Design:

The fourth step in designing a service robot is detailed design. This involves creating detailed 3D computer- aided design (CAD) models of the robot, incorporating all components and subsystems. It also involves choosing appropriate materials for the robot's construction and selecting sensors, actuators, microcontrollers, and other hardware components that meet the technical requirements.

Detailed Design

3D Modeling : 3D modeling is a powerful tool that can be used to design service robots. It allows engineers and designers to visualize the robot in detail and to test its design before it is built. This can help to identify potential problems and to make necessary modifications early on.

3D Modeling

3d Modeling Enhances Service Robot Design:

Improved accuracy and precision: 3D modeling allows for very accurate and precise designs. This is important for service robots, which often need to perform complex and delicate tasks.

Reduced development time and cost: 3D modeling can help to reduce the development time and cost of service robots. This is because it allows engineers to test and validate their designs before they are built.

Improved communication and collaboration: 3D models can be easily shared with other engineers and designers, which can improve communication and collaboration. This is important for complex projects involving multiple stakeholders.

Utilizing 3d Modeling For Service Robot Design Applications:

Creating a prototype: 3D models can be used to create prototypes of service robots. This allows engineers to test the robot's design and functionality before it is built.

Simulating robot performance: 3D models can be used to simulate the performance of service robots in different environments. This can help engineers to identify potential problems and to make necessary modifications.

Generating technical drawings and manufacturing instructions: 3D models can be used to generate technical drawings and manufacturing instructions for service robots. This can help to ensure that the robot is built accurately and precisely.

Material Selection : Material selection is an important aspect of service robot design. The choice of materials will affect the robot's strength, weight, cost, durability, and aesthetics.

Factors to Consider When Choosing Materials for Service Robots:

Strength: Service robots need to be strong enough to withstand the demands of their intended environment and tasks. For example, a service robot that is used in a manufacturing environment will need to be made of materials that are resistant to impact and wear.

Weight: Service robots need to be lightweight enough to be mobile and efficient. However, they also need to be heavy enough to be stable and to have the necessary payload capacity. **Cost**: The cost of materials is an important consideration for any product, including service robots. Engineers need to select materials that are affordable and that meet the required performance specifications.

Durability: Service robots need to be durable and able to withstand repeated use and exposure to the environment. For example, a service robot that is used in a healthcare environment will need to be made of materials that are easy to clean and disinfect.

Aesthetics: Service robots should have a pleasing appearance that is appropriate for their intended environment. For example, a service robot that is used in a hotel lobby will need to have a more polished and professional appearance than a service robot that is used in a warehouse.

Some common materials that are used in service robot design include:

Metals: Metals such as aluminum, steel, and titanium are strong and durable, making them ideal for applications where high performance and reliability are required. However, metals can also be heavy and expensive.

Plastics: Plastics are lightweight and relatively inexpensive, making them a good choice for many service robot applications. However, plastics can be less durable than metals and can be susceptible to wear and tear.

Composites: Composites are materials that are made of two or more different materials. For example, a composite material could be made of carbon fiber and resin. Composites offer a good balance of strength, weight, and cost.

The best materials for a particular service robot will depend on its specific requirements. Engineers need to carefully consider all of the factors involved before selecting materials for their robot design.

Common Materials that are Used in Service Robot Design

Enhancing Service Robot Performance Through Material Selection:

Using Lightweight Materials To Improve Mobility And Efficiency: For example, a service robot that is used to deliver food in a restaurant could be made of lightweight materials such as aluminum and carbon fiber. This would make the robot easier to move around and more efficient in terms of energy consumption.

Using Durable Materials To Improve Reliability: For example, a service robot that is used in a manufacturing environment could be made of durable materials such as steel and titanium. This would make the robot more resistant to impact and wear, improving its reliability.

Using Easy-To-Clean Materials To Improve Hygiene: For example, a service robot that is used in a healthcare environment could be made of materials such as plastic and stainless steel. These materials are easy to clean and disinfect, which is important for preventing the spread of infection.

By carefully selecting the right materials, engineers can design service robots that are strong, lightweight, durable, and easy to clean. This can help to improve the performance and reliability of service robots, and make them more suitable for a wider range of applications.

Component Selection : Component selection is a critical aspect of service robot design. The choice of components will affect the robot's performance, reliability, cost, and power consumption.

Key Factors for Choosing Service Robot Components:

Performance: The components selected should be able to meet the robot's performance requirements. For example, a service robot that is used to perform delicate tasks will need to be equipped with high-precision sensors and actuators.

Reliability: The components selected should be reliable and able to withstand the demands of the robot's intended environment. For example, a service robot that is used in a hazardous environment will need to be equipped with components that are resistant to dust, moisture, and extreme temperatures.

Cost: The cost of components is an important consideration for any product, including service robots. Engineers need to select components that are affordable and that meet the required performance and reliability specifications.

Power consumption: Service robots often need to operate for long periods of time on a single battery charge. Therefore, it is important to select components that have low power consumption.

Some common components that are used in service robot design

include: **Sensors:** Sensors are used to collect data about the robot's environment and its own state. Common sensors used in service robots include cameras, LiDAR sensors, sonar sensors, and force sensors.

Actuators: Actuators are used to move the robot and its components. Common actuators used in service robots include electric motors, servos, and pneumatics.

Microcontrollers: Microcontrollers are used to control the robot's operation. They are responsible for processing sensor data, making decisions, and sending commands to the actuators.

Power systems: Power systems are used to provide power to the robot's components. Common power systems used in service robots include batteries and fuel cells.

The best components for a particular service robot will depend on its specific requirements. Engineers need to carefully consider all of the factors involved before selecting components for their robot design.

Component Selection Enhances Service Robot Performance:

Using high-performance sensors and actuators to improve accuracy and precision: For example, a service robot that is used to perform surgery could be equipped with high-precision cameras and robotic arms. This would allow the robot to perform delicate tasks with great accuracy and precision.

Using reliable components to improve safety and uptime: For example, a service robot that is used in a manufacturing environment could be equipped with components that are resistant to dust, moisture, and extreme temperatures. This would improve the safety and reliability of the robot.

Using low-power components to extend battery life: For example, a service robot that is used to deliver food in a restaurant could be equipped with low-power sensors, actuators, and microcontrollers. This would extend the robot's battery life and allow it to operate for longer periods of time on a single charge.

By carefully selecting the right components, engineers can design service robots that are performant, reliable, cost-effective, and power-efficient. This can help to improve the adoption and use of service robots in a wider range of applications.

Creating a Robot: Detailed Design Examples:

Example-1: Design a Service Robot for Elderly Care:

3D Modeling: Create a detailed 3D CAD model of the robot that incorporates components such as a camera for monitoring vital signs, a medication dispenser, and an interactive touchscreen for communication. Ensure the design is user-friendly with large buttons and clear text for elderly users.

Material Selection: Choose lightweight yet durable materials for the robot's construction, such as high-strength plastics and aluminum alloys. These materials should be cost-effective and easy to clean for hygiene purposes.

Component Selection: Select sensors like infrared sensors for obstacle detection and microcontrollers with wireless connectivity to enable remote monitoring by caregivers. Choose actuators for precise medication dispensing and safe movement within the home.

Example-2: Agricultural Robot for Crop Monitoring:

3D Modeling: Create a 3D CAD model of the robot that includes multiple sensors, a data collection system, and a mobility platform suitable for rough terrain.

Material Selection: Choose materials like weather-resistant plastics and aluminum for the robot's body. Ensure the wheels or tracks are made from materials that can handle outdoor conditions.

Component Selection: Select sensors such as cameras, soil moisture sensors, and GPS for data collection. Use a robust microcontroller or an onboard computer for data processing and control. Employ rugged actuators for mobility, such as high-torque electric motors.

Example-3: Warehouse Inventory Management Robot:

3D Modeling: Create a 3D CAD model that accommodates the robot's barcode scanning system, storage compartments, and a navigation system for efficient warehouse navigation. **Material Selection:** Opt for sturdy materials like steel for the robot's frame and shelves, ensuring it can handle the rigors of a warehouse environment.

Component Selection: Select barcode scanners, RFID readers, and microcontrollers with wireless communication capabilities for inventory tracking. Use precision actuators for the robotic arm responsible for picking and storing items.

Example-4: Underwater Exploration Robot:

3D Modeling: Create a 3D CAD model that incorporates high-resolution cameras, sample collection arms, and a propulsion system suitable for deep-sea exploration.

Material Selection: Choose materials like titanium and specialized corrosion-resistant coatings to withstand the corrosive effects of seawater at depth.

Component Selection: Select high-pressure sensors, underwater cameras, and advanced microcontrollers for data processing and control. Propulsion systems can include thrusters or propellers designed for underwater use.

Once the detailed design is complete, the team can begin to prototype and build the robot.

Software Design:

The fifth step in designing a service robot is software design. This involves developing control algorithms that enable the robot to perform its intended tasks, navigate, and interact with its environment. It also involves designing an intuitive user interface (UI) that allows users to interact with the robot effectively.

Service Robot is Software Design

Control Algorithms

Control algorithms are the software that controls the robot's behavior. They are responsible for processing sensor data, making decisions, and sending commands to the actuators. Control algorithms can be complex, especially for robots that are designed to perform complex tasks.

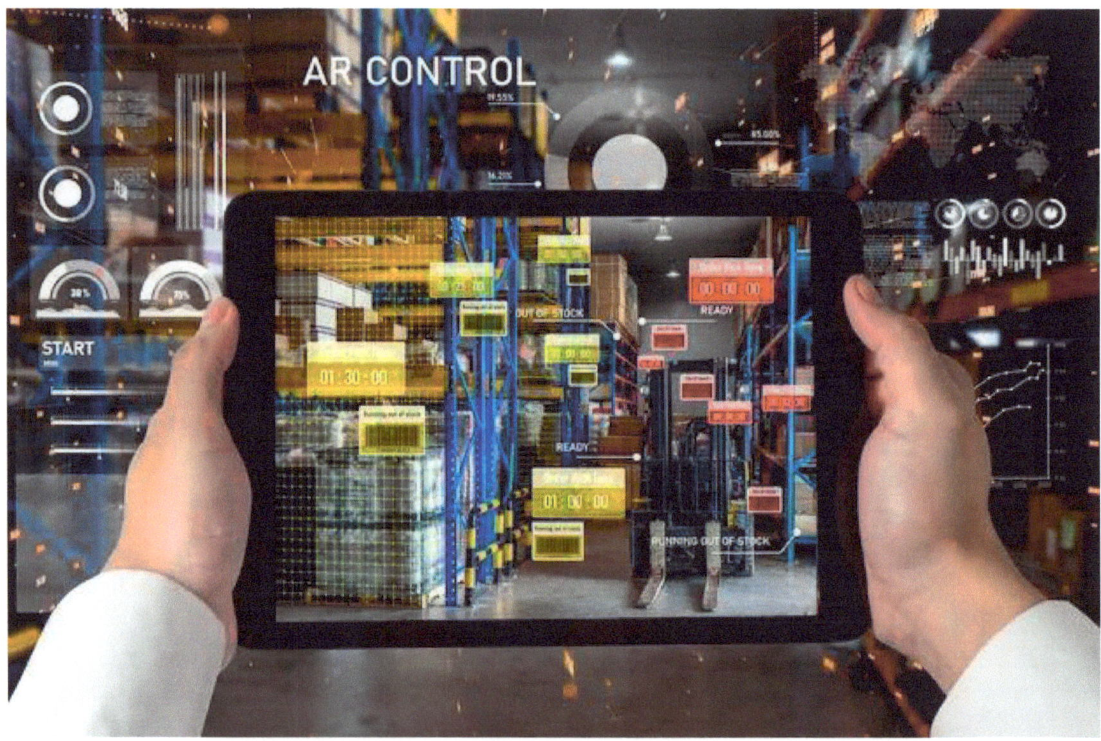

Control Algorithms

Navigation algorithms: Navigation algorithms help the robot to move around its environment safely and efficiently.

Task planning and execution algorithms: Task planning and execution algorithms help the robot to plan and execute its intended tasks.

Perception algorithms: Perception algorithms help the robot to understand its environment and its own state.

Control algorithms for specific tasks: For example, a service robot that is used to deliver food in a restaurant could have control algorithms for tasks such as navigating around people and obstacles, picking up and delivering food, and interacting with customers.

User Interface Design

The user interface (UI) is the way that users interact with the service robot. It can be a physical interface, such as a touchscreen or a button pad, or it can be a virtual interface, such as a mobile app or a website.

The UI should be intuitive and easy to use. It should also be appropriate for the robot's intended environment and users. For example, a service robot that is used in a hospital should have a UI that is easy to use for people with limited mobility.

These are examples of UI elements suitable for service robots:

Touchscreens: Touchscreens are a popular choice for UIs because they are easy to use and understand.

Voice commands: Voice commands can be used to control the robot without having to touch it. This can be useful for people with limited mobility or for robots that are used in noisy environments.

Mobile apps: Mobile apps can be used to control the robot from a distance. This can be useful for robots that are used in large or complex environments.

By carefully designing the control algorithms and user interface, engineers can create service robots that are easy to use and that can perform their intended tasks effectively.

Robot Software Design Examples:

Example-1: Design a Service Robot for Elderly Care:

Control Algorithms: Develop control algorithms that allow the robot to navigate around the home, avoiding obstacles, and reach the elderly person in response to voice commands or touch inputs. Implement algorithms for medication reminders and vital sign monitoring.

User Interface Design: Create an intuitive UI with a touchscreen interface on the robot's body. Users can interact with the robot by tapping icons for medication reminders or accessing vital sign data. Voice recognition software should allow users to issue commands or have conversations with the robot.

Example-2: Agricultural Robot for Crop Monitoring:

Control Algorithms: Create control algorithms that enable the robot to autonomously navigate through fields, collect data from sensors (e.g., soil moisture sensors, cameras), and transmit data to a central server for analysis. Implement path planning algorithms for efficient coverage of agricultural areas.

User Interface Design: Develop a web-based or mobile app for farmers to remotely monitor the robot's activities and view real-time data on crop health and soil conditions. Ensure the interface is user-friendly and provides actionable insights.

Example-3: Warehouse Inventory Management Robot:

Control Algorithms: Design control algorithms for efficient warehouse navigation, shelf scanning, and item picking. Implement algorithms for barcode recognition and inventory database management. Develop path planning algorithms for the robot to navigate through the warehouse efficiently.

User Interface Design: Create a user-friendly web-based interface for warehouse managers to monitor inventory levels, track the robot's progress, and receive alerts for low stock or inventory discrepancies.

Example-4: Underwater Exploration Robot:

Control Algorithms: Develop control algorithms for underwater navigation, depth control, and obstacle avoidance. Implement algorithms for image and data capture in various underwater conditions. Design algorithms for autonomous sample collection.

User Interface Design: Create a specialized control interface for researchers and operators to command the robot remotely. Ensure the interface allows for real-time streaming of underwater imagery and data.

Prototyping:

The sixth step in designing a service robot is prototyping. This involves constructing physical prototypes or mock-ups of the robot's design to test its functionality and appearance. The prototypes can be used to test the robot's mobility, navigation, task performance, and user interface.

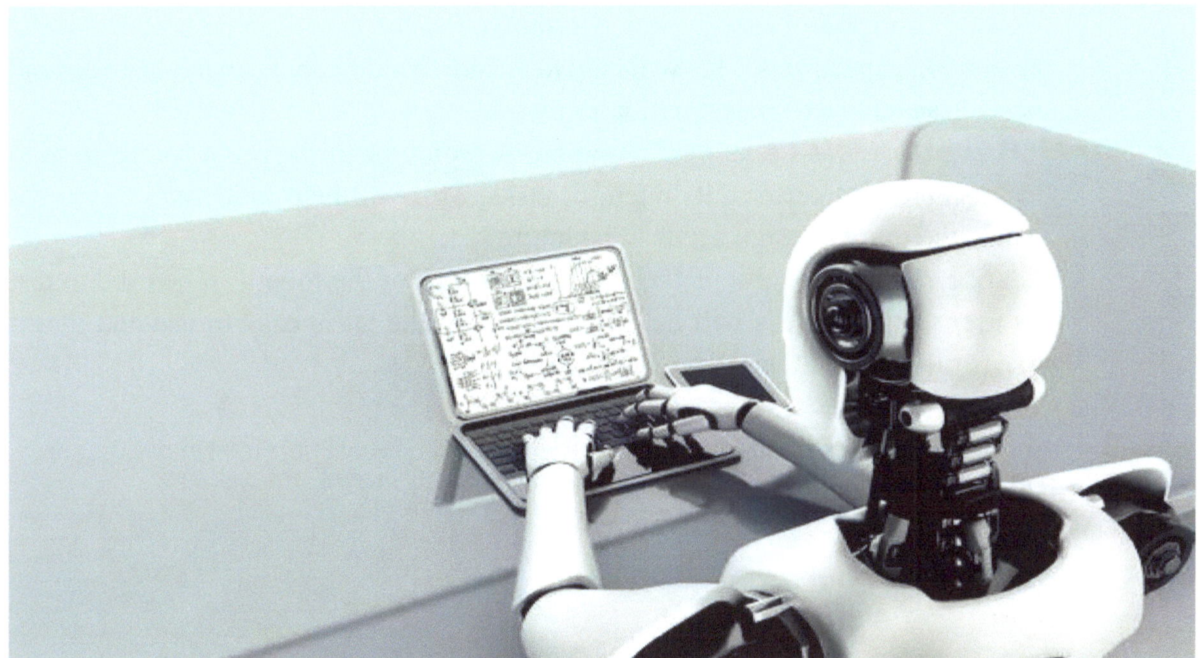

Prototyping

Build Prototypes

There are a variety of ways to build prototypes of service robots. One common approach is to use 3D printing. 3D printing allows engineers to quickly and easily create prototypes of complex shapes and designs. Another common approach is to use off-the-shelf components, such as motors, sensors, and microcontrollers.

Iterative Testing

Once the prototypes have been built, they should be tested rigorously in controlled environments. This will help to identify any design flaws and make necessary adjustments. The testing process should be iterative, with the prototypes being refined and improved over time.

Prototyping is an important step in the service robot design process. It allows engineers to test their designs before they are built, which can save time and money in the long run. Prototyping also helps to ensure that the robot meets the needs of its intended users.

Enhancing Service Robot Design Through Prototyping:

Testing the robot's mobility: A prototype of the robot can be used to test its mobility in different environments. This will help to identify any issues with the robot's design, such as wheels that are too small or motors that are not powerful enough.

Testing the robot's navigation: A prototype of the robot can be used to test its navigation capabilities. This will help to ensure that the robot can navigate around obstacles and reach its destination safely.

Testing the robot's task performance: A prototype of the robot can be used to test its ability to perform its intended tasks. This will help to identify any areas where the robot's design needs to be improved.

Testing the robot's user interface: A prototype of the robot can be used to test its user interface. This will help to ensure that the UI is easy to use and that it meets the needs of the robot's intended users.

By carefully prototyping and testing their designs, engineers can create service robots that are functional, reliable, and easy to use.

Robot Design Prototyping Examples:

Example-1: Design a Service Robot for Elderly Care:

Build Prototypes: Construct physical prototypes of the service robot, starting with a basic frame and adding components like the touchscreen interface, medication dispenser, and vital sign sensors.

Iterative Testing: Test the prototypes in controlled environments simulating elderly homes. Evaluate how well the robot assists with medication reminders and vital sign monitoring. Gather feedback from elderly users to refine the design.

Design a Service Robot for Elderly Care

Example-2: Agricultural Robot for Crop Monitoring:

Build Prototypes: Develop physical prototypes of the agricultural robot, starting with the mobility platform and sensors. Ensure it can navigate fields and collect data.

Iterative Testing: Test the prototypes in real agricultural environments, monitoring their performance in crop monitoring, pest detection, and data collection. Make adjustments to improve accuracy and durability.

Example-3: Warehouse Inventory Management Robot:

Build Prototypes: Construct physical prototypes of the warehouse robot with a focus on the barcode scanning system and mobility. Ensure it can efficiently navigate the warehouse.

Iterative Testing: Test the prototypes in warehouse environments to assess their ability to scan barcodes, update inventory, and navigate through aisles. Adjust the design to optimize performance.

Example-4: Underwater Exploration Robot:

Build Prototypes: Create physical prototypes of the underwater exploration robot, starting with a sealed and pressure-resistant design. Integrate cameras, sample collection arms, and propulsion systems.

Iterative Testing: Conduct tests in controlled underwater environments or simulation tanks to evaluate the robot's ability to explore, capture images, and collect samples. Make necessary modifications for improved functionality.

3.2. Simulation And Testing

Role of Simulation in Service Robotics

Simulation plays a critical role in the development and advancement of service robotics. It involves creating a virtual representation of the real world to simulate the behavior and interactions of robotic systems. The primary objectives of using simulation in service robotics are to model robotic environments, test robot functionalities, validate algorithms, and optimize performance before real-world deployment.

Simulation allows roboticists and engineers to experiment with different scenarios, algorithms, and designs in a controlled virtual environment. It helps in predicting how the robot will behave in various situations, which is vital for refining the robot's behavior, identifying potential issues, and ensuring robustness before physical implementation.

Role of Simulation in Service Robotics

Simulation Basics

a. Software-Based Simulation

Service robotics simulations are predominantly software-based, utilizing specialized platforms and tools to create a digital representation of the robot and its environment.

These simulations use physics engines to model the robot's movements, interactions with the environment, and the effects of various forces and sensors.

b. Visual Representation

Simulations provide a visual representation of the robot, allowing designers and developers to observe and analyze its movements, interactions, and overall performance in different scenarios.

C. Data Generation and Analysis

Simulations generate valuable data that can be analyzed to optimize the robot's behavior and improve its capabilities.

The data collected from simulations helps in making informed design decisions, refining control algorithms, and enhancing the overall efficiency and effectiveness of the robot.

Simulating Robotic Environments

a. Environment Modeling

Simulation involves creating a digital model of the physical environment where the robot will operate, including structures, obstacles, terrain, lighting conditions, and more.
This modeling helps in understanding how the robot will navigate, perceive, and interact with its surroundings.

b. Dynamic Elements

Simulations can incorporate dynamic elements, such as moving objects, changing weather conditions, or human interaction, to test the robot's adaptability and response to real-world unpredictability.

c. Sensor Simulation

Simulating the robot's sensors, such as cameras, LiDAR, and ultrasonic sensors, helps in understanding how the robot perceives its environment and processes sensory data for decision-making.

Advantages of Simulation:

a. Cost-Efficiency

Simulations significantly reduce the costs associated with physical prototypes and real-world testing, saving resources and time during the development process.

b. Rapid Iteration and Prototyping

Simulation enables quick iteration and prototyping, allowing developers to experiment with different designs and algorithms without the need for building and modifying physical robots.

c. Risk Mitigation

Simulating various scenarios helps identify potential risks and challenges early in the development cycle, enabling preemptive adjustments to design and algorithms to mitigate those risks.

d. Safe Testing

Simulation provides a safe environment for testing and validating algorithms and functionalities, ensuring that the robot operates reliably and safely before actual deployment.

Advantages of Simulation

e. Optimized Performance

By fine-tuning the robot's behavior and algorithms through simulation, the performance of the service robot can be optimized for different tasks and environments.In conclusion, simulation is an indispensable tool in the field of service robotics, facilitating the efficient and effective design, development, and testing of robotic systems, ultimately leading to the deployment of reliable and functional service robots in real-world applications.

Simulation Tools for Service Robotics

Service robotics development involves complex systems, making it crucial to test and refine designs before deploying physical robots. Simulation tools provide a virtual environment to model and evaluate robot behavior, saving time and resources in the development process.

Overview of Simulation Software

Definition of Simulation Software: Simulation software for service robotics refers to computer programs that replicate real-world scenarios, allowing developers to test and refine robot designs and algorithms in a controlled virtual environment.

Importance of Simulation: Explain why simulation is vital in service robotics. It helps identify flaws, optimize performance, and reduce the risk associated with deploying robots in unpredictable real-world situations.

Key Features and Capabilities:

Environment Modeling: Describe how simulation tools enable the creation of virtual environments that mimic the physical spaces where service robots will operate. This includes modeling terrain, obstacles, and objects.

Robot Modeling: Explain how these tools allow developers to create detailed virtual representations of service robots, including their hardware and sensors.

Sensor Simulation: Discuss the capability of simulation tools to mimic the behavior of sensors like cameras, lidar, and ultrasonic sensors. Explain how this helps in testing perception algorithms.

Realistic Physics Simulation: Detail how these tools simulate real-world physics, enabling robots to interact with their environments realistically. This includes factors like gravity, friction, and collisions.

Scenario Design:Describe how users can design and customize scenarios for various testing purposes. For instance, testing navigation in a cluttered room or object manipulation tasks.

Data Logging and Analysis: Explain how simulation tools record and analyze data from virtual robot experiments, providing insights into performance and behavior.

Integration with Development: Discuss how simulation tools can be integrated into the broader development process, allowing for rapid prototyping and iterative design improvements.

Comparative Analysis of Simulation Tools:

List of Popular Simulation Tools: Provide a list of widely used simulation tools in the field of service robotics, such as Gazebo, Webots, or ROS-based simulations.

Pros and Cons of Each Tool: Offer a comparative analysis of these tools, highlighting their strengths and weaknesses. Consider factors like ease of use, compatibility with different robot platforms, and community support.

Customization and Extensibility: Discuss the extent to which each tool allows customization and extensibility. Some tools may have a strong ecosystem of plugins and extensions.

Scalability: Consider how well each tool handles scalability, especially when simulating large fleets of robots or complex multi-agent scenarios.

Realism vs. Computation Speed: Highlight the trade-off between realism and computation speed in simulation tools. Some tools prioritize realism but may require powerful hardware, while others focus on faster execution.
Cost and Licensing: Discuss the cost and licensing models associated with each simulation tool. Some may be open source, while others require licenses.
User Community and Support: Mention the size and activity of user communities for each tool, as community support can be valuable for troubleshooting and learning.

Testing Methods for Service Robotics:

In the field of service robotics, rigorous testing is fundamental to ensure the efficiency, safety, and reliability of robots. Testing involves evaluating various aspects of the robot's functionality, performance, and behavior, both in controlled simulated environments and real-world scenarios.

Importance of Testing

Testing in service robotics is critical as it ensures that the robot operates as intended, meeting predefined specifications and safety standards. It helps identify potential issues, defects, or malfunctions, allowing for necessary refinements and improvements. Ultimately, thorough testing enhances user confidence in the robot and its applications.

Types of Testing

Functional Testing:

Description: Evaluates if the robot performs its intended functions accurately and efficiently.
Example: Testing a delivery robot's ability to navigate through various terrains and deliver packages to specified destinations.

Performance Testing:

Description: Measures the speed, accuracy, and efficiency of the robot's movements and tasks.
Example: Analyzing how quickly a service robot can clean a room or respond to user commands.

Usability Testing:

Description: Assesses the ease of use and user-friendliness of the robot's interface and interactions.
Example: Allowing users to interact with a service robot and gathering feedback on their experience and suggestions for improvement.

Safety Testing:

Description: Ensures that the robot operates without causing harm to itself, users, or the environment.
Example: Testing emergency stop mechanisms to halt the robot's operation in case of an obstacle or unsafe situation.

Integration Testing:

Description: Validates the seamless integration of various hardware and software components.
Example: Verifying that the robot's sensors, actuators, and navigation system work harmoniously to avoid collisions.

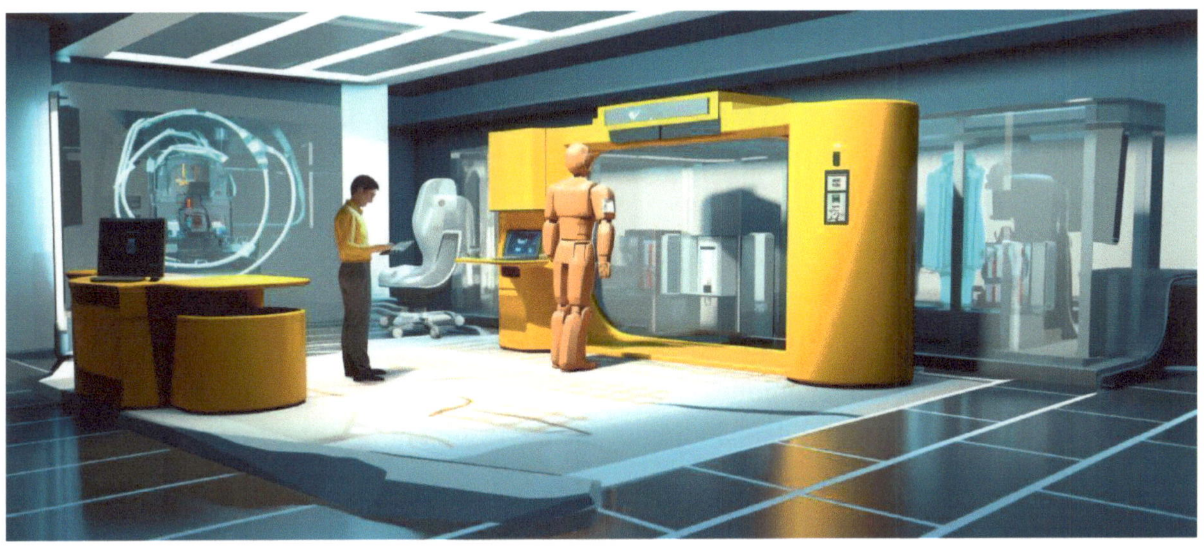

Types of Testing

Load and Stress Testing:

Description: Assesses how the robot performs under maximum load or stressful conditions. **Example**: Evaluating a warehouse robot's ability to handle a high volume of tasks during peak hours.

Service Robot Simulation and Testing:
Modeling the Robot's Behavior:

- The first step in simulation is to create a detailed model of the service robot's behavior. This includes defining how the robot moves, perceives its environment through sensors, and interacts with objects and users.

- Engineers use software tools to develop these models, which may involve mathematical equations, physics simulations, and computer-aided design (CAD) representations.

Virtual Environment Creation:

- A virtual environment is created to simulate the robot's surroundings. This environment can range from a simplified laboratory setup to a highly realistic representation of the real world.
- It includes 3D models of objects, obstacles, and other elements that the robot will encounter during its operation.

Algorithm Testing and Optimization:

- Engineers develop and test the robot's control algorithms in the simulated environment. This includes algorithms for navigation, object recognition, path planning, and decision-making.

- Simulation allows for iterative testing and refinement of these algorithms, making it possible to identify and address issues before physical prototypes are built.

Scenario-based Testing:

- Different scenarios are created to evaluate the robot's performance under various conditions. For example, scenarios may involve navigating through cluttered spaces, recognizing and avoiding obstacles, or interacting with users.
- These scenarios help assess the robot's ability to handle real-world challenges and provide valuable data for algorithm tuning.

Sensor Testing and Fusion:

- The robot's sensors, such as cameras, LiDAR, ultrasonic sensors, and microphones, are tested in the virtual environment to ensure they provide accurate data.
- Sensor fusion techniques, which combine data from multiple sensors to create a comprehensive understanding of the environment, are also tested and optimized.

Failure Mode Analysis:

- Engineers simulate potential failure modes and edge cases to evaluate the robot's robustness. This includes scenarios where sensors malfunction, motors fail, or communication is disrupted.
- Identifying vulnerabilities helps design fail-safes and contingency plans.

Performance Metrics Evaluation:

- Quantitative metrics are defined to measure the robot's performance. These metrics can include navigation accuracy, response time, energy efficiency, and task completion rates.
- Performance benchmarks are established to compare different versions of the robot and track improvements over time.

Human Interaction Testing:

If the robot will interact with humans, human-robot interaction (HRI) is tested in a simulated environment. This assesses the robot's ability to understand human commands, provide appropriate responses, and maintain a natural and safe interaction.

Iterative Refinement:

The simulation and testing phase is iterative. Engineers make adjustments to the robot's design, algorithms, and hardware based on the insights gained from simulation results. These iterations continue until the robot's performance meets the desired standards and safety requirements.

Documentation and Reporting:

Comprehensive documentation of simulation results, findings, and improvements is essential. This documentation guides the next phases of development, including manufacturing and real-world testing.

Real-world vs. Simulated Testing:

Testing service robots can be done in both real-world and simulated environments, each with its advantages and purposes.

Real-world Testing:

Description: Involves testing the robot in actual operating conditions that closely resemble its intended environment.

Advantages:

Provides real and accurate data on how the robot performs in practical scenarios.
Allows for testing unforeseen variables and challenges that may not be replicated in simulations.

Example: Testing a healthcare assistant robot in a hospital to assess its interaction with patients and medical staff.

Simulated Testing:

Description: Involves creating a virtual environment to test the robot's behavior and performance in a controlled, simulated setting.

Advantages:

Cost-effective and allows for rapid testing and iteration.
Enables testing in diverse scenarios that might be challenging to recreate in the real world.

Example: Simulating a smart home environment to test how a domestic service robot interacts with appliances and handles various household tasks.

By utilizing a combination of both real-world and simulated testing, engineers and developers can comprehensively evaluate a service robot's capabilities, ensuring it meets the required standards of functionality, safety, and efficiency before deployment in practical settings.

3.3. Manufacturing And Assembly:

Importance of Manufacturing and Assembly Processes

Manufacturing and assembly are critical stages in the production of service robotics. These processes transform conceptualized designs into tangible, functional robots ready for deployment. The efficiency and precision of manufacturing and assembly significantly impact the robot's performance, safety, and overall cost.

Prototyping Methods and Their Significance in Manufacturing

Prototyping allows for iterative testing and refining of designs before mass production. It helps identify design flaws, improve functionality, and optimize manufacturing processes. Rapid prototyping methods like 3D printing provide quick, cost-effective ways to create physical prototypes for evaluation and validation.

Material Selection and Sourcing

Materials for service robotics need to balance strength, durability, weight, and cost. Common materials include:

Metals (aluminum, steel): For structural components and frames due to their strength.

Plastics (polycarbonate, ABS): Used for casings and non-structural components due to their lightweight and versatile properties.

Composites: Combine the properties of both metals and plastics for specific applications.

Rubber and elastomers: For grips, wheels, and flexible components.

Procurement and Sourcing Strategies for Manufacturing

Strategic procurement involves identifying reliable suppliers, negotiating contracts, ensuring quality standards, and optimizing costs. Sourcing strategies include single or multiple sourcing, just-in-time (JIT) inventory management, and establishing long-term partnerships with suppliers.

Manufacturing Processes

Injection Molding and 3D Printing for Components

Injection Molding: Ideal for mass-producing components with high precision and consistency.
3D Printing (Additive Manufacturing): Valuable for prototyping and low-

volume production, allowing complex geometries and customization.

PCB Assembly and Integration

Printed Circuit Board (PCB) assembly involves soldering electronic components to the PCB, including sensors, actuators, and control systems. This step is crucial for the electronic functionality of the robot.

Integration of Sensors, Actuators, and Other Components

Integration involves assembling various components like sensors (e.g., cameras, proximity sensors), actuators (e.g., motors, servos), and control systems into a cohesive robotic system.

Quality Control and Testing
Quality Assurance During Manufacturing

Quality control ensures that components and products meet predefined standards. Techniques include visual inspection, functional testing, and statistical analysis to detect defects and inconsistencies.

Testing Methodologies to Ensure Performance and Safety

Testing verifies the robot's performance, safety features, and compliance with specifications. Functional testing, stress testing, and simulated real-world scenarios help ensure the robot functions as intended.

Assembly Line Setup and Optimization Designing Efficient Assembly Lines

Efficient assembly lines follow logical sequences, minimizing wasted time and effort. Lean manufacturing principles are often employed to streamline processes and eliminate non-value-added activities.

Automation in Assembly for Improved Efficiency

Automation, including robotic arms and automated guided vehicles (AGVs), enhances efficiency, speed, and precision in assembly. It reduces human error and labor costs while improving throughput and product consistency.

By incorporating these aspects into the manufacturing and assembly of service robotics, companies can produce reliable, efficient, and safe robots that meet the needs of various applications and industries.

3.4. Operation And Maintenance Of Service Robots:

After manufacturing and assembly, service robots are ready for operation. This phase involves setting up the robot's software, calibrating sensors, and ensuring all systems function correctly. Maintenance is an ongoing process. Regular check-ups, software updates, and component replacements are necessary to keep the robot in good working condition. Predictive maintenance techniques, utilizing data from sensors and usage patterns, can help prevent breakdowns.

Key Steps for Operating and Maintaining Service Robots

Initialization and Setup:

When a service robot is first deployed or powered on, it needs to go through an initialization process. This includes checking hardware components, sensors, and software systems to ensure everything is functioning correctly.

The robot's initial configuration, such as Wi-Fi connectivity and user settings, is established during this step.

Software Configuration:

Service robots rely on software for various functions, including navigation, object recognition, and human-robot interaction. Regular software updates and patches are essential to improve performance, fix bugs, and enhance security. These updates may be downloaded and installed automatically or manually.

Sensor Calibration:

Sensors play a vital role in a service robot's ability to perceive its environment. Calibration ensures that sensors provide accurate data.Calibration may involve adjusting the parameters of cameras, lidar, ultrasonic sensors, or other perception devices.

Safety Checks:

Ensuring the safety of both the robot and its users is paramount. Periodic safety checks are conducted to verify that the robot's safety features, such as collision avoidance and emergency stop mechanisms, are operational.In some cases, robots may undergo safety certifications to comply with industry standards.

Battery Management:

Many service robots are battery-powered. Monitoring and managing the robot's battery system is essential to prevent unexpected shutdowns. This includes regular charging, optimizing power consumption, and replacing batteries when they no longer hold a charge effectively.

Data Management:

Service robots generate and collect various data, including sensor data, operational logs, and user interactions. Data storage and management are crucial for performance analysis, debugging, and machine learning-based improvements.

Routine Maintenance:

Routine maintenance tasks may include cleaning the robot's exterior and sensors to ensure they function correctly. Additionally, moving parts, such as wheels or joints, might require lubrication or replacement over time.

Diagnostics and Troubleshooting:

When the robot encounters issues or malfunctions, a diagnostic process is initiated. This involves running tests and analyzing data to identify the root cause of the problem. Troubleshooting may lead to software adjustments or hardware repairs.

User Training and Support:

Users and operators of service robots may require training to understand how to interact with and control the robot effectively. A support system should be in place to assist users with any questions or issues that may arise during operation.

Predictive Maintenance:

Advanced service robots can benefit from predictive maintenance. Data collected from sensors and usage patterns can be analyzed to predict when components might fail or require replacement.This proactive approach helps prevent unexpected downtime and reduces maintenance costs.

Documentation and Record Keeping:

Keeping detailed records of maintenance activities, software updates, and any modifications made to the robot is essential for traceability and compliance.

Regulatory Compliance:

Depending on the application and location, service robots may need to adhere to specific regulations and standards. Compliance with safety and privacy regulations is critical.

Operating Environments
Indoor vs. Outdoor Environments:

Service robots are designed to operate in specific environments. Indoor robots are typically tailored for structured, controlled settings, while outdoor robots require robustness to handle varied terrains and weather conditions.

Operating Environments

Safety Regulations and Compliance:

Adhering to safety regulations is paramount to ensure safe operation in various environments. Compliance with industry standards and governmental guidelines ensures the robot's design and operation meet specified safety criteria.

Operation Procedures

Start-up and Shutdown Processes:

Defined procedures for starting up and shutting down the robot safely are crucial to prevent damage and ensure optimal performance. This involves initializing systems, checking sensors, and securing the robot after use.

Navigation and Path Planning:

Efficient navigation is achieved through algorithms that calculate optimal paths for the robot to reach destinations, avoiding obstacles and following specified trajectories.

Task Execution and Interaction with Humans:

Service robots must be programmed to execute tasks efficiently and interact with humans in a friendly and intuitive manner. This includes understanding human commands, providing necessary information, and completing tasks accurately.

Maintenance and Servicing

Preventive Maintenance: Regular, scheduled maintenance is essential to prevent failures and prolong the robot's operational life. This involves inspecting, cleaning, and replacing components as needed.

Troubleshooting and Diagnostics: In the event of malfunctions or issues, systematic troubleshooting and diagnostics are performed to identify the problem and implement solutions, minimizing downtime.

Upgrades and Software Updates: Continuous improvements are achieved through regular updates to both hardware and software. Upgrades enhance performance, add new features, and address security vulnerabilities.

Safety and Risk Management

Risk Assessment and Mitigation: Identifying potential risks and implementing mitigation strategies is crucial to ensure the safety of the robot, its operators, and the environment it operates in.
Emergency Procedures: Established emergency protocols guide actions in unexpected or hazardous situations, prioritizing safety and minimizing harm to humans and the robot.
Human-Robot Interaction Safety: Ensuring that the robot is designed to interact with humans in a safe manner, considering factors like speed, force, and proximity to avoid accidents or injuries.

Data Management and Analysis

Data Collection and Storage: Managing the vast amount of data collected by the robot, storing it securely, and making it accessible for analysis and decision-making purposes.
Data Analysis for Performance Improvement: Analyzing collected data helps identify patterns, inefficiencies, and areas for improvement in the robot's performance and behavior.
Privacy and Ethical Considerations: Addressing ethical concerns related to data privacy and ensuring that collected data is used responsibly and in accordance with privacy regulations.

Remote Monitoring and Control

Remote Operation Capabilities: Enabling operators to monitor and control the robot remotely, enhancing operational flexibility and efficiency.
Telemetry and Feedback Systems: Implementing telemetry systems to provide real-time feedback on the robot's performance, aiding operators in making informed decisions and adjustments.

This elaboration provides a comprehensive understanding of the operational and maintenance aspects involved in the design, development, and utilization of service robots.

3.5. Deployment Strategies

Deploying service robots effectively is crucial. The choice of deployment strategy depends on the robot's intended application

Individual Ownership: Some service robots are designed for personal use, like robot vacuum cleaners or personal assistants.

: Service robots can be deployed in businesses, hospitals, hotels, and retail environments to perform specific tasks, such as customer service or cleaning.

Public Spaces: Some service robots are deployed in public areas like airports or train stations to provide information or assistance.

Industrial Deployment: In industrial settings, robots are used for tasks like manufacturing, warehouse automation, and inspection.

Factors like cost, safety, regulations, and user acceptance influence deployment decisions.Service robots must be integrated into the existing infrastructure and workflows, and user training and support may be required.

Deployment strategies for service robots are crucial for determining how and where these robots will be used to achieve their intended goals. Here are the key steps involved in developing and executing a deployment strategy:

Deployment strategies

Needs Assessment: Before deploying a service robot, a thorough understanding of the specific needs and requirements of the target users or environment is essential. This involves:

- Conducting surveys, interviews, or consultations to gather data on what tasks the robot should perform.
- Identifying pain points and areas where automation or assistance from a robot would be beneficial.
- Analyzing user preferences, limitations, and expectations to tailor the robot's functionalities accordingly.

Target Environment Analysis: Understanding the environment where the service robot will operate is crucial for effective deployment. This entails:

- Evaluating the physical layout, obstacles, and conditions in the intended environment.
- Assessing lighting, temperature, noise levels, and other environmental factors that might affect the robot's performance.
- Identifying potential hazards or challenges that the robot may encounter and planning for their mitigation.

Integration Planning: Integrating the service robot seamlessly into the existing ecosystem involves careful planning:

- Ensuring compatibility and integration with other technologies and systems present in the environment.
- Developing interfaces and communication protocols to enable interaction with the robot.
- Planning for data exchange and integration with backend systems for efficient operations.

Training and User Acceptance: Preparing users for interaction with the service robot is essential for a successful deployment:

- Designing comprehensive training programs to educate users about the robot's capabilities, limitations, and proper usage.
- Addressing any concerns or skepticism users might have by fostering a positive perception of the robot's value and benefits.
- Encouraging active user participation and gathering feedback during the training phase to make necessary adjustments.

Pilot Deployment: Before a full-scale rollout, a pilot deployment is usually conducted to assess the robot's real-world performance and make necessary refinements:

- Selecting a representative sample of users or locations to deploy and evaluate the robot's functionality.
- Monitoring the robot's performance, gathering user feedback, and identifying areas for improvement.
- Using the data and insights gained during the pilot phase to optimize the robot's design, software, and operational procedures.

Scalable Deployment: Once the pilot phase is successful, scaling the deployment involves a more extensive implementation:

- Expanding the deployment to cover a broader user base or a larger geographical area.
- Ensuring the infrastructure can support the increased load and adapting the system for higher performance and reliability.
- Implementing a strategy for managing increased maintenance needs and addressing any unforeseen challenges that arise with scale.

Adaptation and Expansion: Continuous adaptation and expansion are vital to ensure the service robot remains effective and relevant over time:

- Regularly updating the robot's software and hardware to enhance its capabilities and address emerging needs.
- Analyzing user feedback and usage data to identify areas for improvement and expansion of functionalities.
- Strategically planning for future upgrades, new features, or integration with evolving technologies to keep the robot competitive and beneficial.
- These deployment strategies collectively ensure a successful integration of service robots into various environments, optimizing their performance, user acceptance, and overall effectiveness.

3.6. Ethical Considerations in Service Robotics

Service robotics, which encompasses a wide range of autonomous and semi-autonomous machines designed to perform tasks that benefit humans, has seen significant advancements in recent years. These robots are increasingly being integrated into various sectors, including healthcare, manufacturing, hospitality, and domestic settings, to improve efficiency, safety, and overall quality of life. However, with the proliferation of service robots, there come ethical considerations that demand careful examination.

Ethical Considerations in Service Robotics

Ethical Concerns in Service Robotics

Privacy and Data Security

One of the foremost ethical concerns in service robotics is the protection of privacy and data security. Many service robots are equipped with sensors, cameras, and microphones to interact with and assist humans. While these features are essential for their functionality, they also raise concerns about potential surveillance, data breaches, and unauthorized access to personal information. Balancing the benefits of service robots with individuals' right to privacy is a critical challenge.

Autonomy and Accountability

Service robots are designed to operate autonomously to a certain extent, making decisions and taking actions without direct human intervention. This autonomy raises questions about accountability when things go wrong. Who is responsible when a service robot makes a mistake or causes harm? Determining liability and establishing clear lines of responsibility is a complex issue that requires careful legal and ethical consideration.

Ethical Concerns in Service Robotics

Impact on Employment

The widespread adoption of service robots also has implications for the job market. While these robots can enhance productivity and efficiency, they may lead to job displacement in certain industries. Ethical considerations arise concerning the potential loss of livelihoods and the responsibility of society and policymakers to address this issue by providing opportunities for retraining and upskilling.

Human-Robot Relationships

As service robots become more integrated into our daily lives, they have the potential to form emotional bonds with humans. This raises questions about the ethical implications of these relationships. Should robots be designed to mimic human emotions, and to what extent should humans rely on robots for companionship or emotional support? The development of guidelines for the ethical design and use of emotionally engaging robots is an ongoing challenge.

Bias and Fairness

Service robots often rely on artificial intelligence algorithms for decision-making, which can inadvertently perpetuate bias and discrimination. For example, in healthcare settings, robots may provide different levels of care to individuals based on their demographic characteristics. Ensuring that service robots are programmed and trained to be fair and unbiased is a critical ethical concern.

Ethical Frameworks for Service Robotics

The integration of service robots into our daily lives brings about numerous opportunities and challenges, and addressing the ethical considerations associated with this technology is paramount. To navigate this complex landscape, the development of robust ethical frameworks is essential. These frameworks should serve as guiding principles that inform the design, deployment, and regulation of service robots.

Interdisciplinary collaboration is a key ingredient in crafting these frameworks. Roboticists, ethicists, policymakers, and various stakeholders must come together to contribute their expertise and perspectives. This collaboration ensures that the ethical frameworks are not only technically feasible but also morally sound. It helps strike a delicate balance between technological advancement and ethical responsibility.

Ethical Frameworks for Service Robotics

The promise of service robotics is significant, as these machines can enhance our lives and tackle pressing societal issues. However, the ethical considerations underscore the need for a careful and comprehensive approach. Ethical principles must be at the forefront of development to ensure that service robots benefit society in ways that align with our values and respect individual rights.

In conclusion, as we continue to embrace service robots, we must prioritize ethical principles to guide their evolution. This approach will not only harness the potential benefits of this technology but also safeguard against potential pitfalls, ensuring that service robotics align with the values and ethical standards of our society.

Chapter 4: Fundamental Concepts In Service Robotics

4.1. Introduction to Fundamental Concepts Overview Of Robotics In Service Applications

The rapid evolution of robotics has spurred a transformative revolution in recent years, transcending its traditional role in industrial settings to encompass a diverse range of applications within the service sector. This comprehensive overview delves deeply into the multifaceted role of robotics in service applications, shedding light on its profound significance, remarkable versatility, and far-reaching impact on modern society. Explore the innovative ways in which robotics is shaping our world and driving advancements across various domains, fundamentally changing the way we interact with technology and navigate the complexities of contemporary living.

The Versatility Of Service Robots:

An exceptional trait of service robotics lies in its remarkable versatility. These robots showcase a remarkable ability to seamlessly adapt to a wide array of environments and carry out a vast spectrum of tasks. Their adaptability knows no bounds, and they are an indispensable presence across various industries. From enhancing healthcare services to transforming the hospitality sector, revolutionizing agriculture, optimizing transportation and logistics, aiding in education, and offering valuable assistance in household chores, service robots are true multi-taskers, showcasing their flexibility and importance across different domains.

Human-Robot Interaction (HRI):

Human-Robot Interaction (HRI) stands as a fundamental and integral aspect of service robotics. It encompasses the ways in which humans and robots interact, communicate, and collaborate within diverse environments. Service robots are meticulously designed to comprehend and respond to human gestures, speech, and actions, effectively becoming valuable collaborators in a wide array of human-centric settings and scenarios. The seamless and intuitive interaction between humans and robots is key to maximizing the potential and benefits of service robotics, enabling a harmonious and productive coexistence for improved quality of life and efficiency.

Importance Of Fundamental Principles In Service Robotics

The importance of fundamental principles in service robotics cannot be overstated. These principles serve as the foundation upon which the development, design, and operation of service robots are built. Here are several key reasons highlighting their significance:

Safety

Fundamental principles underpin the safety of service robots. Ensuring that robots operate without causing harm to humans or themselves is paramount. Principles related to risk assessment, emergency stop mechanisms, and fail-safe designs are essential for safe operation in various service environments.

> **Risk Assessment:** Conduct a thorough risk assessment to identify and mitigate all potential hazards associated with the robot's design, operation, and environment.
>
> **Emergency Stop Mechanisms:** Implement emergency stop mechanisms that can be easily activated in the event of a hazard.
>
> **Fail-Safe Design:** Design the robot with fail-safe features to minimize the risk of injury or damage in the event of a failure.
>
> **Human-Robot Interaction Design:** Design the robot and its user interface in a way that minimizes the risk of human error and maximizes safety.
>
> **Sensor Fusion:** Use sensor fusion to integrate data from multiple sensors to create a comprehensive and accurate understanding of the robot's environment.
>
> **Machine Learning:** Use machine learning algorithms to train the robot to recognize and avoid hazards, and to adapt to changing condition

In addition to these general safety parameters, there are also specific safety considerations that may need to be taken into account depending on the type of service robot being built. For example, a service robot that is designed to operate in a healthcare environment may need to be designed with additional safety features to protect patients and staff.It is important to note that safety is a top priority when designing and building service robots. By carefully considering all of the relevant safety parameters, developers can help to ensure that their robots are safe for humans and other robots to interact with.

Efficiency

Principles of efficiency guide the design of service robots to perform tasks with maximum productivity. Efficient algorithms, motion planning, and control systems are critical for optimizing resource utilization and minimizing energy consumption. Notausschalter und Notfall-Stopp-Systeme, die es ermöglichen, den Roboter im Notfall sofort abzuschalten.

> **Algorithm Efficiency:** Utilize efficient algorithms for navigation, perception, and task execution to minimize computation time and resource usage.
>
> **Motion Planning:** Implement optimal motion planning algorithms to ensure robots take the most efficient paths while avoiding obstacles.
>
> **Control Systems:** Design control systems that are responsive and precise to minimize unnecessary movements and maximize accuracy in task execution.
>
> **Energy Efficiency:** Incorporate energy-efficient components and systems to prolong battery life and reduce the robot's overall energy consumption.
>
> **Sensors:** Choose sensors that provide accurate data while consuming minimal power. Optimize sensor data processing for efficiency.
>
> **Resource Utilization:** Efficiently allocate resources like computing power, memory, and communication bandwidth to various robot subsystems.

Efficiency is a critical aspect of service robot design, as it directly impacts productivity, operational costs, and the overall effectiveness of these robots in various applications.

Principles in Service Robotics

Reliability

Fundamental principles contribute to the reliability of service robots. Robots must operate consistently and predictably, especially in critical applications like healthcare and manufacturing. Principles of fault tolerance and redundancy help ensure continued operation even in the presence of hardware or software failures.

Fault Tolerance:
Robustness against hardware or software failures.
The ability to detect faults and respond appropriately, such as switching to backup systems or initiating safe shutdown procedures.

Redundancy:
Duplication of critical components or systems to provide backup in case of failure. Redundant sensors, actuators, or communication channels to maintain functionality.

System Diagnostics:
Continuous monitoring of robot components and subsystems for signs of malfunction. Real-time diagnostics to identify and isolate faults.

Predictive Maintenance:
Utilizing sensor data and predictive analytics to schedule maintenance proactively, reducing downtime and preventing unexpected failures.

Sensing and Perception Reliability:

Testing and Validation: Rigorous testing and validation procedures, including stress testing and failure mode analysis, to identify and address reliability issues during development.

These reliability parameters are critical for building service robots that can operate consistently, predictably, and safely in various real-world scenarios. Adhering to these principles helps ensure the robot's performance and maintain its functionality even in the presence of failures or adverse conditions.

Autonomy: Service robots often need to operate autonomously in dynamic and unstructured environments. Fundamental principles of perception, decision-making, and control enable robots to navigate, interact, and adapt to changing conditions without constant human intervention.

Perception: The robot must be able to perceive its environment using sensors such as cameras, lidar, and ultrasonic sensors. This includes being able to identify objects, people, and obstacles, as well as estimating the distance and orientation of objects.

Decision-Making: The robot must be able to make decisions about how to navigate its environment and interact with objects. This includes decisions about how to avoid obstacles, reach goals, and plan paths.

Control: The robot must be able to control its movement and actuators in order to execute its decisions. This includes controlling motors, grippers, and other actuators.

Adaptability

Service robots interact with diverse environments and users, and adaptability is crucial. Principles of machine learning, computer vision, and sensor fusion enable robots to learn and adapt to new tasks, recognize objects, and respond to various user needs.

Environment Interaction:

Ability to navigate and operate in different types of environments (e.g., indoor, outdoor, structured, unstructured).

User Interaction:

Capability to interact with diverse users, including understanding and responding to their commands and requests.

Task Learning:

Incorporation of machine learning algorithms to enable the robot to learn and perform new tasks.

Object Recognition:

Utilization of computer vision techniques to identify and recognize objects in the robot's surroundings.

User Needs Response:

Capacity to respond effectively to various user needs and preferences.

Sensor Fusion:

Integration of multiple sensors (e.g., cameras, LiDAR, ultrasonic sensors) to provide comprehensive environmental awareness.

Adaptive Control Systems:

Implementation of control systems that can dynamically adjust robot behavior based on changing circumstances.

Learning Algorithms:

Integration of algorithms that allow the robot to continuously learn and improve its performance over time.

Feedback Mechanisms:

Establishment of mechanisms to receive and process feedback from users and the environment for better adaptation.

Fault Tolerance:

Robustness in handling unexpected situations or malfunctions, ensuring the robot can recover gracefully.

Scalability:

Ability to adapt to different scales of operation, from small-scale tasks to larger and more complex operations.

User-Centered Design

Principles of user-centered design ensure that service robots are user-friendly and intuitive. Understanding human needs, preferences, and limitations is fundamental to creating robots that can effectively assist, support, or collaborate with humans.

User Feedback Loop:

Establish a mechanism for users to provide feedback and suggestions for continuous improvement.

Cultural Sensitivity:

Take into account cultural differences and norms that may affect user interactions with the robot.

Long-term User Engagement:

Design strategies to maintain user engagement and satisfaction over time.

Human-Robot Collaboration:

Ensure that the robot can work collaboratively with humans, understanding social cues and etiquette

Aesthetics and Design Appeal

Consider the robot's physical appearance and design to make it more appealing and relatable to users. These parameters collectively contribute to a user-centered design approach, ensuring that service robots are not only functional but also genuinely helpful and user-friendly.

Human Autonomy:

Respecting and upholding the autonomy and decision-making abilities of individuals when interacting with the robot.

Avoidance of Deception:

Ensuring that the robot does not engage in deceptive practices or provide false information.

Long-term Impacts:

Considering the potential societal, economic, and environmental impacts of widespread deployment of service robots.

Continuous Monitoring and Evaluation:

Implementing mechanisms for ongoing assessment of the robot's performance, ethics, and impact on society.

Human-Robot Interaction Ethics

Promoting positive and respectful interactions between humans and robots, taking into account cultural and social norms.

Stakeholder Engagement:

Involving relevant stakeholders, including users, experts, and affected communities, in the development and decision-making process.

Compliance with Legal and Regulatory Frameworks:

Adhering to applicable laws, regulations, and industry standards related to robotics, AI, and data protection.

These ethical considerations are essential for ensuring that service robots are developed and deployed in a manner that aligns with human values, rights, and societal well-being.

Regulatory Compliance

Many regions and industries have specific regulations and standards governing the deployment of robots, especially in healthcare and transportation. Fundamental principles play a crucial role in ensuring that robots comply with these regulations, fostering acceptance and trust.

Safety Standards: Ensure compliance with safety standards such as ISO 13482 (for personal care robots) and ISO 10218 (for industrial robots). Safety should be a top priority in robot design.

Data Privacy: Address data privacy regulations like GDPR (General Data Protection Regulation) to protect user information, especially in healthcare applications where sensitive data may be involved.

HIPAA Compliance: If the robot deals with healthcare data in the United States, comply with the Health Insurance Portability and Accountability Act (HIPAA) to safeguard patient information.

Accessibility Standards: Adhere to accessibility guidelines (e.g., WCAG) to ensure that the robot can be used by individuals with disabilities.

Ethical Considerations: Consider ethical guidelines for AI and robotics, such as those provided by organizations like the IEEE (Institute of Electrical and Electronics Engineers) and the ACM (Association for Computing Machinery).

Liability and Insurance: Address liability issues and ensure appropriate insurance coverage in case of accidents or malfunctions, especially in transportation where accidents can have serious consequences.

Certifications: Seek relevant certifications such as CE marking (for European markets) or FDA approval (for medical robots) to demonstrate compliance with industry-specific requirements.

Security Standards: Implement cybersecurity measures to protect against data breaches and hacking, especially in critical applications like autonomous vehicles.

Environmental Regulations: Comply with environmental regulations for the disposal of robot components, including batteries and electronics, to minimize the impact on the environment.

Human-Robot Interaction Guidelines: Follow guidelines for human-robot interaction to ensure that the robot's behavior is predictable, understandable, and safe for users.

Licensing and Permits: Ensure compliance with local licensing and permitting requirements for operating robots in public spaces, such as sidewalks or roads.

Noise Regulations: If the robot generates noise, comply with noise pollution regulations to minimize disturbance to the environment and people.

Sustainability

As service robots become more prevalent, principles of sustainability are essential. Designing robots with energy efficiency and recyclability in mind contributes to reducing their environmental impact and long- term viability.

Energy Efficiency:
Use of low-power components and systems.
Optimized power management to minimize energy consumption during operation.

Recyclability:
Designing robots with easily disassembled and separable parts.
Selecting materials that can be recycled or repurposed at the end of the robot's lifecycle.

Materials Selection:

Prioritizing eco-friendly and sustainable materials in the construction of robots. Avoiding the use of hazardous or non-recyclable materials.

Manufacturing Process:
Implementing eco-friendly manufacturing processes to reduce waste and emissions. Considering local sourcing and production to minimize transportation-related environmental impact.

Durability and Longevity:
Designing robots to withstand prolonged use and environmental stressors, reducing the need for frequent replacements.

End-of-Life Disposal:
Providing clear instructions for proper disposal or recycling of the robot's components. Ensuring compliance with local regulations for electronic waste disposal.

Transport Efficiency:
Optimizing packaging and transportation methods to reduce carbon emissions during shipping.

Repairability and Upgradability:
Designing robots with easily replaceable or upgradable parts to extend their lifespan. Providing accessible repair manuals and support for maintenance.

Software and Firmware Updates:
Enabling over-the-air updates to improve functionality and performance without the need for physical hardware changes.

User Education:
Providing users with information on sustainable practices, including energy-saving tips and proper disposal methods.

Ethical Sourcing:
Ensuring that raw materials used in the production of the robot are sourced from responsible and ethical suppliers.

Kinematics and Dynamics in Service Robotics

1.Kinematics:

Kinematics and dynamics form the bedrock of service robotics, enabling a profound grasp of robot motion and functionality across diverse contexts. Kinematics, the first pillar, dissects motion independently of force or torque dynamics. It zeroes in on the precise depiction of a robot's components – often its joints or end- effectors – delineating their spatial position, velocity, and acceleration during motion. These insights are pivotal in crafting robots that navigate and manipulate their surroundings with precision and efficiency.

Key Aspects of Kinematics:

Forward Kinematics: This aspect calculates the position and orientation of a robot's end- effector (e.g., gripper or tool) based on the joint angles or joint trajectories. It enables robots to determine where they are in space.

Inverse Kinematics: Inverse kinematics, on the other hand, determines the joint angles or joint trajectories required to achieve a desired end-effector position and orientation. It's essential for planning and controlling robot motions, especially for tasks like reaching and grasping objects.

Workspace Analysis: Kinematics analysis helps define the reachable workspace of a robot, which is crucial for task planning. It identifies the regions in space where the robot can operate effectively.

Singularities: Kinematics also helps identify singularities, which are configurations where a robot's manipulator may lose degrees of freedom. Avoiding singularities is important for safe and reliable robot operation.

Compelling Real-Time Kinematics Examples

Self-driving delivery robots use kinematics to navigate through crowded sidewalks and streets. The robots use sensors to detect their surroundings, and then use kinematic algorithms to calculate a path that will avoid obstacles and safely reach their destination.

For example, a self-driving delivery robot may use kinematics to calculate a path around a pedestrian crossing the street. The robot would first use its sensors to detect the pedestrian's location and speed. Then, the robot would use kinematic algorithms to calculate a path that would allow it to safely pass the pedestrian. Kinematics is essential for the safe and efficient operation of self-driving delivery robots. By using kinematics to calculate a path that avoids obstacles and safely reaches their destination, self-driving delivery robots can help to reduce traffic congestion and improve air quality.

Few More Examples:

Robotic arms in factories use kinematics to precisely position and move parts. The robots use sensors to detect the location of the parts, and then use kinematic algorithms to calculate the movement of the robot's joints that will allow it to reach the part and perform the desired operation.

Robotic arm may use kinematics to precisely position a circuit board on an assembly line. The robot would first use its sensors to detect the location of the circuit board. Then, the robot would use kinematic algorithms to calculate the movement of the robot's joints that would allow it to reach the circuit board and place it on the assembly line with the required precision.

Kinematics is essential for the accurate and efficient operation of robotic arms. By using kinematics to precisely position and move parts, robotic arms can help to improve the quality and consistency of manufactured products.

Kinematics

2.Dynamics:

Dynamics within the realm of service robotics delves into the intricate interplay of forces, torques, and accelerations that orchestrate a robot's movement. It scrutinizes the intricate choreography of a robot's interaction with its surroundings, taking into account the omnipresent influences of gravity, inertia, friction, and external forces. By comprehending the dynamic forces at play, robotics engineers can design robots capable of maneuvering through intricate environments, manipulating objects with finesse, and maintaining stability amidst the myriad external factors that shape their operational landscapes.

Key Aspects of Dynamics:

Forward Dynamics: This aspect involves calculating the robot's motion (e.g., joint accelerations) given the applied forces or torques. It's essential for simulating and controlling the robot's behavior.

Inverse Dynamics: Inverse dynamics, conversely, calculate the forces or torques required at each joint to achieve a desired motion or trajectory. It plays a crucial role in controlling and optimizing robot movements, such as path following and manipulation tasks.

Collision Avoidance: Understanding the dynamics of a robot is crucial for collision detection and avoidance. By considering the dynamics, robots can plan and execute motions that avoid obstacles and ensure safety.

Efficiency and Energy Consumption: Dynamics analysis helps in optimizing robot movements to minimize energy consumption, especially in applications where power efficiency is essential, such as mobile robots and drones.

Control Design: Dynamics is at the core of control system design for robots. It helps engineers develop control algorithms that regulate the robot's behavior, maintain stability, and achieve precise control of its motion.

Dynamics

A Persuasive Instance of Dynamics in Delivery Robots:

Delivery robots are becoming increasingly common in our cities. They are used to deliver packages, food, and other goods to people's homes and businesses. Delivery robots need to be able to navigate complex and dynamic environments, such as sidewalks, streets, and crosswalks. They also need to be able to avoid collisions with people, objects, and other vehicles.

Dynamics is essential for the safe and efficient operation of delivery robots. Delivery robots need to be able to control their speed and acceleration to avoid collisions and to maintain their stability. They also need to be able to compensate for external forces, such as wind and bumps in the road.

One way that delivery robots use dynamics is to control their speed and acceleration. Delivery robots use sensors to detect their surroundings and to identify potential hazards. The robot's controller then uses this information to calculate the safest and most efficient path to the destination. The controller also takes into account the robot's own dynamics, such as its mass and inertia, to determine how the robot should accelerate and decelerate.

Another way that delivery robots use dynamics is to maintain their stability. Delivery robots have a center of gravity, which is the point at which the weight of the robot is evenly distributed. If the robot's center of gravity is not over its base of support, the robot will tip over. Delivery robots use dynamics to control their center of gravity and to maintain their stability. For example, a delivery robot may lean back slightly when it is accelerating and forward slightly when it is decelerating. This helps to keep the robot's center of gravity over its base of support.

Dynamics is also important for delivery robots to compensate for external forces. For example, a delivery robot may encounter wind resistance as it is moving down the street. The robot's controller can use dynamics to calculate the amount of force that is required to overcome the wind resistance. The controller then applies the necessary force to the robot's motors to keep the robot on track.

Dynamics is essential for the safe and efficient operation of delivery robots. By understanding and using dynamics, delivery robots can navigate complex and dynamic environments, avoid collisions, and maintain their stability.

A Practical Scenario of Dynamic Collision Avoidance for a Delivery Robot

A delivery robot is driving down the street when it detects a pedestrian crossing in front of it. The robot's controller calculates the safest and most efficient way to avoid a collision. The controller takes into account the robot's own dynamics, such as its mass and inertia, as well as the dynamics of the pedestrian, such as their speed and direction of travel.

The controller determines that the safest way to avoid a collision is to decelerate the robot and to swerve to the right. The controller calculates the amount of force that is required to decelerate the robot and to swerve to the right. The controller then applies the necessary force to the robot's motors.

The robot decelerates and swerves to the right, avoiding a collision with the pedestrian. The robot then continues on its way to its destination.

This is just one example of how dynamics is used in service robotics. Dynamics is essential for the safe and efficient operation of many different types of service robots, including delivery robots, cleaning robots, and security robots.

In summary, kinematics and dynamics are fundamental concepts in service robotics that govern how robots move, interact with their environment, and respond to control inputs. A solid understanding of these principles is crucial for designing, controlling, and optimizing the performance of service robots across a wide range of applications, from manufacturing and healthcare to logistics and autonomous vehicles.

Robot Kinematics:

Understanding Robot Motion in Service Robotics

Robot kinematics is a fundamental concept in service robotics that focuses on understanding and describing the motion of robotic systems without considering the forces and torques involved. It plays a crucial role in tasks such as planning robot movements, calculating joint angles, and ensuring that robots can interact with their environment effectively. This overview explores the key aspects of robot kinematics and its significance in service robotics.

Components of Robot Kinematics:

Joint Space: Robot kinematics often start with joint space, which represents the configuration of a robot's joints. Each joint's position and orientation contribute to defining the robot's overall posture. The joint space is described using variables such as joint angles or joint coordinates.

End-Effector Pose: The end-effector is the part of the robot responsible for interacting with the environment. In kinematics, determining the end-effector's position and orientation in space is a crucial objective. This is known as forward kinematics.

Forward Kinematics: Forward kinematics calculates the position and orientation of the robot's end-effector based on the joint variables. It helps answer questions like "Where is the robot's hand located?" Given the joint angles or joint coordinates.

Inverse Kinematics: Inverse kinematics, on the other hand, seeks to answer the question "What joint angles are required to place the end-effector in a specific position and orientation?" This aspect of kinematics is essential for planning and controlling robot motions.

Applications and Importance in Service Robotics:

Path Planning and Trajectory Generation: Robot kinematics is central to path planning, where robots determine a sequence of positions and orientations to reach a target location. Trajectory generation relies on kinematics to create smooth and feasible motion paths.

Object Manipulation: In tasks like pick-and-place operations or object manipulation, robot kinematics is essential. Knowing how to position the end-effector accurately is critical for grasping objects and performing tasks with precision.

Collision Avoidance: Understanding robot kinematics helps in collision avoidance. Robots can calculate safe paths by considering their physical dimensions and constraints, thus avoiding obstacles in their path.

Simulation and Control: Kinematic models are used for simulating robot behavior and developing control algorithms. By understanding how joint angles affect end-effector position, engineers can design control strategies for tasks like robot arm movement or mobile robot navigation.

Human-Robot Interaction: Kinematics is crucial for ensuring safe and intuitive human-robot interaction. Robots need to move in ways that humans can predict and understand, making accurate kinematic modeling vital for collaborative settings.

Virtual Reality and Teleoperation: In teleoperation scenarios or virtual reality applications, robot kinematics is used to translate the operator's motions into corresponding robot movements, enabling remote control and immersive experiences.

In conclusion, robot kinematics is a fundamental concept that underpins many aspects of service robotics. It enables robots to understand and control their movements, interact with their surroundings, and collaborate with humans effectively. A strong grasp of robot kinematics is essential for developing and deploying service robots across a wide range of applications, enhancing their precision, safety, and versatility.

Robot Kinematics

4.2 Robot Kinematics: Understanding Robot Motion

Robot Dynamics: Forces, Torques, and Control

Robot dynamics is a crucial aspect of service robotics that deals with understanding the forces, torques, and control strategies involved in a robot's motion and interaction with its environment. This knowledge is essential for ensuring that robots can perform tasks accurately, safely, and efficiently. In this overview, we'll explore the key components of robot dynamics and their significance in service robotics.

1. Forces and Torques:

Forces: Forces in robot dynamics refer to the pushes or pulls that act on different parts of a robot. These forces can result from various sources, including gravity, contact with objects, or external loads. Understanding the forces acting on a robot is essential for maintaining stability and balance.

Torques: Torques are rotational forces that cause a change in the angular motion of a robot's joints or links. They play a critical role in controlling a robot's orientation and are generated by actuators such as motors and gears.

2. Inertia and Mass Distribution:

Inertia: Inertia is a measure of an object's resistance to changes in its state of motion. In robot dynamics, inertia refers to how mass is distributed throughout a robot's structure. Accurate knowledge of inertia properties is crucial for predicting how a robot will respond to applied forces and torques.

3. Control Strategies:

Feedback Control: Feedback control systems use sensors to measure a robot's current state (e.g., position, velocity, or force) and compare it to a desired state. Control algorithms then adjust the robot's actions to minimize the error and achieve the desired outcome. Feedback control is fundamental for tasks like path following and maintaining stability.

Proportional-Integral-Derivative (PID) Control: PID control is a widely used feedback control method that adjusts control inputs (e.g., motor speeds) based on proportional, integral, and derivative terms. It is effective for achieving accurate and stable control in many robotic applications.

Model-Based Control: Model-based control relies on mathematical models of a robot's dynamics to predict its behavior and optimize control inputs. These models are used to generate control strategies that consider the robot's dynamics and constraints.

4. Interaction with the Environment:

Gravitational Effects: Robot dynamics must account for the effects of gravity on different parts of the robot. This is critical for tasks involving lifting, balancing, or carrying objects. **Contact Forces:** When robots interact with objects or surfaces, they experience contact forces. Understanding these forces is vital for tasks like manipulation, walking, or climbing.

5. Stability and Robustness:

Stability Analysis: Robot dynamics are used to analyze the stability of robot motions. Stability analysis ensures that robots can maintain balance and control, even in the presence of external disturbances.
Robust Control: Robust control strategies are designed to handle uncertainties in a robot's dynamics and disturbances in the environment. These strategies help ensure reliable and predictable robot behavior.

In conclusion, robot dynamics is a fundamental aspect of service robotics that involves understanding and controlling the forces, torques, and control strategies that govern a robot's motion. This knowledge is crucial for designing robots that can perform tasks accurately, safely, and efficiently in various real-world applications, from manufacturing and healthcare to autonomous vehicles and beyond.

Inverse Kinematics And Trajectory Planning In Robotics

Inverse kinematics and trajectory planning are essential concepts in robotics, particularly for manipulators and mobile robots. They play a fundamental role in determining how a robot should move its joints or

Key Aspects Of Inverse Kinematics:

Solving for Joint Angles: Inverse kinematics seeks to find the joint angles that satisfy the desired position and orientation of the end-effector. This is often a nonlinear and complex mathematical problem, especially for robots with many degrees of freedom.

Importance in Manipulation: Inverse kinematics is crucial for robotic manipulators (e.g., robot arms) as it allows them to reach, grasp, and manipulate objects precisely. It's widely used in tasks like pick-and-place, assembly, and 3D printing.

Real-Time Calculations: In practical applications, robots need to solve inverse kinematics in real- time to respond to changing environments or user commands. Efficient algorithms are essential for achieving real-time performance.

Trajectory Planning:

Trajectory planning involves generating a sequence of positions and orientations for a robot's end- effector or mobile base to follow over time. It ensures smooth and feasible robot motions from an initial state to a goal state.

Key Aspects Of Trajectory Planning:

Path Planning vs. Trajectory Planning: Path planning focuses on finding a collision-free path from the start to the goal, while trajectory planning refines this path into a smooth and continuous motion trajectory.

Interpolation Techniques: Trajectory planning often uses interpolation methods to generate waypoints between the start and goal configurations. Common interpolation techniques include linear, cubic, and spline interpolation.

Time Parameterization: Assigning time durations to trajectory segments ensures that the robot moves at a controlled velocity and acceleration, avoiding abrupt changes in motion that could lead to instability or inefficiency.

Applications And Significance:

Robotic Manipulation: Inverse kinematics and trajectory planning are essential for robotic arms to perform tasks like grasping objects, painting, welding, or surgical procedures.

Mobile Robotics: Trajectory planning is crucial for mobile robots (e.g., autonomous cars, drones) to navigate through environments, follow paths, and reach specific GPS waypoints safely and efficiently.

Human-Robot Interaction: In applications involving human-robot collaboration or teleoperation, smooth and predictable robot motions are essential for user comfort and safety.

3D Printing and Additive Manufacturing: Precise toolpath planning using inverse kinematics and trajectory planning is essential for 3D printers to create complex objects with accuracy.

Animation and Gaming: In computer graphics and animation, these concepts are used to control the motion of virtual characters and objects.

In summary, inverse kinematics and trajectory planning are fundamental concepts in robotics that enable robots to achieve precise and controlled movements in various applications. Whether it's a robot arm assembling products on an assembly line or an autonomous vehicle navigating city streets, these concepts are at the heart of creating efficient and reliable robotic motion.

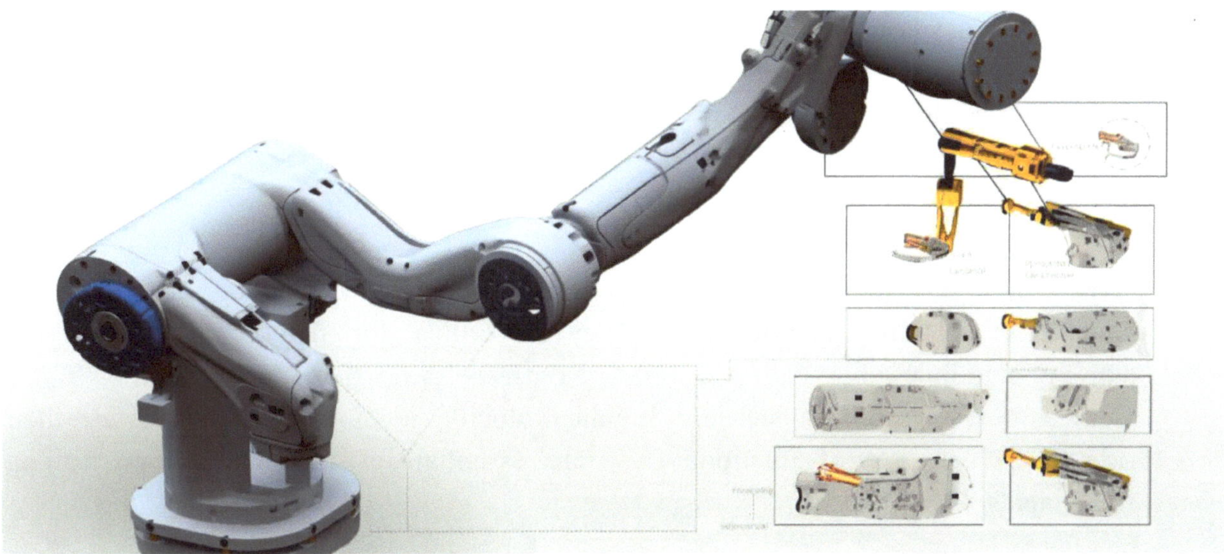

Inverse Kinematics

4.3. Sensing And Perception

Sensors are critical components in robotics, providing robots with the ability to perceive and interact with their environment. The selection of appropriate sensors is essential to ensure that robots can effectively sense, process, and respond to the surrounding world. In this overview, we'll explore different types of sensors commonly used in robotics and the considerations involved in their selection.

1. Vision Sensors:

Cameras: Cameras are perhaps the most versatile sensors in robotics. They capture visual information, allowing robots to recognize objects, navigate, and perform tasks requiring visual perception. Types of cameras include RGB cameras, depth cameras, and stereo cameras.

Lidar (Light Detection and Ranging): Lidar sensors emit laser beams and measure the time it takes for them to bounce back. This technology creates detailed 3D maps of the environment, making lidar valuable for mapping, localization, and obstacle avoidance.

Depth Sensors: These sensors measure the distance between the sensor and objects in the environment, providing depth information. Examples include structured light sensors and time-of- flight cameras.

2. Proximity and Contact Sensors:

Ultrasonic Sensors: Ultrasonic sensors emit high-frequency sound waves and measure the time it takes for them to bounce back. They are useful for detecting nearby objects and obstacles.

Infrared Sensors: Infrared sensors can detect the presence of objects based on their heat signatures. They are commonly used in obstacle detection and proximity sensing.

Tactile Sensors: Tactile sensors provide information about physical contact with objects and surfaces. They are used in robot grippers and end-effectors to grasp objects safely and with precision.

3. Inertial Sensors:

Accelerometers: Accelerometers measure linear acceleration, allowing robots to detect changes in velocity and orientation. They are critical for maintaining balance and stability in mobile robots.

Gyroscopes: Gyroscopes measure angular velocity, helping robots determine their orientation and rate of rotation. They are essential for tracking and controlling movements.

4. Force and Torque Sensors:

Force Sensors: These sensors measure forces applied to robot end-effectors or components. They are used in tasks requiring force feedback, such as assembly and manipulation.

Torque Sensors: Torque sensors measure the rotational force or torque applied to robot joints or components. They are vital for tasks like precise control of robot arms.

5. Environmental Sensors:

Temperature Sensors: Temperature sensors monitor environmental conditions, which is valuable for applications like agriculture, where maintaining specific temperature ranges is essential.

Humidity Sensors: Humidity sensors measure moisture levels in the environment, important in applications like climate control and greenhouse management.

Gas Sensors: Gas sensors detect the presence and concentration of specific gases, useful in monitoring air quality and safety in industrial settings.

6. Selection Considerations:

Task Requirements: The choice of sensors should align with the robot's intended tasks. For example, a robot designed for autonomous driving may require a combination of cameras, lidar, and radar sensors.

Cost and Budget: Sensor selection should consider the available budget. High-end sensors may offer superior performance but could be cost-prohibitive for some applications.

Environmental Conditions: Consider the operating environment, including factors like lighting conditions, temperature variations, and exposure to dust or moisture. Sensors should be robust enough to handle these conditions.

Accuracy and Precision: Different sensors offer varying levels of accuracy and precision. The required level of accuracy depends on the specific application.

Integration and Compatibility: Sensors should be compatible with the robot's control system and software. Integration should be feasible without significant technical challenges.

Redundancy: In critical applications, redundancy in sensor selection can enhance safety and reliability. Redundant sensors can provide backup in case of sensor failure.

In conclusion, selecting the right sensors for a robot is a crucial decision that directly impacts its capabilities and performance. By carefully considering the requirements of the robot's tasks, environmental conditions, and budget constraints, engineers and developers can make informed choices that enable robots to perceive and interact with the world effectively.

Vision Systems And Object Recognition in Robotics

Vision systems and object recognition are pivotal components of robotics that enable machines to perceive, interpret, and interact with the visual world. These technologies empower robots to recognize objects, navigate environments, and perform tasks in a wide range of applications. In this overview, we'll delve into vision systems, object recognition, and their significance in the field of robotics.

1. Vision Systems

Vision systems in robotics refer to the integration of cameras, image sensors, and associated hardware and software to capture, process, and analyze visual information from the robot's environment. These systems are used to replicate the human sense of sight and enable robots to make sense of the world around them.

Key Aspects of Vision Systems:

Camera Types: Robots employ various types of cameras, including RGB cameras for color information, depth cameras for 3D perception, and specialized cameras like fisheye or thermal cameras for specific applications.

Image Processing: Image processing algorithms enhance and extract meaningful information from images. This includes techniques for noise reduction, image segmentation, feature extraction, and image stitching.

Computer Vision: Computer vision is the field of artificial intelligence that focuses on teaching machines to interpret and understand visual data. It encompasses object detection, image classification, facial recognition, and more.

2. Object Recognition:

Object recognition is a subset of computer vision that enables robots to identify and categorize objects within images or the robot's field of view. It involves the extraction of relevant features and patterns from visual data, which are then used to recognize and differentiate objects.

Key Aspects Of Object Recognition:

Feature Extraction: Object recognition algorithms extract distinctive features from images, such as edges, corners, or textures, which are used to describe objects and facilitate their identification.

Deep Learning: Deep learning techniques, particularly convolutional neural networks (CNNs), have revolutionized object recognition. CNNs can automatically learn and recognize objects from large datasets, achieving state-of-the-art accuracy.

Object Localization: In addition to recognizing objects, robots often need to determine their spatial location within the scene. Object localization techniques help robots pinpoint the precise position and orientation of recognized objects.

Applications And Significance:

Robotic Vision: Vision systems are fundamental for robots to navigate and interact in complex environments. They enable robots to detect obstacles, follow paths, and avoid collisions.

Object Manipulation: Object recognition is essential for robots to grasp, manipulate, and interact with objects in a wide range of applications, from manufacturing and logistics to healthcare and household assistance.

Autonomous Vehicles: Vision systems are key components in autonomous vehicles, helping them detect road signs, pedestrians, and other vehicles for safe and efficient navigation.

Surveillance and Security: Robots equipped with vision systems can monitor and analyze security camera feeds, recognize intruders or suspicious activities, and enhance surveillance capabilities.

Medical Imaging: In healthcare, robots with vision systems aid in medical imaging and diagnostics, supporting doctors in tasks like surgery, radiology, and patient monitoring.

Agriculture and Precision Farming: Vision systems help automate tasks like crop monitoring, fruit picking, and weed detection in agriculture, improving productivity and reducing resource use.

Retail and Inventory Management: Robots in retail environments use object recognition to manage inventory, assist customers, and perform stock-taking tasks.

In conclusion, vision systems and object recognition are integral to the advancement of robotics, enabling robots to perceive and interact with their surroundings effectively. These technologies enhance robot autonomy, improve safety, and expand the range of applications where robots can be deployed, ultimately driving innovation in various industries.

Lidar And Depth Sensing

Lidar and depth sensing are two technologies that are used to measure the distance between an object and a sensor. Lidar uses light to measure distance, while depth sensing uses a variety of other technologies, such as stereo vision, structured light, and time-of-flight.

Lidar

Lidar works by sending out pulses of laser light and measuring the time it takes for the light to reflect back to the sensor. The distance to the object is calculated by multiplying the time of flight by the speed of light. Lidar sensors can typically measure distances of up to several hundred meters with high accuracy.

Depth sensing

Depth sensing uses a variety of different technologies to measure distance. Stereo vision uses two cameras to look at an object from different angles and then calculates the distance to the object by comparing the images. Structured light uses a projector to cast a pattern of light onto an object and then measures the distortion of the pattern to calculate the distance to the object. Time-of-flight depth sensors send out pulses of light and measure the time it takes for the light to reflect back to the sensor.

Lidar And Depth Sensing Are Used In A Wide Variety of Applications, Including:

Robotics: Lidar and depth sensing are used by robots to navigate and avoid obstacles.
Self-driving cars: Lidar and depth sensing are used by self-driving cars to perceive their surroundings and make decisions about how to navigate.
Augmented reality and virtual reality: Lidar and depth sensing are used in augmented reality and virtual reality applications to track the user's position and orientation.
Security and surveillance: Lidar and depth sensing are used in security and surveillance systems to detect intruders and monitor areas of interest.
Industrial automation: Lidar and depth sensing are used in industrial automation applications to measure distances and perform tasks such as picking and packing items.

Lidar And Depth Sensing Each Have Their Own Advantages And Disadvantages

Lidar

Advantages: Lidar sensors are very accurate and can measure distances in low-light and even no-light conditions.
Disadvantages: Lidar sensors can be expensive and bulky.
Depth sensing

Advantages: Depth sensing sensors are typically less expensive and smaller than lidar sensors.
Disadvantages: Depth sensing sensors can be less accurate than lidar sensors and may not work well in low-light or no-light conditions.

Overall, lidar and depth sensing are two powerful technologies that can be used to measure distance with high accuracy. The best technology for a particular application will depend on the specific requirements of the application.

Sensor Fusion Techniques

Sensor fusion techniques combine data from multiple sensors to provide a more accurate and reliable estimate of the state of a system or environment. This is because each sensor has its own strengths and weaknesses, and by combining data from multiple sensors, it is possible to overcome the limitations of each individual sensor.

There are a variety of different sensor fusion techniques, but they can generally be classified into two categories: centralized fusion and decentralized fusion.

Centralized Fusion

In centralized fusion, all of the sensor data is sent to a central processing unit, where it is fused together to produce a single output. This approach is typically used in applications where the sensors are located close together and where there is a high bandwidth connection between the sensors and the central processing unit.

Decentralized Fusion

In decentralized fusion, the sensor data is fused locally at each sensor node. The fused data is then sent to other sensor nodes or to a central processing unit. This approach is typically used in applications where the sensors are distributed over a large area or where there is a limited bandwidth connection between the sensors and the central processing unit.

Some Common Sensor Fusion Techniques Include:

Kalman Filtering: The Kalman filter is a recursive algorithm that can be used to estimate the state of a system from a sequence of noisy measurements. The Kalman filter is widely used in a variety of applications, including robotics, navigation, and signal processing.

Particle Filtering: The particle filter is a Monte Carlo algorithm that can be used to estimate the state of a system from a sequence of noisy measurements. The particle filter is well-suited for tracking objects in dynamic environments.

Bayesian Estimation: Bayesian estimation is a statistical approach to estimating the probability distribution of a parameter given a set of observations. Bayesian estimation can be used to fuse sensor data in a variety of ways.

Sensor Fusion is Used in a Wide Variety of Applications, Including:

Robotics: Sensor fusion is used by robots to navigate, avoid obstacles, and interact with their environment.
Self-Driving Cars: Sensor fusion is used by self-driving cars to perceive their surroundings and make decisions about how to navigate.
Augmented Reality And Virtual Reality: Sensor fusion is used in augmented reality and virtual reality applications to track the user's position and orientation.
Security And Surveillance: Sensor fusion is used in security and surveillance systems to detect intruders and monitor areas of interest.
Industrial Automation: Sensor fusion is used in industrial automation applications to improve the accuracy and efficiency of tasks such as picking and packing items.

Conclusion

Sensor fusion is a powerful technique that can be used to improve the accuracy and reliability of measurements from multiple sensors. Sensor fusion is used in a wide variety of applications, including robotics, self-driving cars, augmented reality and virtual reality, security and surveillance, and industrial automation.

Perception Algorithms For Mapping And Localization

Perception algorithms for mapping and localization are used to build and maintain a map of the environment, as well as to determine the robot's pose within the map. These algorithms are essential for robots to be able to navigate autonomously.

There are a variety of different perception algorithms for mapping and localization, but they can generally be classified into two categories: Feature-based and sensor-fusion-based.

Feature-Based Algorithms

Feature-Based Algorithms: build a map of the environment by identifying and tracking features, such as corners, edges, and planes. These features can be detected using a variety of different sensors, such as cameras, lidar, and sonar. Once the features have been detected, they can be used to build a map of the environment using a variety of different techniques, such as simultaneous localization and mapping (SLAM).
Sensor-fusion-based algorithms

Sensor-Fusion-Based: algorithms combine data from multiple sensors to build and maintain a map of the environment. This approach is typically more robust than feature-based algorithms, as it is not as sensitive to noise or occlusions. Sensor-fusion based algorithms are also better suited for dynamic environments, where the features of the environment may change over time.

Perception Algorithms for Mapping and Localization

Some Common Perception Algorithms For Mapping And Localization Include:

Particle Filtering SLAM: Particle filtering SLAM is a feature-based algorithm that uses particle filtering to track the robot's pose and build a map of the environment.

Lidar SLAM: Lidar SLAM is a feature-based algorithm that uses lidar data to track the robot's pose and build a map of the environment.

Visual SLAM: Visual SLAM is a feature-based algorithm that uses camera data to track the robot's pose and build a map of the environment.

Extended Kalman filter (EKF) SLAM: EKF SLAM is a sensor-fusion based algorithm that uses an extended Kalman filter to track the robot's pose and build a map of the environment.

Graph-Based SLAM: Graph-based SLAM is a sensor-fusion based algorithm that uses a graph to represent the map of the environment.

Perception Algorithms For Mapping And Localization Are Used In A Variety Of Applications, Including:

Robotics: Perception algorithms for mapping and localization are used by robots to navigate autonomously, such as mobile robots, self-driving cars, and drones.

Augmented Reality And Virtual Reality: Perception algorithms for mapping and localization are used in augmented reality and virtual reality applications to track the user's position and orientation. **Security And Surveillance:** Perception algorithms for mapping and localization are used in security and surveillance systems to detect intruders and monitor areas of interest.

Industrial Automation: Perception algorithms for mapping and localization are used in industrial automation applications to track the movement of objects and robots.

Conclusion

Perception algorithms for mapping and localization are essential for robots to be able to navigate autonomously. These algorithms are used in a variety of applications, including robotics, augmented reality and virtual reality, security and surveillance, and industrial automation.

Control Systems

A control system is a system that manages, commands, directs, or regulates the behavior of other devices or systems. Control systems are used in a wide variety of applications, including robotics, automation, and manufacturing.

In the realm of robotics, control systems are instrumental in shaping how robots interact with their environment and execute tasks. There exist two primary categories of control systems: open-loop and closed-loop.

Open-Loop Control System:

An open-loop control system, in the context of robotics, functions like a basic set of instructions without the capacity for real-time adjustments. It lacks the ability to measure and provide feedback on the robot's actions. As a result, open-loop control systems are typically deployed for rudimentary and straightforward tasks, where real-time error correction or adaptation to changing conditions is unnecessary. Examples in service robotics include simple actions such as turning a light on or off, where precise feedback isn't a critical factor.

Closed -Loop Control System:

Conversely, closed-loop control systems are fundamental to the capabilities and precision of service robots. These systems integrate feedback mechanisms, enabling the robot to constantly measure and assess its actions, then adjust in response to deviations from the desired outcome. In the robotics context, closed-loop control is invaluable for executing complex and dynamic tasks. For instance, it allows a robot to control the temperature of a room, navigate through cluttered environments, or regulate the speed and precision of its movements. In service robotics, the choice between open- loop and closed-loop control systems significantly influences a robot's capacity to excel in a wide range of applications, from basic tasks to intricate, real-world challenges.

Control Systems Are Typically Made Up Of Three Main Components:

Sensor: The sensor measures the output of the system.
Controller: The controller compares the measured output to the desired output and generates a control signal.
Actuator: The actuator applies the control signal to the system to change the output.

Control systems can be either analog or digital. Analog control systems use continuous signals, while digital control systems use discrete signals. Digital control systems are becoming more common, as they are more precise and reliable than analog control systems.

Control Systems

Control Systems Are Used In A Wide Variety Of Applications, Including:

Robotics: Control systems are used to control the movement and behavior of robots.
Automation: Control systems are used to automate tasks in factories and other industrial settings. Manufacturing: Control systems are used to control the manufacturing process, such as assembling products and welding parts.
Transportation: Control systems are used to control traffic lights, aircraft autopilots, and other transportation systems.
Power Generation And Distribution: Control systems are used to control power plants and electrical grids.
Environmental Control: Control systems are used to control the temperature, humidity, and other environmental conditions in buildings and other facilities.

Control systems are an essential part of many modern technologies. They allow us to automate tasks, improve efficiency, and create more complex and sophisticated systems.

4.4 Robot Control Architecture

Robot control architecture refers to the overall framework or structure that governs how a robot's control system is organized and functions. It's a crucial aspect of robotics that defines how various components and modules work together to ensure the robot operates effectively and efficiently. There are several different control architectures in robotics, and the choice of architecture depends on the specific application and requirements. Here is a general overview of robot control architecture:

Hierarchical Control: This architecture is structured in a hierarchical manner, with different levels of control. The top level is responsible for high-level tasks and decision-making, while lower levels manage more detailed control aspects, such as motion planning, trajectory generation, and feedback control.

Behavior-Based Control: In this architecture, the robot's behavior is decomposed into multiple behaviors or modules, each responsible for a specific task or function. These behaviors operate independently and may react to sensor inputs. The overall behavior of the robot emerges from the interactions of these modules.

Reactive Control: Reactive control systems focus on quick, real-time responses to sensor data. They don't typically involve high-level planning or complex decision-making. Instead, they rely on pre-defined rules or algorithms to react to immediate environmental changes.

Deliberative Control: Deliberative control architectures emphasize planning and decision-making. They involve high-level reasoning and long-term goal setting. These systems typically use models of the environment and take into account a robot's capabilities and limitations.

Hybrid Control: Hybrid control architectures combine elements of both reactive and deliberative control. They aim to strike a balance between real-time responsiveness and long-term planning, adapting to different situations as needed.

Centralized vs. Distributed Control: Depending on the application, a robot's control architecture may centralize control in a single processing unit or distribute control across multiple processors or embedded controllers. Distributed control can enhance redundancy and fault tolerance.

Closed-Loop Control: Most robot control architectures employ closed-loop control, where sensor feedback continuously informs and adjusts the robot's actions to maintain desired states or trajectories. Closed-loop control is essential for tasks requiring precision and adaptation to changing conditions.

Open-Loop Control: In some specific applications, open-loop control is used for simple, repetitive tasks. Open-loop control does not rely on sensor feedback, making it less adaptable but suitable for tasks where environmental variations are minimal.

The choice of control architecture depends on the robot's purpose, its sensory and computational capabilities, and the complexity of the tasks it needs to perform. In practice, many modern robots use a combination of these architectures to handle different aspects of control, creating a versatile and adaptable robotic system.

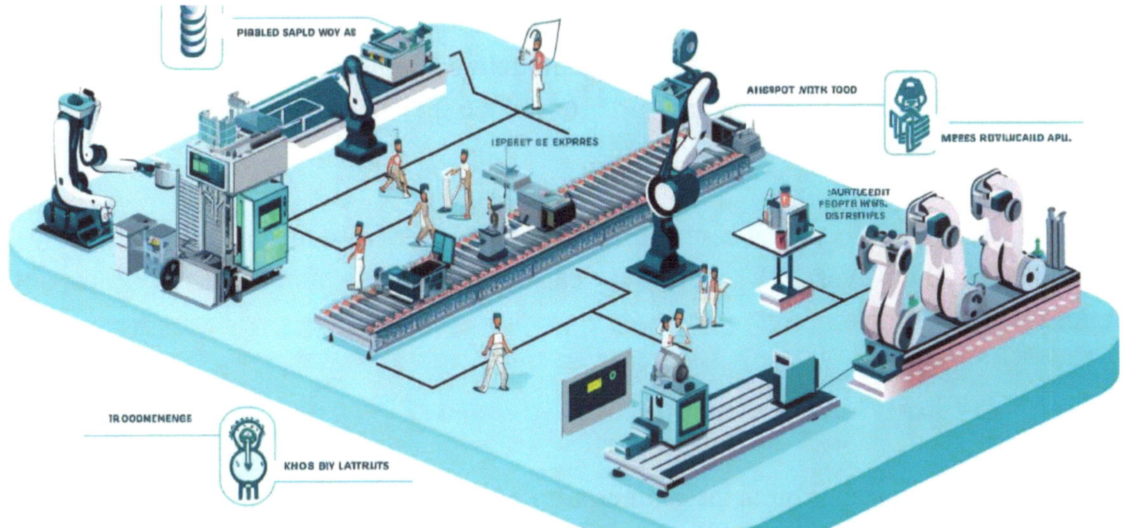

Robot Control Architecture

Feedback Control and PID Controllers

In a closed-loop control system, feedback is used to continuously measure and compare the actual output of the system to the desired or reference output. The difference, known as the error, is then used to adjust the system's inputs or control actions to minimize the error and maintain the desired output. This feedback loop allows the system to adapt to changing conditions, disturbances, and uncertainties, making it an essential choice for tasks that require precision and stability.

Feedback control is a type of control system that uses feedback to control the output of a system. Feedback is a signal that is generated by measuring the output of the system and comparing it to the desired output. The feedback signal is then used to adjust the input to the system in order to reduce the error between the actual output and the desired output.

A PID (Proportional-Integral-Derivative) controller is a specific type of feedback controller commonly used in various applications, including robotics, automation

PID controllers are a type of feedback controller that uses a **proportional-integral-derivative (PID)** algorithm to control the output of a system. PID controllers are one of the most common types of feedback controllers, and they are used in a wide variety of applications.

The PID Algorithm Uses Three Terms To Calculate The Control Signal:

Proportional term: The proportional term is proportional to the error between the actual output and the desired output.
Integral term: The integral term is proportional to the accumulated error over time.

Derivative term: The derivative term is proportional to the rate of change of the error. The PID algorithm combines these three terms to calculate a control signal that is used to adjust the input to the system in order to reduce the error between the actual output and the desired output.

PID controllers are very effective at controlling a wide variety of systems, and they are relatively easy to implement. However, PID controllers can be difficult to tune, and they may not be suitable for all applications.

Utilizing a PID Controller For Room Temperature Regulation:
An Example

- The sensor measures the temperature of the room.
- The controller compares the measured temperature to the desired temperature.
- The controller calculates a control signal based on the error between the actual temperature and the desired temperature.
- The actuator applies the control signal to the system, such as turning on or off a heater or air conditioner.
- The sensor measures the temperature of the room again.
- The controller repeats steps 2-5 until the desired temperature is reached.

PID Controllers Are Used In A Wide Variety Of Applications, Including:

Robotics: PID controllers are used to control the movement and behavior of robots.
Automation: PID controllers are used to automate tasks in factories and other industrial settings. **Manufacturing:** PID controllers are used to control the manufacturing process, such as assembling products and welding parts.
Transportation: PID controllers are used to control traffic lights, aircraft autopilots, and other transportation systems.
Power Generation and Distribution: PID controllers are used to control power plants and electrical grids.
Environmental Control: PID controllers are used to control the temperature, humidity, and other environmental conditions in buildings and other facilities.

PID controllers are an essential part of many modern technologies. They allow us to automate tasks, improve efficiency, and create more complex and sophisticated systems.

Model-Based Control

Model-based control (MBC) is a control system design methodology that uses a model of the system being controlled to design the controller. The model is used to predict the behavior of the system under different control inputs, and this information is used to design a controller that will achieve the desired performance.

MBC Has A Number Of Advantages Over Traditional Control System Design Methods:

Improved Performance: MBC can achieve better performance than traditional control system design methods, such as PID control. This is because MBC uses a model of the system to design the controller, which allows the controller to take into account the dynamics of the system.

Reduced Design Time: MBC can reduce the design time for complex control systems. This is because MBC can be used to simulate the performance of the control system before it is implemented, which can help to identify and correct any potential problems.

Improved Robustness: MBC can improve the robustness of control systems to disturbances and uncertainties. This is because MBC uses a model of the system to design the controller, which allows the controller to compensate for disturbances and uncertainties.

MBC Is Used In A Wide Variety Of Applications, Including:

Robotics: MBC is used to control the movement and behavior of robots.

Automation: MBC is used to automate tasks in factories and other industrial settings.

Manufacturing: MBC is used to control the manufacturing process, such as assembling products and welding parts.

Transportation: MBC is used to control traffic lights, aircraft autopilots, and other transportation systems.

Power generation and distribution: MBC is used to control power plants and electrical grids. **Environmental Control:** MBC is used to control the temperature, humidity, and other environmental conditions in buildings and other facilities.

MBC is a powerful tool for designing control systems for complex and dynamic systems. MBC can improve the performance, robustness, and design time of control systems.

Couple Of Examples Of How Model-Based Control (MBC) Is Applied in Robotics:

Autonomous Navigation: In robotics, MBC can be employed to design controllers for autonomous navigation systems. For instance, a mobile robot equipped with MBC can use a model of its environment to plan and execute precise movements, ensuring it avoids obstacles and reaches its destination efficiently.

Manipulator Arm Control: For robots with manipulator arms used in manufacturing or pick-and- place tasks, MBC can optimize the control of these arms. By using a model of the robot's kinematics and dynamics, MBC can calculate the necessary joint angles and forces to grasp and manipulate objects with accuracy and stability.

Quadcopter Flight: Drones and quadcopters benefit from MBC in controlling their flight dynamics. By utilizing a model of the quadcopter's physical properties and environmental conditions, MBC can enhance stability, maneuverability, and response to external disturbances.

Humanoid Robot Locomotion: Humanoid robots designed for walking or running require advanced control to maintain balance and execute complex movements. MBC can be instrumental in creating controllers that factor in the robot's kinematics and dynamics to achieve human-like locomotion.

MBC plays a vital role in enhancing the capabilities and performance of robotic systems across various applications, making them more capable, efficient, and adaptable to their environments.

MBC is a complex topic, but it is a powerful tool for designing control systems for complex and dynamic systems. MBC can improve the performance, robustness, and design time of control systems.

Adaptive And Learning Control

Adaptive and learning control are two related control system design methodologies that use feedback to control the output of a system and adapt to changing conditions.

Adaptive control systems use feedback to estimate the parameters of the system being controlled and then use this information to adjust the controller to achieve the desired performance. Adaptive control systems are particularly useful for controlling systems with unknown or time-varying parameters.

Learning control systems use feedback to learn the relationship between the control inputs and the system outputs. This information is then used to design a controller that will achieve the desired performance. Learning control systems are particularly useful for controlling systems with complex and non-linear dynamics.

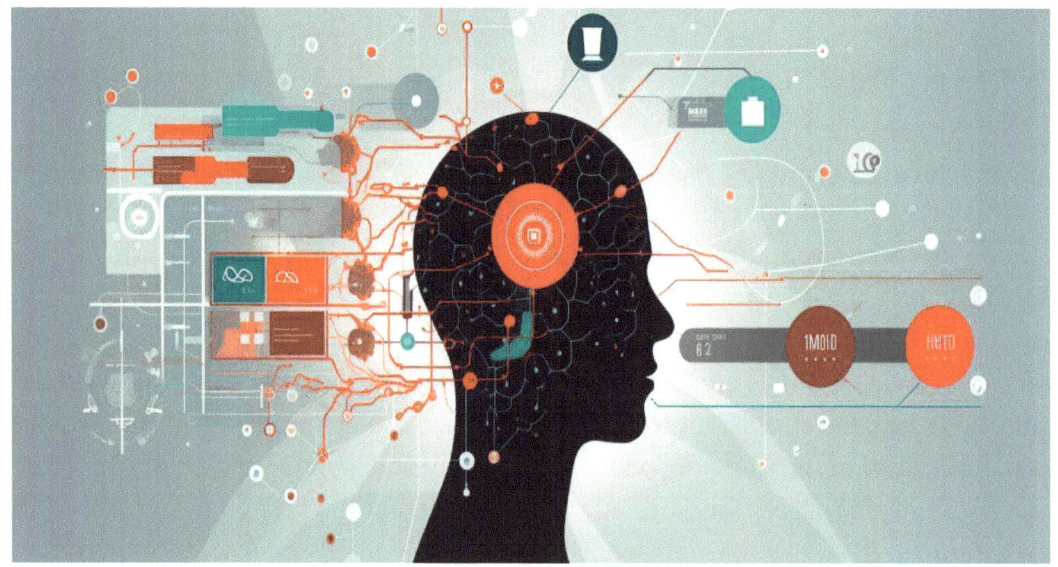

Adaptive And Learning Control Systems Are Used In A Wide Variety Of Applications, Including:

Robotics: Adaptive and learning control systems are used to control the movement and behavior of robots.

Automation: Adaptive and learning control systems are used to automate tasks in factories and other industrial settings.

Manufacturing: Adaptive and learning control systems are used to control the manufacturing process, such as assembling products and welding parts.

Transportation: Adaptive and learning control systems are used to control traffic lights, aircraft autopilots, and other transportation systems.

Power Generation And Distribution: Adaptive and learning control systems are used to control power plants and electrical grids.

Environmental Control: Adaptive and learning control systems are used to control the temperature, humidity, and other environmental conditions in buildings and other facilities.

Certainly, Here Are A Couple Of Examples Of How Adaptive Control Systems Are Applied In Robotics:

Robotic Grasping: In robotics, adaptive control systems can be used to improve the grasp stability and success rate of robotic hands. These systems adjust the grip strength and finger positions based on feedback from sensors to effectively handle objects with varying shapes and surface properties.

Legged Robot Locomotion: For legged robots, adaptive control is valuable in maintaining balance and adapting to changing terrains. These systems continuously adjust joint angles and gait patterns to prevent falls and ensure stable locomotion on uneven surfaces.

Sensor Calibration: Robots equipped with various sensors, such as cameras and LiDAR, may use adaptive control to continuously calibrate sensor parameters. This ensures accurate perception and helps the robot adapt to changes in lighting conditions and sensor degradation.

Collision Avoidance: Adaptive control in robotics can be employed to enhance collision avoidance strategies. Robots can adapt their trajectories and behaviors in real-time to avoid obstacles and operate safely in dynamic environments.

Adaptive control systems play a crucial role in making robots more versatile, capable of handling uncertainty and variability in their operating environments, and improving their overall performance in real- world scenarios.

Learning Control Systems :

Learning control systems find several applications in robotics to improve the adaptability and performance of robots. Here are some use cases:

Learning control systems

Skill Acquisition: Learning control systems are employed to enable robots to acquire and improve their skills over time. They can learn from past experiences and gradually refine their movements, such as grasping objects more precisely or executing tasks with greater accuracy.

Path Planning And Navigation: In complex environments, robots can use learning control to enhance their path planning and navigation capabilities. They learn from previous navigation experiences to make better decisions in real-time, avoiding obstacles and finding optimal routes.

Object Recognition: Learning control can be used in vision systems to improve object recognition. Robots can learn from a database of images or data to identify objects with greater accuracy, even in varying lighting or background conditions.

Human-Robot Interaction: In collaborative and human-robot interaction scenarios, learning control enables robots to adapt their behavior based on human cues and preferences. They can learn to respond more effectively to human gestures, speech, and actions.

Grasping And Manipulation: Learning control systems help robots refine their grasping and manipulation strategies. By learning from trial and error, robots can adapt their grasp force, finger positions, and manipulation techniques to handle a wider range of objects.

Task Optimization: Robots can use learning control to optimize their task execution. For example, they can learn to weld joints with more precision or assemble products with fewer errors by continuously improving their control strategies.

Error Correction: Learning control systems allow robots to detect and correct errors during operations. They can learn to identify issues and apply corrective actions in real-time, reducing the need for human intervention.

Adaptation To Environmental Changes: Learning control enables robots to adapt to changes in their working environment, such as temperature variations, surface conditions, or the presence of new obstacles.

These applications demonstrate how learning control systems in robotics contribute to increased adaptability, improved performance, and a wider range of capabilities for robots in various tasks and environments.

Navigation And Path Planning

Navigation and path planning are two essential tasks for service robots. Navigation is the process of determining the robot's current location and orientation, and then moving the robot to a desired goal location. Path planning is the process of finding a collision-free path from the robot's current location to the desired goal location.

There are a variety of different navigation and path planning algorithms that can be used for service robots. The choice of algorithm will depend on the specific requirements of the application, such as the environment in which the robot will be operating, the accuracy and speed required, and the computational resources available.

Some Common Navigation And Path Planning Algorithms For Service Robots Include:

Dead reckoning: Dead reckoning is a simple navigation algorithm that uses the robot's odometry readings to estimate its current location. Dead reckoning is relatively accurate over short distances, but it can become inaccurate over long distances due to errors in the odometry readings.

Landmark-based navigation: Landmark-based navigation is a navigation algorithm that uses landmarks in the environment to determine the robot's current location.

Landmarks can be natural features, such as walls and doorways, or artificial features, such as beacons or QR codes. Landmark- based navigation is more accurate than dead reckoning, but it requires a map of the environment with the landmarks identified.

Simultaneous localization and mapping (SLAM): SLAM is a navigation algorithm that builds a map of the environment while simultaneously tracking the robot's location. SLAM is a more complex algorithm than dead reckoning or landmark-based navigation, but it can be used to navigate in unknown or dynamic environments.

Path planning: Path planning algorithms find a collision-free path from the robot's current location to the desired goal location. There are a variety of different path planning algorithms, such as Dijkstra's algorithm, A* algorithm, and Rapidly-exploring Random Tree (RRT). The choice of path planning algorithm will depend on the specific requirements of the application, such as the complexity of the environment and the time constraints.

Navigation and path planning are essential tasks for service robots. By carefully selecting the appropriate navigation and path-planning algorithms, service robots can be designed to navigate autonomously in a variety of different environments.

Applications Of Navigation And Path Planning In Service Robots

Delivery robots: Delivery robots use navigation and path planning to navigate to customers' homes and businesses to deliver packages.

Cleaning robots: Cleaning robots use navigation and path planning to navigate around homes and businesses to clean floors and other surfaces.

Security robots: Security robots use navigation and path planning to patrol areas and detect intruders.

Agricultural robots: Agricultural robots use navigation and path planning to navigate farms and perform tasks such as harvesting crops and spraying pesticides.

Navigation and path planning are essential technologies for the development of autonomous service robots. By enabling robots to navigate autonomously, navigation and path planning algorithms can help to improve the efficiency and productivity of service robots and to expand the range of tasks that they can perform.

4.5. Localization Techniques Path Planning Algorithms

As mentioned in my previous response, there are a variety of different path planning algorithms that can be used for service robots. Some common path planning algorithms include:

Dijkstra's Algorithm: Dijkstra's algorithm is a graph search algorithm that finds the shortest path from a source node to a destination node. Dijkstra's algorithm is relatively simple to implement and is efficient for finding shortest paths in small graphs. However, Dijkstra's algorithm can be inefficient for finding shortest paths in large graphs.

A Algorithm: The A algorithm is a graph search algorithm that is similar to Dijkstra's algorithm, but the A algorithm uses a heuristic function to estimate the cost of reaching the destination node. This allows the A* algorithm to find shorter paths than Dijkstra's algorithm in many cases. However, the A* algorithm is more complex to implement than Dijkstra's algorithm and can be slower for finding shortest paths in large graphs.

Rapidly-Exploring Random Tree (RRT): RRT is a randomized path planning algorithm that is well-suited for planning paths in complex and dynamic environments. RRT works by randomly exploring the state space and building a tree of connected states. The RRT algorithm then uses the tree to find a path from the start state to the goal state.

Path Planning Algorithms

In addition to these basic path-planning algorithms, there are a number of more advanced path-planning algorithms that have been developed for specific applications. For example, there are path-planning algorithms that are specifically designed for planning paths for robots in dynamic environments, such as warehouses and factories. There are also path-planning algorithms that are specifically designed for planning paths for robots with limited computational resources, such as robots used in search and rescue operations.

The choice of path-planning algorithm will depend on the specific requirements of the application. For example, if the robot needs to find the shortest path to a goal location as quickly as possible, then the A* algorithm may be a good choice. However, if the robot needs to plan a path in a complex and dynamic environment, then the RRT algorithm may be a better choice.

Path planning algorithms are an essential part of the development of autonomous service robots. By enabling robots to plan and navigate their own paths, path-planning algorithms can help to improve the efficiency, productivity, and safety of service robots.

Obstacle Avoidance Strategies

Obstacle avoidance strategies are techniques that service robots use to avoid obstacles in their environment. There are a variety of different obstacle avoidance strategies that can be used, and the best strategy for a particular application will depend on the specific requirements of the application.

Some Common Obstacle Avoidance Strategies Include:

Reactive Obstacle Avoidance: Reactive obstacle avoidance strategies are based on the robot's immediate surroundings. In reactive obstacle avoidance, the robot uses sensors to detect obstacles in its path and then takes evasive action to avoid the obstacles. Reactive obstacle avoidance strategies are relatively simple to implement, but they can be inefficient and unreliable in complex environments.

Predictive Obstacle Avoidance: Predictive obstacle avoidance strategies use a model of the environment to predict the future locations of obstacles. This allows the robot to plan a path that will avoid obstacles before they reach the robot's current position. Predictive obstacle avoidance strategies are more efficient and reliable than reactive obstacle avoidance strategies, but they are also more complex to implement.

Hybrid Obstacle Avoidance: Hybrid obstacle avoidance strategies combine elements of both reactive and predictive obstacle avoidance. This allows the robot to benefit from the advantages of both approaches while minimizing the disadvantages of each approach.

In addition to these basic obstacle-avoidance strategies, there are a number of more advanced obstacle- avoidance strategies that have been developed for specific applications. For example, there are obstacle avoidance strategies that are specifically designed for robots that operate in dynamic environments, such as warehouses and factories. There are also obstacle avoidance strategies that are specifically designed for robots with limited computational resources, such as robots used in search and rescue operations.

The choice of obstacle avoidance strategy will depend on the specific requirements of the application. For example, if the robot needs to avoid obstacles quickly and reliably in a complex environment, then a predictive obstacle avoidance strategy may be a good choice. However, if the robot has limited computational resources, then a reactive obstacle avoidance strategy may be a better choice.

Obstacle avoidance strategies are an essential part of the development of autonomous service robots. By enabling robots to avoid obstacles, obstacle avoidance strategies help to improve the safety and reliability of service robots.

Human-Robot Interaction In Navigation

Human-robot interaction (HRI) in navigation is the study of how humans and robots can interact to navigate in shared spaces. This is an important area of research, as robots are becoming increasingly common in our environment, and we need to be able to interact with them safely and efficiently.

There are a number of different ways that humans and robots can interact in navigation. Some common examples include:

Pedestrian-Robot Interaction: This refers to the interaction between pedestrians and robots in public spaces, such as sidewalks and streets. It is important to design robots that can safely navigate around pedestrians and avoid collisions.

Human-Robot Collaboration: This refers to the interaction between humans and robots in collaborative workspaces, such as factories and warehouses. It is important to design robots that can work safely and efficiently alongside humans.

Human-Robot Assistance: This refers to the interaction between humans and robots that are designed to assist humans, such as service robots and assistive robots. It is important to design robots that can understand and respond to human needs, and that can navigate safely in human environments.

There are a number of challenges associated with HRI in navigation. One challenge is that humans and robots have different navigation styles and goals. For example, humans may tend to take the most direct path to their destination, while robots may take a more roundabout path to avoid obstacles. Another challenge is that humans and robots may have different communication styles. For example, humans may use natural language to communicate, while robots may use machine-readable codes.

Despite these challenges, there has been significant progress in the field of HRI in navigation in recent years. Researchers are developing new algorithms and technologies that allow robots to better understand and respond to human needs, and to navigate safely in human environments.

Some Examples Of How HRI in Navigation Is Being Used Today:

Self-driving cars: Self-driving cars use HRI to navigate around pedestrians and other vehicles. For example, self-driving cars may use cameras and sensors to detect pedestrians crossing the street, and then slow down or stop to avoid a collision.

 Delivery robots use HRI to navigate around people and other obstacles in public spaces. For example, delivery robots may use sensors to detect pedestrians and then slow down or stop to avoid a collision.

Human-Robot Interaction in Navigation

Cleaning Robots: Cleaning robots use HRI to navigate around furniture and other objects in a room or building. For example, cleaning robots may use sensors to detect obstacles and then change their course to avoid them.

Security Robots: Security robots use HRI to navigate around people and other obstacles while patrolling an area. For example, security robots may use sensors to detect people and then slow down or stop to avoid a collision.

HRI in navigation is a rapidly developing field with the potential to revolutionize the way we interact with robots. By developing new algorithms and technologies that allow robots to better understand and respond to human needs, and to navigate safely in human environments, we can create robots that can assist us in a wide range of tasks, from delivering packages to patrolling our streets.

Human-Centered Design

Human-centered design (HCD) is a problem-solving approach that puts people at the center of every stage of the design process. HCD is used to create products, services, and systems that are useful, usable, and accessible to everyone.

HCD Is A Five-Step Process:

Empathize: The first step is to understand the needs, wants, and pain points of the people who will be using the product or service. This is done through interviews, surveys, and observations.

Define: Once you have a good understanding of the users, you can start to define the problem that you are trying to solve. This involves working with the users to identify the specific needs that the product or service should address.

Ideate: Once you have a clear understanding of the problem, you can start to generate ideas for solutions. This is a brainstorming process where all ideas are welcome, no matter how crazy they may seem.

Prototype: Once you have a few ideas, you can start to create prototypes. Prototypes are low-fidelity versions of the product or service that you can use to test your ideas with users and get feedback.

Test: Once you have a prototype that you are happy with, you can start to test it with users in a more realistic setting. This will help you to identify any potential problems with the product or service and make necessary changes before you launch it.

HCD is an iterative process, meaning that you may need to go back and forth between the different steps as you learn more about the users and the problem you are trying to solve.

HCD is important because it helps to ensure that the products and services that we create are actually useful and usable for the people who will be using them. HCD also helps to reduce the risk of failure, as it allows us to identify and address any potential problems early in the design process.

Examples Of Contemporary Applications for Human-Centered Design (HCD):

Self-driving cars: HCD is being used to design self-driving cars that are safe and easy for people to use. For example, HCD is being used to design self-driving cars that can communicate with pedestrians and other drivers, and that can safely navigate around obstacles.

Medical devices: HCD is being used to design medical devices that are easy for patients and healthcare professionals to use. For example, HCD is being used to design insulin pumps that are easy for diabetics to use, and that can communicate with other medical devices, such as blood glucose monitors.

Public transportation systems: HCD is being used to design public transportation systems that are easy for everyone to use, including people with disabilities. For example, HCD is being used to design bus stops and train stations that are accessible to everyone, and that provide clear and concise information about schedules and routes.

HCD is a powerful tool that can be used to create products and services that are truly useful and usable for everyone. By putting people at the center of the design process, we can create products and services that make a positive impact on the world.

4.6 User-Centered Design Principles

Ethical Considerations In Service Robotics

Service robots are becoming increasingly common in our lives, and it is important to consider the ethical implications of their use.

Ethical Considerations in Service Robotics

Some Of The Key Ethical Considerations In Service Robotics:

Safety: Service robots must be designed and operated in a safe manner. This means that they must be able to avoid collisions with people and objects, and they must not pose any other safety hazards. **Privacy:** Service robots may collect personal data about the people they interact with. It is important to ensure that this data is collected and used in a responsible and ethical manner.

Autonomy: Service robots may be able to make their own decisions to some extent. It is important to ensure that these decisions are made in a way that is aligned with human values and interests.

Transparency: It is important to be transparent about the use of service robots. This means informing people about the capabilities of the robots and how they are being used.

Accountability: There must be a clear system of accountability for the actions of service robots. This means that there must be someone who can be held responsible if a robot causes harm.

In addition to these general ethical considerations, there are also a number of specific ethical issues that may arise in the development and use of service robots. For example, there may be concerns about the use of service robots to replace human workers, or about the use of service robots to monitor and control people.

It is important to have a public conversation about the ethical implications of service robotics. This will help to ensure that service robots are developed and used in a responsible and ethical manner.

Some Specific Examples Of Ethical Considerations In Service Robotics:

Delivery Robots: Delivery robots may be used to deliver packages to people's homes. This raises concerns about privacy, as the robots may be able to see and hear what is happening inside people's homes. It is important to ensure that delivery robots are designed with privacy in mind, and that people are informed about how their data is being collected and used.

Cleaning Robots: Cleaning robots may be used to clean homes and businesses. This raises concerns about safety, as the robots may have access to dangerous cleaning chemicals. It is important to ensure that cleaning robots are designed with safety in mind and that they are not able to access dangerous chemicals.

Security Robots: Security robots may be used to patrol public spaces and monitor people's activities. This raises concerns about civil liberties, as the robots may be able to collect data about people without their consent. It is important to ensure that security robots are used in a responsible and ethical manner, and that people are informed about their right to privacy.

It is important to note that the ethical implications of service robotics are still being debated. There is no one-size-fits-all answer to the question of how to develop and use service robots in a responsible and ethical manner. However, it is important to be aware of the ethical issues involved, and to have a public conversation about these issues.

Human-Robot Collaboration And Safety

Human-robot collaboration (HRC) is a rapidly growing field that is transforming the way we work. HRC allows humans and robots to work together to perform tasks that would be difficult or impossible for either to do alone.

HRC Can Offer A Number Of Benefits, Including:

- Increased productivity
- Improved quality
- Reduced costs
- Enhanced safety
- Improved ergonomics

However, HRC also presents a number of safety challenges. Humans and robots have different capabilities and limitations, and they may not always be able to anticipate each other's actions. This can lead to accidents and injuries.

Some Of The Key Safety Challenges In HRC:

Collision Avoidance: Robots must be able to avoid collisions with humans and other objects in the shared workspace.

Force Limiting: Robots must be able to limit the force they apply to humans and objects in the shared workspace.

Awareness: Humans and robots must be aware of each other's location and intentions in the shared workspace.

Communication: Humans and robots must be able to communicate effectively with each other.

Trust: Humans must trust the robots they are working with.

There Are A Number Of Ways To Mitigate The Safety Risks in HRC. These Include:

Risk Assessment: Before deploying HRC systems, it is important to conduct a risk assessment to identify and evaluate the potential hazards.

Engineering Controls: Engineering controls, such as safety barriers and sensors, can be used to reduce the risk of accidents.

Administrative Controls: Administrative controls, such as training and procedures, can also be used to reduce the risk of accidents.

It is important to note that safety is a critical consideration in HRC. By carefully designing and implementing HRC systems, and by taking appropriate safety measures, we can minimize the risk of accidents and injuries.

Human-Robot Collaboration and Safety

Some Examples Of How Safety Is Being Addressed In HRC:

Collaborative Robots (Cobots): Cobots are designed to work safely alongside humans. They have features such as force limiting and collision avoidance that help to prevent accidents.

Safety Sensors: Safety sensors can be used to detect the presence of humans and other objects in the shared workspace. If a hazard is detected, the robot can be stopped automatically.

Safeguarding Devices: Safeguarding devices, such as safety cages and interlocked gates, can be used to prevent humans from entering the robot's workspace while it is in operation.

Training: It is important to train workers on the safe operation of HRC systems. Workers should be aware of the potential hazards and how to avoid them.

HRC is a promising technology with the potential to revolutionize the way we work. By addressing the safety challenges in HRC, we can create safe and productive work environments where humans and robots can collaborate to achieve their goals.

User Experience (UX) In Human-Robot Interaction

User experience (UX) in human-robot interaction (HRI) is the design of robots and the interactions between robots and humans to be positive, meaningful, and efficient. It is important to consider UX in HRI because robots are becoming increasingly common in our lives, and we need to be able to interact with them in a way that is easy, enjoyable, and productive.

User Experience (UX) in Human-Robot Interaction

There Are A Number Of Factors That Contribute To Good UX In HRI, Including:

Usability: Robots should be easy to use and understand. This means that they should have clear and concise controls, and they should provide feedback to the user to indicate their actions and status.

Accessibility: Robots should be accessible to everyone, including people with disabilities. This means that they should have controls that are easy to reach and operate, and they should provide feedback in multiple modalities, such as visual, auditory, and tactile feedback.

Trust: Users must trust the robots they are interacting with. This means that the robots must be reliable and predictable, and they must be able to handle errors gracefully.

Delight: Robots should be enjoyable to interact with. This means that they should have a friendly and engaging personality, and they should be able to surprise and delight users.

There Are A Number Of Ways To Design Robots For Good UX. Some Of These Include:

User-Centered Design: User-centered design is a design approach that puts the user at the center of the design process. This means involving users in the design process to understand their needs and wants, and to get their feedback on prototypes.

Iterative Design: Iterative design is a design approach that involves designing, building, and testing prototypes repeatedly. This allows designers to refine the design and get feedback from users early and often.

Multidisciplinary Design: Designing robots for good UX requires a multidisciplinary team that includes designers, engineers, and social scientists. This team can work together to create robots that are both functional and user-friendly.

UX is an important consideration in the design and development of robots. By designing robots for good UX, we can create robots that are easy to use, trust, and enjoy interacting with. This will help to ensure that robots are widely accepted and used in society.

Contemporary Considerations Of UX In Human-Robot Interaction:

Social Robots: Social robots are designed to interact with humans in a social way. They often have human-like features and behaviors, such as faces, voices, and the ability to move. Social robots are being used in a variety of settings, such as healthcare, education, and customer service. When designing social robots, it is important to consider the emotions and feelings of the users. For example, social robots should be designed to be respectful and avoid offending users.

Assistive Robots: Assistive robots are designed to help people with disabilities or mobility issues. They can help with tasks such as cooking, cleaning, and getting around. When designing assistive robots, it is important to consider the needs of the users and to make sure that the robots are easy to use and control.

Service Robots: Service robots are designed to perform tasks that are repetitive, dangerous, or difficult for humans to do. They are often used in industrial and commercial settings. When designing service robots, it is important to consider the safety of the users and to make sure that the robots are able to operate safely in human environments.

UX is an essential consideration in the design and development of robots. By designing robots for good UX, we can create robots that are easy to use, trust, and enjoy interacting with. This will help to ensure that robots are widely accepted and used in society

Chapter 5: Advanced Sensing and Perception

5.1. Overview Of Sensing And Perception In Service Robots

Sensing and perception are essential for service robots to navigate and interact with their environment in a safe and efficient manner. Sensing refers to the process of acquiring data about the environment through sensors. Perception refers to the process of interpreting and understanding the sensed data.

Service Robots Typically Use A Variety Of Sensors To Collect Data About Their Environment. Some Common Sensors Include:

Cameras: Cameras are used to collect visual data about the environment. This data can be used to identify objects, track their movement, and estimate their distance.

LiDAR: LiDAR (Light Detection and Ranging) sensors emit laser pulses and measure the time it takes for the pulses to reflect back to the sensor. This data can be used to create a 3D map of the environment.

Radar: Radar sensors emit radio waves and measure the time it takes for the waves to reflect back to the sensor. This data can be used to detect objects and track their movement.

Sonar: Sonar sensors emit sound waves and measure the time it takes for the waves to reflect back to the sensor. This data can be used to detect objects and measure their distance.

Force Sensors: Force sensors measure the amount of force that is applied to them. This data can be used to detect collisions and to estimate the weight of objects.

Service robots use a variety of algorithms to process and interpret the sensed data. These algorithms can be used to identify objects, track their movement, estimate their distance, and create a map of the environment. Once the robot has a map of the environment, it can use this information to plan its path to the destination, avoid obstacles, and interact with objects in a safe and controlled manner.

Utilizing Sensing And Perception In Service Robots: Applications And Advancements

Delivery Robots: Delivery robots use sensing and perception to navigate through crowded sidewalks and streets. They use sensors to detect obstacles, such as people and vehicles, and to plan a path around them.

Cleaning Robots: Cleaning robots use sensing and perception to navigate around furniture and other objects in a room or building. They use sensors to detect obstacles and avoid collisions.

Security Robots: Security robots use sensing and perception to patrol an area and to detect intruders. They use sensors to detect objects and people and to track their movement.

Assistive Robots: Assistive robots use sensing and perception to help people with disabilities or mobility issues. They use sensors to detect the user's needs and to provide assistance in a safe and controlled manner.

Sensing and perception are essential for service robots to operate safely and efficiently in the real world. By developing new and innovative sensing and perception technologies, researchers are enabling service robots to perform a wide range of tasks that can improve our lives.

Sensing and Perception are used in Service Robots

Multimodal Sensing Systems

Multimodal sensing systems are systems that use two or more different types of sensors to collect data about the environment. This data can then be combined and fused to create a more comprehensive and accurate understanding of the environment than could be achieved using any single sensor type.

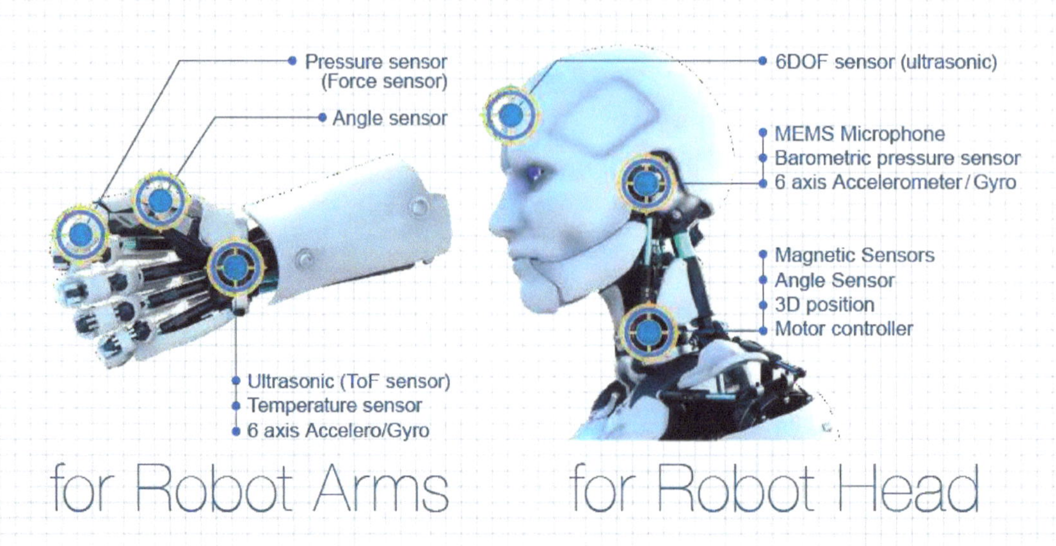

Multimodal Sensing Systems

Multimodal sensing systems are used in a variety of applications, including service robotics, self-driving cars, and medical imaging. In service robotics, multimodal sensing systems are used to help robots navigate and interact with their environment in a safe and efficient manner. For example, a delivery robot might use a combination of cameras, LiDAR, and radar sensors to detect obstacles and plan its path to the destination.

One of the key advantages of multimodal sensing systems is that they can overcome the limitations of individual sensor types. For example, cameras can be used to identify objects in well-lit environments, but they can be less effective in low-light or challenging weather conditions. LiDAR sensors can be used to create a 3D map of the environment, but they can be expensive and power-hungry. By combining cameras and LiDAR sensors, robots can create a more comprehensive and accurate understanding of their environment, even in challenging conditions.

Another advantage of multimodal sensing systems is that they can be used to verify and cross-validate sensor data. For example, a robot might use a camera to identify a pedestrian crossing the street. The robot could then use LiDAR to measure the distance to the pedestrian and to confirm that it is safe to cross. By combining data from multiple sensors, robots can make more informed decisions and reduce the risk of errors.

Multimodal sensing systems are becoming increasingly important in service robotics as robots are deployed in more complex and challenging environments. By using multimodal sensing systems, robots can navigate and interact with their environment in a more safe and efficient manner.

Multimodal sensing systems

These Are Some Examples Of How Multimodal Sensing Systems Are Used In Service Robots:

Delivery Robots: Delivery robots use multimodal sensing systems to navigate through crowded sidewalks and streets. They might use a combination of cameras, LiDAR, and radar sensors to detect obstacles, such as people and vehicles, and to plan a path around them.

Cleaning Robots: Cleaning robots use multimodal sensing systems to navigate around furniture and other objects in a room or building. They might use a combination of cameras and ultrasonic sensors to detect obstacles and avoid collisions.

Security Robots: Security robots use multimodal sensing systems to patrol an area and to detect intruders. They might use a combination of cameras, LiDAR, and thermal imaging sensors to detect objects and people and to track their movement.

Assistive Robots: Assistive robots use multimodal sensing systems to help people with disabilities or mobility issues. They might use a combination of cameras, infrared sensors, and force sensors to detect the user's needs and to provide assistance in a safe and controlled manner.

Multimodal sensing systems are an essential part of the development of service robots. By enabling robots to perceive their environment more accurately and comprehensively, multimodal sensing systems can help robots perform a wider range of tasks and to operate more safely and efficiently in the real world.

Integration Of Vision, Lidar, Sonar, And Other Sensors

The integration of vision, LiDAR, sonar, and other sensors is an essential part of the development of autonomous service robots. By combining data from multiple sensor types, robots can create a more comprehensive and accurate understanding of their environment, which allows them to navigate and interact with the environment in a safe and efficient manner.

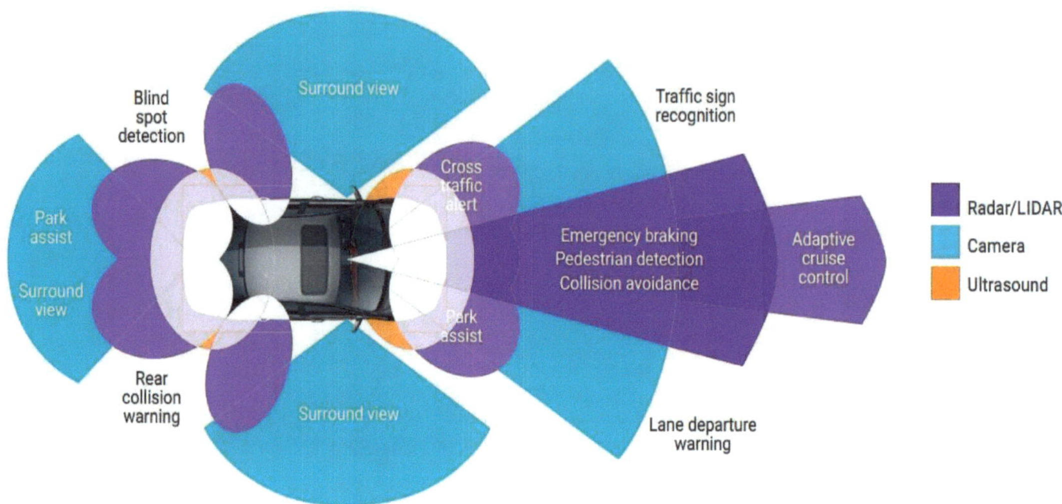

Integration of Vision, Lidar, Sonar, and Other Sensors

There are a number of different ways to integrate vision, LiDAR, sonar, and other sensors. One common approach is to use a sensor fusion algorithm. Sensor fusion algorithms combine data from multiple sensors to create a unified representation of the environment. This unified representation can then be used by the robot's navigation and planning algorithms to make decisions about how to move and interact with the environment.

Another approach to integrating vision, LiDAR, sonar, and other sensors is to use a multi-modal perception system. Multi-modal perception systems use different sensor types to complement each other's strengths and weaknesses. For example, a multi-modal perception system might use cameras to identify objects in the environment and LiDAR to measure their distance. The combined data from the cameras and LiDAR can then be used to create a more accurate and comprehensive understanding of the environment.

The integration of vision, LiDAR, sonar, and other sensors is a complex task, but it is essential for the development of autonomous service robots. By combining data from multiple sensor types, robots can create a more comprehensive and accurate understanding of their environment, which allows them to navigate and interact with the environment in a safe and efficient manner.

Sensor Integration In Service Robots: Enhancing Capabilities And Versatility

Delivery Robots: Delivery robots use a combination of vision, LiDAR, and radar sensors to navigate through crowded sidewalks and streets. They use vision sensors to identify objects and people, LiDAR sensors to measure distances, and radar sensors to detect moving objects. This combination of sensors allows delivery robots to safely navigate through complex and dynamic environments.

Cleaning Robots: Cleaning robots use a combination of vision and ultrasonic sensors to navigate around furniture and other objects in a room or building. They use vision sensors to identify objects and ultrasonic sensors to measure distances. This combination of sensors allows cleaning robots to safely navigate around obstacles and clean floors and surfaces thoroughly.

Security Robots: Security robots use a combination of vision, LiDAR, and thermal imaging sensors to patrol an area and detect intruders. They use vision sensors to identify objects and people, LiDAR sensors to measure distances, and thermal imaging sensors to detect heat signatures. This combination of sensors allows security robots to detect intruders even in low-light or challenging weather conditions.

Assistive Robots: Assistive robots use a combination of vision, infrared sensors, and force sensors to help people with disabilities or mobility issues. They use vision sensors to identify objects and people, infrared sensors to detect obstacles, and force sensors to measure the amount of force being applied. This combination of sensors allows assistive robots to safely navigate around obstacles and provide assistance to users in a safe and controlled manner.

The integration of vision, LiDAR, sonar, and other sensors is a rapidly developing field with the potential to revolutionize the way we interact with robots. By developing new and innovative sensor fusion and multi-modal perception algorithms, researchers are enabling service robots to perform a wider range of tasks and to operate more safely and efficiently in the real world.

Fusion Techniques For Enhancing Perception

Fusion techniques for enhancing perception in service robots combine data from multiple sensors to create a more comprehensive and accurate understanding of the environment.

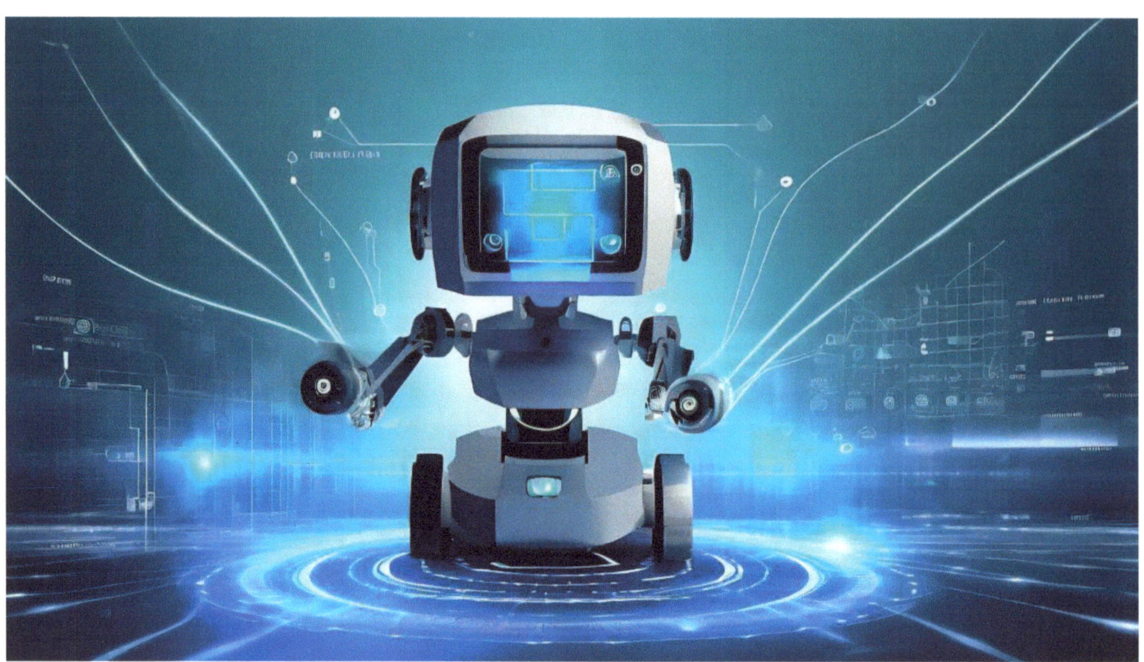

Fusion Techniques for Enhancing Perception

This can be done in a variety of ways, but some common fusion techniques are listed below.

Common Fusion Techniques:

Weighted Average: This is a simple but effective fusion technique that combines the data from multiple sensors by taking a weighted average. The weights are typically determined based on the reliability and accuracy of each sensor.

Kalman Filter: The Kalman filter is a more sophisticated fusion technique that uses a statistical model to track the state of the environment over time. The Kalman filter can combine data from multiple sensors to improve the accuracy of the state estimate.

Particle Filter: The particle filter is another sophisticated fusion technique that uses a set of weighted particles to represent the state of the environment. The particle filter can combine data from multiple sensors to update the weights of the particles and to create a more accurate representation of the state of the environment.

In addition to these general fusion techniques, there are also a number of specialized fusion techniques that have been developed for specific applications. For example, there are fusion techniques that are specifically designed for combining data from vision and LiDAR sensors.

Sensor Fusion Techniques: Enhancing Perception And Reliability In Service Robots

Improve the Accuracy of Object Detection: By combining data from multiple sensors, fusion techniques can help robots to more accurately detect objects in their environment. This is especially important in challenging environments, such as low-light or crowded conditions.

Reduce Noise: Fusion techniques can be used to reduce noise in the data from individual sensors. This can improve the overall accuracy and reliability of the robot's perception system.

Increase the Range of Perception: Fusion techniques can be used to increase the range of the robot's perception system. For example, a robot might use a fusion technique to combine data from a camera and a LiDAR sensor to see further away.

Improve the Robustness of Perception: Fusion techniques can be used to improve the robustness of the robot's perception system to errors in individual sensors. For example, if a camera sensor fails, a robot can still perceive its environment using a fusion technique to combine data from other sensors.

Fusion techniques are an essential part of the development of autonomous service robots. By combining data from multiple sensors, fusion techniques can help robots to create a more comprehensive and accurate understanding of their environment, which allows them to navigate and interact with the environment in a safe and efficient manner.

Transformative Fusion Techniques: Powering The Capabilities Of Service Robots

Delivery Robots: Delivery robots use fusion techniques to combine data from cameras, LiDAR, and radar sensors to create a more comprehensive and accurate understanding of their environment. This allows them to safely navigate through crowded sidewalks and streets, avoid collisions with objects and other vehicles, and identify and deliver packages to the correct addresses.

Cleaning Robots: Cleaning robots use fusion techniques to combine data from vision and ultrasonic sensors to create a more comprehensive and accurate understanding of their environment. This allows them to safely navigate around furniture and other objects in a room or building, and to thoroughly clean floors and surfaces.

Security Robots: Security robots use fusion techniques to combine data from vision, LiDAR, and thermal imaging sensors to create a more comprehensive and accurate understanding of their environment. This allows them to detect intruders even in low-light or challenging weather conditions, and to track their movements.

Assistive Robots: Assistive robots use fusion techniques to combine data from vision, infrared sensors, and force sensors to create a more comprehensive and accurate understanding of their environment. This allows them to safely navigate around obstacles and to provide assistance to users in a safe and controlled manner.

Fusion techniques are a rapidly developing field with the potential to revolutionize the way we interact with robots. By developing new and innovative fusion algorithms, researchers are enabling service robots to perform a wider range of tasks and to operate more safely and efficiently in the real world.

Environmental Mapping And Localization

Environmental mapping and localization are essential components of various fields, including robotics, autonomous vehicles, and augmented reality applications. These technologies allow machines to understand and navigate their surroundings effectively. Here, we'll delve into these concepts in detail:

Environmental Mapping: Environmental mapping involves creating a representation of the physical world using sensors and other data sources. This representation can be two-dimensional (2D) or three-dimensional (3D) and may include information about obstacles, landmarks, and other relevant features. There are several approaches to environmental mapping:

2D Occupancy Grid Mapping: In this method, the environment is divided into a grid of cells. Each cell is assigned a probability indicating the likelihood of it being occupied by an obstacle. Laser range finders and sonar sensors are commonly used for this purpose.

3D Mapping: This extends the concept to three dimensions. It is crucial for applications like autonomous flying drones or robots operating in complex, multi-level environments.

Simultaneous Localization and Mapping (SLAM): SLAM is a technique used to create a map of an unknown environment while simultaneously keeping track of the agent's location within that environment. It's widely employed in robotics and autonomous systems.

Semantic Mapping: This involves not only mapping the physical layout but also assigning semantic labels to different parts of the environment, such as identifying objects like tables, chairs, walls, etc.

Localization: Localization is the process of determining a device's or a robot's position within a mapped environment. This is usually done using various sensors and techniques.

Global Positioning System (GPS): GPS is a satellite-based navigation system that provides precise location information on Earth.

Inertial Measurement Unit (IMU): IMUs use a combination of accelerometers and gyroscopes to measure an object's linear and angular motion, which can be used for estimating position.

Visual Odometry: This technique uses cameras to estimate an agent's position by tracking features in the environment as it moves.

Lidar and Radar: These sensors use laser or radio waves, respectively, to measure distances to objects in the environment. This information can be used for localization.

This involves using fixed beacons with known positions that can communicate with the device or robot to provide positional information.

Integration: The synergy between mapping and localization is crucial for applications that require accurate spatial awareness. For instance, in autonomous vehicles, environmental mapping helps create detailed maps of roads and landmarks, while localization ensures the vehicle knows exactly where it is on those maps. This combination is also vital in robotics for tasks like navigation, exploration, and manipulation.

By effectively combining environmental mapping and localization, machines can operate autonomously in a wide range of environments, enabling them to perform tasks that would otherwise be challenging or impossible for humans to accomplish manually.

Environmental Mapping and Localizatio

5.2 Simultaneous Localization And Mapping (Slam) Algorithms

Simultaneous localization and mapping (SLAM) algorithms are used by robots to build a map of their environment while simultaneously determining their location within that map. This is a challenging problem, as the robot does not have any prior knowledge of the environment and must rely on its sensors to gather information.

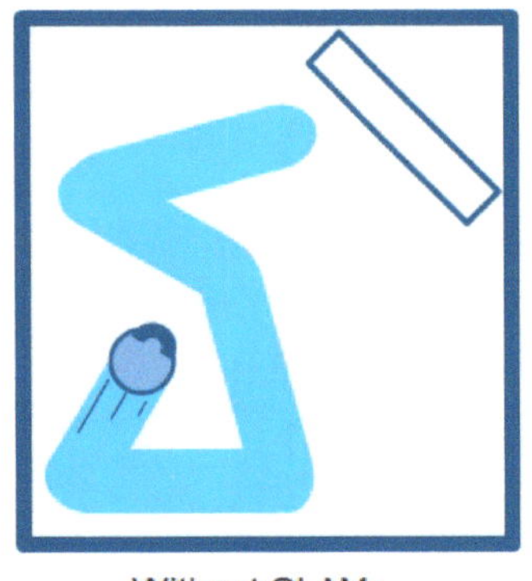

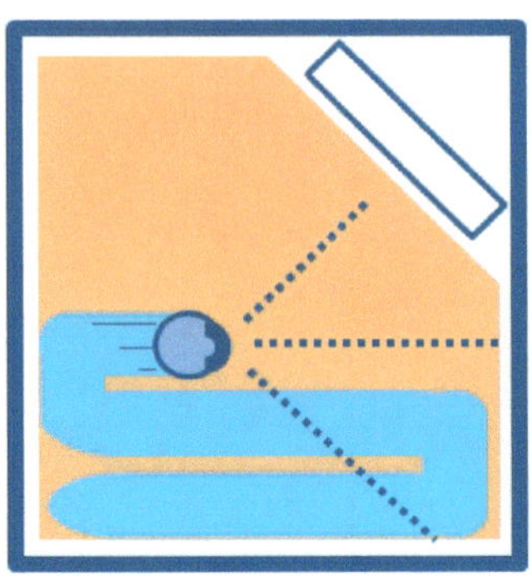

Simultaneous localization and mapping

There Are A Number Of Different Slam Algorithms, But They All Work By Following A Similar Basic Principle:

The robot collects data from its sensors. This data can include images, LiDAR scans, sonar measurements, and other types of data.

The robot uses the sensor data to build a map of the environment. This map can be represented in a variety of ways, such as a grid map or a topological map.

The robot uses the map to determine its location within the environment. This is done by matching the sensor data to the map.

The robot repeats steps 1-3 until it has built a complete map of the environment and knows its location within that map.

Slam Algorithms Are Used In A Variety Of Robotics Applications, Including:

Autonomous Navigation: SLAM algorithms allow robots to navigate autonomously in unknown environments. For example, SLAM algorithms are used by self-driving cars and delivery robots to navigate through streets and sidewalks.

Mapping and Exploration: SLAM algorithms can be used to create maps of unknown environments. This can be useful for tasks such as exploring disaster zones or mapping new planets.

> **Search and Rescue:** SLAM algorithms can be used to help robots search for and rescue people in disaster zones or other dangerous environments.

SLAM algorithms are an essential part of the development of autonomous robots. By allowing robots to build maps of their environment and to determine their location within those maps, SLAM algorithms enable robots to perform a wide range of tasks that would be impossible without them.

These Are The Examples Of How Slam Algorithms Are Being Used In Service Robots:

> **Delivery Robots:** Delivery robots use SLAM algorithms to build maps of their environment and to determine their location within those maps. This allows them to navigate autonomously through crowded sidewalks and streets, and to avoid collisions with objects and other vehicles.

> **Cleaning Robots:** Cleaning robots use SLAM algorithms to build maps of the homes or businesses that they are cleaning. This allows them to efficiently clean all of the surfaces in the environment, and to avoid cleaning the same areas multiple times.

> **Security Robots:** Security robots use SLAM algorithms to build maps of the areas that they are patrolling. This allows them to detect intruders even in low-light or challenging weather conditions, and to track their movements.

> **Assistive Robots:** Assistive robots use SLAM algorithms to build maps of the homes or businesses that they are assisting in. This allows them to navigate around obstacles and to provide assistance to users in a safe and   controlled manner.

SLAM algorithms are a rapidly developing field with the potential to revolutionize the way we interact with robots. By developing new and innovative SLAM algorithms, researchers are enabling service robots to perform a wider range of tasks and to operate more safely and efficiently in the real world.

Dynamic Map Updating For Navigation

Dynamic map updating for navigation is the process of keeping a map of the environment up-to-date in real-time. This is important for mobile robots, such as delivery robots, cleaning robots, and security robots, that need to navigate in dynamic environments.

There are a number of different approaches to dynamic map updating. One common approach is to use a sensor fusion algorithm to combine data from multiple sensors to create a more comprehensive and accurate understanding of the environment. The sensor fusion algorithm can then be used to update the map in real- time.

Another approach to dynamic map updating is to use a machine learning algorithm to learn the dynamics of the environment. The machine learning algorithm can then be used to predict changes in the environment and to update the map accordingly.

Dynamic map updating is a challenging problem, but it is essential for mobile robots to operate safely and efficiently in dynamic environments. By developing new and innovative dynamic map updating algorithms, researchers are enabling mobile robots to perform a wider range of tasks and to operate in more complex and challenging environments.

Dynamic Map Updating for Navigation

Exploring Applications Of Dynamic Map Updating In Service Robots:

Delivery Robots: Delivery robots use dynamic map updating to keep track of changes in the environment, such as traffic congestion and pedestrians crossing the street. This allows them to safely navigate through crowded sidewalks and streets, and to avoid collisions with objects and other vehicles.

Cleaning Robots: Cleaning robots use dynamic map updating to keep track of changes in the environment, such as furniture being moved or new obstacles appearing. This allows them to safely navigate around obstacles and to thoroughly clean floors and surfaces.

Security Robots: Security robots use dynamic map updating to keep track of changes in the environment, such as people entering or leaving the area that they are patrolling. This allows them to detect intruders even in low-light or challenging weather conditions, and to track their movements.

Assistive Robots: Assistive robots use dynamic map updating to keep track of changes in the environment, such as doors opening and closing or people moving around. This allows them to safely navigate around obstacles and to provide assistance to users in a safe and controlled manner.

Dynamic map updating is a rapidly developing field with the potential to revolutionize the way we interact with robots. By developing new and innovative dynamic map updating algorithms, researchers are enabling service robots to perform a wider range of tasks and to operate more safely and efficiently in the real world.

Object Recognition And Manipulation

Object recognition and manipulation are two essential tasks for service robots. Object recognition is the ability of a robot to identify and classify objects in its environment. Object manipulation is the ability of a robot to grasp and move objects.

Object recognition is a challenging task, as there are many different types of objects in the world and they can be presented in a variety of different ways.

For Example, A Robot May Need To Be Able To Recognize A Cup Regardless Of Its Size, Color, Or Orientation.

There are a number of different approaches to object recognition. One common approach is to use machine learning. Machine learning algorithms can be trained on a dataset of images and labels to learn to identify objects in new images.

Another approach to object recognition is to use deep learning. Deep learning algorithms are a type of machine learning algorithm that is particularly well-suited for object recognition tasks. Deep learning algorithms can be trained on large datasets of images to learn to identify objects with high accuracy.

Object manipulation is also a challenging task, as robots need to be able to grasp and move objects without damaging them. Additionally, robots need to be able to manipulate objects in a safe and controlled manner to avoid collisions with people and other objects.

There are a number of different approaches to object manipulation. One common approach is to use a gripper. Grippers are devices that can be used to grasp objects. Grippers come in a variety of different shapes and sizes, and they can be designed to grasp specific types of objects.

Another approach to object manipulation is to use a robotic arm. Robotic arms are multi-jointed manipulators that can be used to move objects in a variety of ways. Robotic arms are often used in industrial applications, but they are also becoming increasingly common in service robotics applications.

Object recognition and manipulation are essential tasks for service robots. By developing new and innovative object recognition and manipulation algorithms and technologies, researchers are enabling service robots to perform a wider range of tasks and to operate more safely and efficiently in the real world.

Object Detection And Recognition Techniques

Object detection and recognition techniques are used by robots to identify and locate objects in their environment. These techniques are essential for robots to perform a wide range of tasks, such as navigation, manipulation, and interaction with humans.

There Are Two Main Categories Of Object Detection And Recognition Techniques:

Traditional Techniques: These techniques rely on hand-crafted features, such as color, texture, and shape, to identify objects. Traditional techniques are relatively simple to implement, but they can be less accurate than deep learning techniques, especially in challenging environments.

Deep Learning Techniques: Deep learning techniques use neural networks to learn features from data. Neural networks can be trained on large datasets of images to learn to identify objects with high accuracy. Deep learning techniques are more complex to implement than traditional techniques, but they are generally more accurate and robust.

Some Common Object Detection And Recognition Techniques Include:

Histogram of Oriented Gradients (HOG): HOG is a traditional technique that extracts features from images by computing the distribution of oriented gradients. HOG features are robust to changes in illumination and pose, making them well-suited for object detection tasks.

Support Vector Machines (SVMs): SVMs are machine learning algorithms that can be used to classify objects based on their HOG features. SVMs are trained on a dataset of labeled images to learn to distinguish between different classes of objects.

Convolutional Neural Networks (CNNs): CNNs are a type of deep learning algorithm that is well- suited for object detection and recognition tasks. CNNs learn to identify objects by extracting features from images using a series of convolutional layers.

You Only Look Once (YOLO): YOLO is a real-time object detection algorithm that uses a single neural network to detect and classify objects in an image. YOLO is one of the fastest and most accurate object detection algorithms available.

Object detection and recognition techniques are constantly evolving, and new techniques are being developed all the time. As object detection and recognition techniques become more accurate and efficient, they will enable robots to perform a wider range of tasks and to operate more safely and efficiently in the real world.

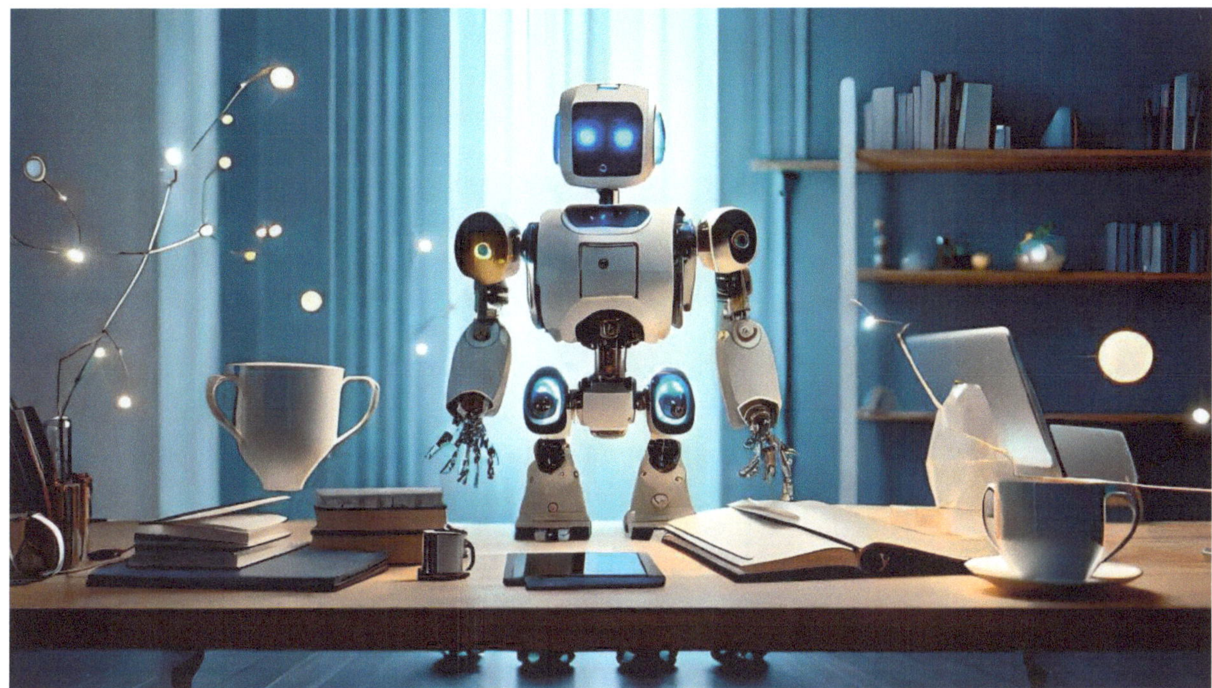

Object detection and recognition techniques

Object detection and recognition techniques are essential for the development of service robots. By developing new and innovative object detection and recognition techniques, researchers are enabling service robots to perform a wider range of tasks and to operate more safely and efficiently in the real world.

Object Recognition And Manipulation Applications In Service Robots

Delivery Robots: Delivery robots use object recognition to identify and classify objects, such as packages and mailboxes. They use object manipulation to grasp and move objects to their destination.

Cleaning Robots: Cleaning robots use object recognition to identify and classify objects, such as furniture, trash cans, and dirty surfaces. They use object manipulation to move objects around and to clean surfaces.

Security Robots: Security robots use object recognition to identify and classify objects, such as people, vehicles, and weapons. They use object manipulation to open and close doors, to pick up and disarm objects, and to subdue intruders.

Assistive Robots: Assistive robots use object recognition to identify and classify objects, such as food, drinks, and medications. They use object manipulation to grasp and move objects to assist users with tasks such as eating, drinking, and taking medication.

Object recognition and manipulation are rapidly developing fields with the potential to revolutionize the way we interact with robots. By developing new and innovative object recognition and manipulation algorithms and technologies, researchers are enabling service robots to perform a wider range of tasks and to operate more safely and efficiently in the real world.

Grasping And Manipulation Strategies

Grasping and manipulation strategies are the methods that robots use to grasp and move objects. These strategies are essential for robots to perform a wide range of tasks, such as delivery, cleaning, and assembly.

There are a number of different grasping and manipulation strategies that robots can use. The choice of strategy depends on a number of factors, such as the type of object being grasped, the desired accuracy and precision of the manipulation, and the constraints of the robot's hardware.

Some Common Grasping And Manipulation Strategies Include:

Power grasping: Power grasping is a simple and effective grasping strategy that uses the robot's fingers to apply a large amount of force to the object. Power grasping is often used for grasping objects that are heavy or difficult to grip.

Precision grasping: Precision grasping is a more delicate grasping strategy that uses the robot's fingers to apply a small amount of force to the object with high precision. Precision grasping is often used for grasping objects that are fragile or delicate.

In-hand manipulation: In-hand manipulation is a strategy that robots use to manipulate objects that are already grasped. In-hand manipulation can be used to reorient objects, to move them to a different location, or to perform other tasks.

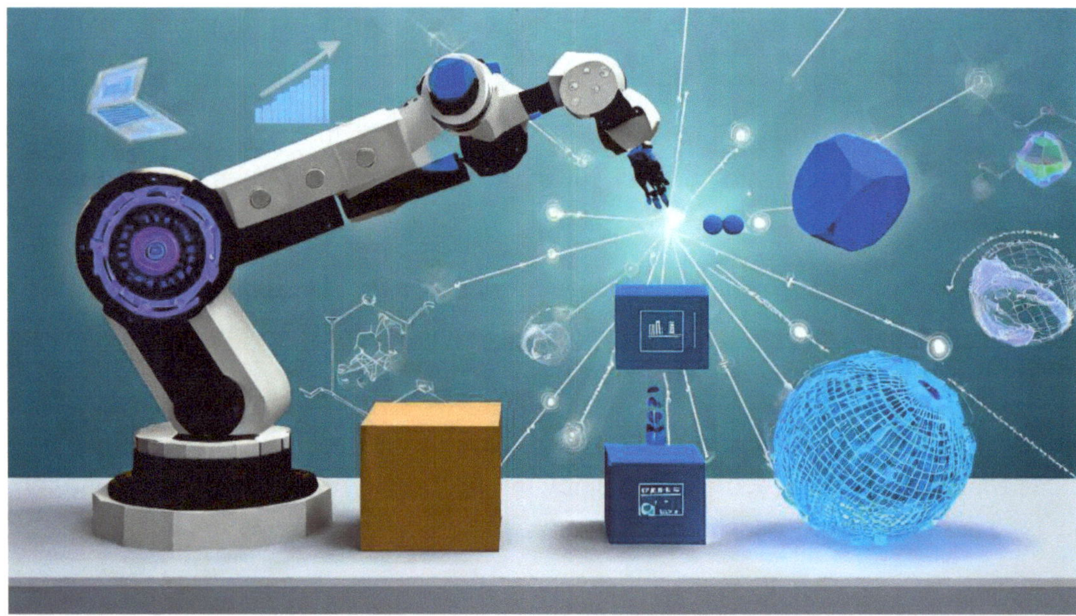

Grasping and Manipulation Strategies

Robots can use a variety of sensors to assist with grasping and manipulation tasks. For example,

robots can use cameras to identify and locate objects, and they can use force sensors to measure the amount of force being applied to an object.

Grasping and manipulation strategies are constantly evolving, and new strategies are being developed all the time. As grasping and manipulation strategies become more sophisticated, they will enable robots to perform a wider range of tasks and to operate more safely and efficiently in the real world.

Applications Of Grasping And Manipulation Strategies In Service Robots

Delivery Robots: Delivery robots use grasping and manipulation strategies to pick up and deliver packages. They typically use power grasping to pick up packages, and they may use precision grasping or in-hand manipulation to place packages in specific locations, such as on a doorstep or in a mailbox.

Cleaning Robots: Cleaning robots use grasping and manipulation strategies to pick up and move objects, such as trash and dirty dishes. They typically use power grasping to pick up objects, and they may use precision grasping or in-hand manipulation to place objects in specific locations, such as a trash can or a sink.

Security Robots: Security robots use grasping and manipulation strategies to open and close doors, to pick up and disarm objects, and to subdue intruders. They typically use power grasping for these tasks.

Assistive Robots: Assistive robots use grasping and manipulation strategies to help people with disabilities or mobility issues with tasks such as eating, drinking, and taking medication. They typically use precision grasping for these tasks.

Grasping and manipulation strategies are essential for the development of service robots. By developing new and innovative grasping and manipulation strategies, researchers are enabling service robots to perform a wider range of tasks and to operate more safely and efficiently in the real world.

Human-Robot Interaction

Human-robot interaction (HRI) is the study of interactions between humans and robots. HRI is a multidisciplinary field that draws on contributions from human-computer interaction, artificial intelligence, robotics, natural language processing, design, and psychology.

The goal of HRI is to develop robots that can interact with humans in a safe, efficient, and natural way. This requires robots to be able to understand and respond to human language, gestures, and facial expressions. Robots also need to be able to plan and execute actions in a way that is safe and predictable for humans.

HRI is becoming increasingly important as robots are deployed in a wider range of applications, such as healthcare, manufacturing, and customer service. For example, HRI is essential for the development of robots that can assist with tasks such as surgery, rehabilitation, and product assembly. HRI is also important for the development of robots that can provide customer service in stores and restaurants.

There are a number of challenges that need to be addressed in order to develop effective HRI systems. One challenge is that humans and robots have very different cognitive and physical abilities. This can make it difficult for robots to understand human intentions and to respond in a way that is appropriate.

Another challenge is that human expectations of robots are constantly evolving. As robots become more sophisticated, people expect them to be able to interact with them in more natural and intuitive ways.

Despite these challenges, HRI is a rapidly developing field with the potential to revolutionize the way we interact with machines. By developing new and innovative HRI systems, researchers are enabling robots to perform a wider range of tasks and to operate more safely and efficiently in the real world.

Some Examples Of How HRI Is Being Used In Service Robots:

Delivery Robots: Delivery robots use HRI to interact with customers and to navigate through crowded sidewalks and streets. They may use voice commands or touchscreens to communicate with customers, and they may use sensors to avoid obstacles and to interact with other robots.

Cleaning Robots: Cleaning robots use HRI to understand and respond to human commands. They may use voice commands or remote controls to receive instructions from users, and they may use sensors to avoid obstacles and clean surfaces efficiently.

Security Robots: Security robots use HRI to interact with people and to detect intruders. They may use voice commands or cameras to identify and greet people, and they may use sensors to detect suspicious activity.

Assistive Robots: Assistive robots use HRI to understand and respond to the needs of users. They may use voice commands or cameras to identify and respond to user requests, and they may use sensors to safely navigate around obstacles and to provide assistance to users.

HRI is an essential part of the development of service robots. By developing new and innovative HRI systems, researchers are enabling service robots to perform a wider range of tasks and to operate more safely and efficiently in the real world

5.3 Natural Language Processing For Communication

Natural language processing (NLP) is a field of computer science that deals with the interaction between computers and human (natural) languages. It's a subfield of artificial intelligence that deals with the ability of computers to understand and process human language, including speech and text.

Nlp Is Used For A Variety Of Tasks, Including:

Machine Translation: Translating text from one language to another.

Speech recognition: Converting spoken language into text.

Text Analysis: Extracting meaning from text, such as sentiment analysis, topic extraction, and question answering.

Natural Language Generation: Generating human-like text, such as creative text formats and code.

NLP can be used to improve communication between humans and robots in a number of ways.

For Example, NLP Can Be Used To:

Enable Robots to Understand and Respond to Natural Language Commands: This allows humans to interact with robots in a more natural and intuitive way.

Generate Natural Language Responses: This allows robots to communicate with humans in a more natural and engaging way.

Translate Languages: This allows robots to communicate with people from different cultures. **Analyze Text:** This allows robots to extract information from text, such as the sentiment of a customer review or the steps involved in a recipe.

Generate Text: This allows robots to create content, such as news articles, blog posts, and poems.

Natural Language Processing for Communication

Utilizing NLP For Enhanced Communication In Service Robots: Illustrative Instances:

Delivery Robots: Delivery robots use NLP to understand and respond to customer instructions, such as "Deliver this package to 123 Main Street." They also use NLP to generate natural language responses, such as "I'm on my way to deliver your package."

Cleaning Robots: Cleaning robots use NLP to understand and respond to human commands, such as "Clean the kitchen." They also use NLP to generate natural language responses, such as "I'm starting to clean the kitchen now."

Security Robots: Security robots use NLP to interact with people and to detect intruders. They may use NLP to understand and respond to human commands, such as "Identify yourself." They may also use NLP to generate natural language responses, such as "You are not authorized to enter this area."

Assistive Robots: Assistive robots use NLP to understand and respond to the needs of users. They may use NLP to understand and respond to user commands, such as "Help me get dressed." They may also use NLP to generate natural language responses, such as "I'm here to help you get dressed."

NLP is a rapidly developing field with the potential to revolutionize the way we interact with robots. By developing new and innovative NLP techniques, researchers are enabling service robots to communicate with humans in a more natural and effective way. Overall, NLP is a powerful tool that can be used to improve communication between humans and robots. By enabling robots to understand and respond to natural language, NLP can make robots more accessible and useful to people from all walks of life.

Gesture And Emotion Recognition

Gesture and emotion recognition are two important areas of research in human-robot interaction (HRI). Gesture recognition allows robots to understand human body language, while emotion recognition allows robots to understand human emotions. Both of these technologies are essential for developing robots that can interact with humans in a natural and effective way.

Gesture and Emotion Recognition

Gesture Recognition Can Be Used For A Variety Of Tasks, Including

Controlling Robots: Gestures can be used to control robots, such as commanding a robot to move forward, backward, or turn.

Communicating with Robots: Gestures can be used to communicate with robots, such as giving a thumbs up to indicate approval or a thumbs down to indicate disapproval.

Interpreting Human Intentions: Gestures can be used to interpret human intentions, such as understanding whether someone is asking for help or giving directions.

Emotion Recognition Can Be Used For A Variety Of Tasks, Including

Adapting to Human Emotions: Robots can use emotion recognition to adapt their behavior to human emotions, such as speaking more gently to someone who is feeling sad or speaking more enthusiastically to someone who is feeling happy.

Providing Social Support: Robots can use emotion recognition to provide social support to humans, such as comforting someone who is feeling sad or encouraging someone who is feeling overwhelmed.

Detecting Dangerous Situations: Robots can use emotion recognition to detect dangerous situations, such as when someone is feeling angry or aggressive.

Gesture and emotion recognition are still in their early stages of development, but they have the potential to revolutionize the way we interact with robots. By enabling robots to understand human body language and emotions, gesture and emotion recognition can make robots more accessible and useful to people from all walks of life.

How Service Robots Harness Gesture And Emotion Recognition

Delivery Robots: Delivery robots can use gesture and emotion recognition to understand and respond to customer instructions and reactions. For example, a delivery robot could use gesture recognition to understand when a customer is waving it over or when a customer is giving it a thumbs up or thumbs down. It could also use emotion recognition to understand whether a customer is feeling happy, sad, or angry.

Cleaning Robots: Cleaning robots can use gesture and emotion recognition to understand and respond to human commands and preferences. For example, a cleaning robot could use gesture recognition to understand when a person wants it to clean a specific area or when a person wants it to stop cleaning. It could also use emotion recognition to understand whether a person is feeling comfortable or uncomfortable with its presence.

Security Robots: Security robots can use gesture and emotion recognition to detect intruders and to assess the safety of a situation. For example, a security robot could use gesture recognition to identify people who are entering a restricted area or who are making threatening gestures. It could also use emotion recognition to identify people who are feeling angry or aggressive.

Assistive Robots: Assistive robots can use gesture and emotion recognition to provide personalized assistance to users. For example, an assistive robot could use gesture recognition to understand when a user wants it to help them with a task or when they want it to move out of the way. It could also use emotion recognition to identify when a user is feeling frustrated or overwhelmed and to adjust their behavior accordingly.

Gesture and emotion recognition are rapidly developing fields with the potential to revolutionize the way we interact with robots. By developing new and innovative gesture and emotion recognition techniques, researchers are enabling service robots to perform a wider range of tasks and to operate more safely and efficiently in the real world.

Socially Assistive Robot Behavior

Socially assistive robot behavior (SARB) refers to the ways in which robots can use their social intelligence to assist humans in a variety of ways. SARB encompasses a wide range of behaviors, such as:

Providing Companionship: SARs can provide companionship to people who are lonely or isolated. They can do this by engaging in conversation, playing games, or simply being present.

   Offering Emotional support: SARs can offer emotional support to people who are feeling stressed, anxious, or depressed. They can do this by listening to people's concerns, offering words of encouragement, or simply being a shoulder to cry on.

   Promoting Cognitive Engagement: SARs can promote cognitive engagement in people with dementia or other cognitive impairments. They can do this by playing games, engaging in puzzles, or simply stimulating conversation.

   Assisting with Physical Activities: SARs can assist with physical activities in people with mobility impairments. They can do this by helping people to walk, stand up, or reach objects.

   Providing Educational Support: SARs can provide educational support to children and adults alike. They can do this by teaching new skills, helping with homework, or providing feedback on creative work.

SARB is a rapidly developing field with the potential to revolutionize the way we interact with robots. By developing new and innovative SARB techniques, researchers are enabling robots to provide a wider range of assistance to people from all walks of life.

Service Beyond Companionship: Socially Assistive Robots In Various Sectors

A socially Assistive Robot (SAR) is used in a nursing home to provide companionship to elderly residents. The SAR can engage in conversation, play games, and simply be present with residents. This can help to reduce loneliness and isolation and improve the overall well-being of residents.

A SAR is used in a hospital to provide emotional support to children who are undergoing chemotherapy. The SAR can listen to children's concerns, offer words of encouragement, and simply be a shoulder to cry on. This can help to reduce stress and anxiety and improve the quality of life for children with cancer.

A SAR is used in a school to promote cognitive engagement in students with autism. The SAR can play games, engage in puzzles, and simply stimulate conversation with students. This can help to improve students' attention, memory, and problem-solving skills.

A SAR is used in a rehabilitation center to assist with physical activities in patients with stroke. The SAR can help patients to walk, stand up, and reach objects. This can help patients to regain their mobility and independence.

A SAR is used in a museum to provide educational support to visitors. The SAR can teach visitors about the exhibits, answer questions, and provide feedback on creative work. This can help visitors to learn more about the museum's collection and have a more engaging experience.

SARB is a promising new field with the potential to make a significant positive impact on the lives of people from all walks of life. By developing new and innovative SARB techniques, researchers are enabling robots to provide a wider range of assistance and support to people in need.

Machine Learning And Deep Learning In Perception

Machine learning (ML) and deep learning (DL) are two types of artificial intelligence (AI) that are well- suited for perception tasks. Perception is the ability to understand the environment through sensory data, such as vision, hearing, and touch.

ML algorithms can be trained to learn patterns in data and to make predictions based on those patterns. For example, an ML algorithm can be trained to recognize different types of objects in images. DL algorithms are a type of ML algorithm that uses artificial neural networks to learn complex patterns in data. Neural networks are inspired by the structure of the human brain, and they are able to learn patterns that are difficult or impossible to learn with traditional ML algorithms.

Ml And Dl Are Used In A Variety Of Perception Tasks, Including:

Image Recognition: ML and DL algorithms can be used to recognize objects, faces, and scenes in images. This technology is used in a variety of applications, such as self-driving cars, security systems, and social media.

Natural Language Processing (NLP): ML and DL algorithms can be used to understand and respond to human language. This technology is used in a variety of applications, such as machine translation, chatbots, and voice assistants.

Speech Recognition: ML and DL algorithms can be used to convert speech into text. This technology is used in a variety of applications, such as voice dictation, transcription, and voice assistants.

Sensor Fusion: ML and DL algorithms can be used to fuse data from multiple sensors to create a more comprehensive and accurate understanding of the environment. This technology is used in a variety of applications, such as robotics, self-driving cars, and augmented reality.

ML and DL are rapidly developing fields, and new applications are being developed all the time. As ML and DL algorithms become more sophisticated and accurate, they will be able to perform a wider range of perception tasks and to operate in more challenging environments.

Machine Learning and Deep Learning in Perception

The Cognitive Edge: Ml And Dl In Service Robot Development

Delivery Robots: Delivery robots use ML and DL to recognize objects, such as packages and mailboxes, in their environment. They also use ML and DL to avoid obstacles and navigate safely through crowded sidewalks and streets.

Cleaning Robots: Cleaning robots use ML and DL to recognize objects, such as furniture and trash cans, in their environment. They also use ML and DL to avoid obstacles and clean surfaces efficiently.

Security Robots: Security robots use ML and DL to recognize objects, such as people, vehicles, and weapons, in their environment. They also use ML and DL to detect intruders and to track their movements.

Assistive Robots: Assistive robots use ML and DL to recognize objects, such as food, drinks, and medications, in their environment. They also use ML and DL to avoid obstacles and to assist users with tasks such as eating, drinking, and taking medication.

ML and DL are essential for the development of service robots. By enabling robots to perceive their environment in a more accurate and comprehensive way, ML and DL are enabling service robots to perform a wider range of tasks and to operate more safely and efficiently in the real world.

Deep Neural Networks For Object Recognition

Deep neural networks (DNNs) are a type of machine learning algorithm that is well-suited for object recognition tasks. DNNs are inspired by the structure of the human brain, and they are able to learn complex patterns in data. DNNs for object recognition are typically trained on large datasets of images and labels. The DNN learns to extract features from the images and to associate those features with the corresponding labels. Once the DNN is trained, it can be used to recognize objects in new images.

Dnns For Object Recognition Are Used In A Variety Of Applications, Including:

Self-Driving Cars: DNNs are used in self-driving cars to recognize objects on the road, such as other vehicles, pedestrians, and traffic signs.

Security Systems: DNNs are used in security systems to detect intruders and other suspicious activity.

Social Media: DNNs are used in social media platforms to recognize objects and people in photos and videos.

Service Robots: DNNs are used in service robots to recognize objects in their environment, such as packages, trash cans, and people.

DNNs for object recognition are still under development, but they have made significant progress in recent years. DNNs are now able to recognize objects with high accuracy, even in challenging conditions, such as low light or cluttered environments.

DNNs In Action: Object Recognition In Service Robotic

Delivery Robots: Delivery robots use DNNs to recognize packages and mailboxes, as well as obstacles to avoid. This enables them to navigate safely through crowded sidewalks and streets.

Cleaning Robots: Cleaning robots use DNNs to recognize furniture, trash cans, and other objects in their environment. This enables them to avoid obstacles and to clean surfaces efficiently.

Security Robots: Security robots use DNNs to recognize people, vehicles, and weapons. This enables them to detect intruders and to track their movements.

Assistive Robots: Assistive robots use DNNs to recognize food, drinks, and medications. This enables them to assist users with tasks such as eating, drinking, and taking medication.

DNNs for object recognition are essential for the development of service robots. By enabling robots to recognize objects in their environment, DNNs are enabling them to perform a wider range of tasks and to operate more safely and efficiently in the real world.

Overall, DNNs are a powerful tool for object recognition. By enabling machines to learn complex patterns in data, DNNs are enabling them to perform tasks that were once thought to be impossible. As DNNs continue to develop, they are likely to revolutionize the way we interact with machines and the world around us.

Reinforcement Learning For Autonomous Decision-Making

Reinforcement learning (RL) is a type of machine learning that enables agents to learn how to behave in an environment by trial and error. RL agents are rewarded for taking actions that lead to desired outcomes and penalized for taking actions that lead to undesired outcomes. Over time, the agent learns to choose actions that maximize its expected reward.

RL is well-suited for autonomous decision-making tasks, as it allows agents to learn how to behave in complex and dynamic environments without being explicitly programmed. RL agents can be used to control a wide range of systems, including robots, self-driving cars, and financial trading algorithms.

Rl Algorithms Are Typically Divided Into Two Categories: value-based and policy-based. Value-based algorithms learn to estimate the value of each state in the environment, while policy-based algorithms learn to directly map states to actions.

RL algorithms have been used to achieve impressive results in a variety of autonomous decision-making tasks. For example, RL agents have been trained to play games at a superhuman level, to control robots to perform complex tasks, and to drive self-driving cars autonomously.

From Learning To Decision-Making: How Rl Drives Service Robots :

Delivery Robots: RL agents can be used to train delivery robots to navigate safely through crowded sidewalks and streets and to deliver packages to customers efficiently.

Cleaning Robots: RL agents can be used to train cleaning robots to clean different types of environments efficiently and to avoid obstacles.

Security Robots: RL agents can be used to train security robots to patrol buildings and to detect and respond to intruders.

Assistive Robots: RL agents can be used to train assistive robots to help people with disabilities or mobility issues with tasks such as eating, dressing, and bathing.

RL is a promising technology for autonomous decision-making in service robots. By enabling robots to learn how to behave in their environment by trial and error, RL is enabling them to perform a wider range of tasks and to operate more safely and efficiently in the real world.

Overall, RL is a powerful tool for autonomous decision-making. By enabling machines to learn how to behave in complex and dynamic environments, RL is enabling them to perform tasks that were once thought to be impossible. As RL continues to develop, it is likely to revolutionize the way we interact with machines and the world around us.

Chapter 6: Human-Robot Interaction (HRI)

6.1. Introduction

Human-Robot Interaction (HRI) is of fundamental significance in service robotics, as it is the key to enabling service robots to safely and effectively interact with humans and their environment. Service robots are designed to operate in a wide range of settings, from homes and hospitals to offices and stores, where they are expected to work alongside humans and provide assistance with a variety of tasks. This requires that service robots be able to understand and respond to human needs, as well as communicate and coordinate with humans in a way that is natural and efficient.

Human-robot interaction (HRI)

Service Robotics: Why Human-Robot Interaction (HRI) Matters

Safety: Service robots need to be designed to be safe for humans to interact with. This means that they need to be able to avoid collisions and prevent injuries. It also means that they need to be able to operate reliably and consistently. HRI is essential for ensuring that service robots can be safely deployed in real-world environments.

Efficiency: Service robots are designed to help humans with tasks, so it is important for them to be able to interact with humans in a way that is efficient and productive. This means that HRI should be designed to minimize the cognitive load on humans and to allow for seamless interaction between humans and robots.

Naturalness: Service robots should be able to interact with humans in a way that is natural and intuitive. This means that they should be able to understand and respond to human gestures, body language, and speech. It also means that they should be able to communicate with humans in a way that is clear and concise.

HRI is also important for ensuring that service robots are accepted and adopted by humans. If humans do not trust or feel comfortable interacting with service robots, then they are less likely to use them. HRI can help to build trust and acceptance by designing service robots to be transparent, predictable, and adaptable.

Service Robots Evolved: Real-Life Applications Of HRI For Improvement

Personal Service Robots: Personal service robots are increasingly being used to assist people with everyday tasks, such as cooking, cleaning, and shopping. These robots typically interact with humans through natural language and gestures.

For example, a personal service robot might be able to understand and respond to voice commands, such as "Please make me a cup of coffee" or "Please help me fold my laundry."

Healthcare Robots: Healthcare robots are being used to assist healthcare professionals with tasks such as patient care, surgery, and rehabilitation. These robots typically interact with humans through touch and voice.

For example, a healthcare robot might be able to help a nurse lift and transfer a patient, or it might be able to provide physical therapy exercises to a patient.

Educational Robots: Educational robots are being used to help children learn and develop new skills. These robots typically interact with humans through play and games.

For example, an educational robot might be able to teach a child about math or science concepts through interactive games and activities.

Overall, HRI is an essential component of service robotics. By designing service robots to interact with humans in a safe, efficient, and natural way, we can ensure that these robots can be used to improve the lives of people in a variety of ways.

6.2. Foundations of HRI in Service Robotics

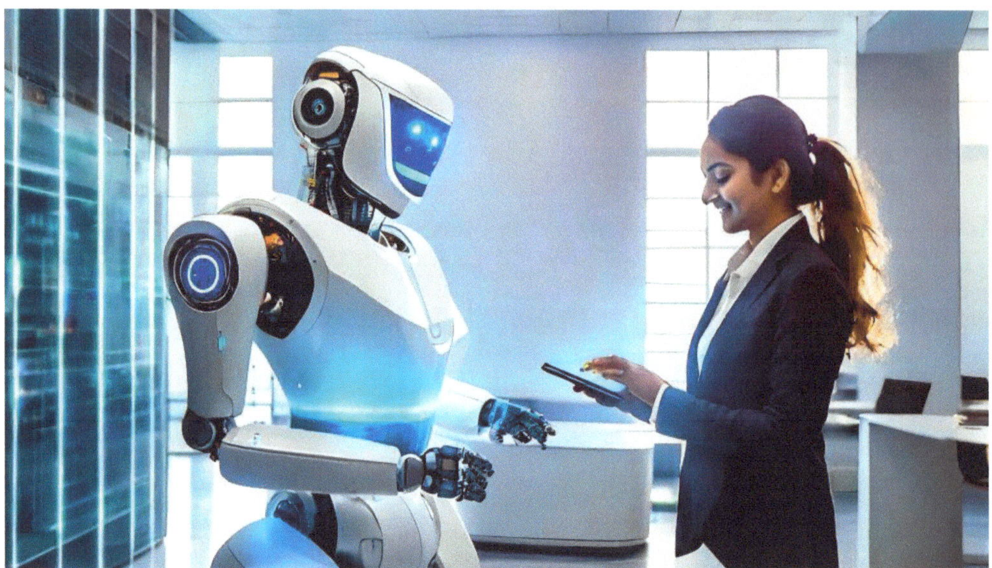

Foundations of HRI in Service Robotics

The Foundations Of Human-Robot Interaction (Hri) In Service Robotics Are Based On A Number Of Different Disciplines, Including:

Artificial Intelligence (AI): AI is used to develop robots that can understand their environment, make decisions, and take actions. This is essential for robots to be able to interact with humans in a safe and effective way.

Machine Learning (ML): ML is a type of AI that allows robots to learn from data. This is important for robots to be able to adapt to their environment and to the people they interact with.

Robotics: Robotics is the field of engineering that deals with the design, construction, operation, and application of robots. This includes the development of robots that can move and manipulate objects in the world.

Human-Computer Interaction (HCI): HCI is the study of how people interact with computers and other technologies. This is important for designing robots that are easy and enjoyable for people to interact with.

Social Psychology: Social psychology is the study of how people think about, influence, and relate to one another. This is important for understanding how people will interact with robots and how robots can be designed to interact with people in a socially acceptable way.

In addition to these disciplines, HRI in service robotics also draws on a number of other fields, such as ethics, law, and public policy. This is because the deployment of service robots raises a number of important ethical, legal, and social issues.

Service Robots And HRI Foundations: Practical Implementations

AI: AI is being used to develop robots that can understand human language, gestures, and facial expressions. This is essential for robots to be able to interact with humans in a natural and intuitive way. For example, AI is being used to develop robots that can provide customer service, assist with healthcare, and provide companionship.

ML: ML is being used to develop robots that can learn from data and adapt to their environment. This is important for robots to be able to safely and effectively perform tasks in a variety of settings. For example, ML is being used to develop robots that can clean homes and offices, deliver packages, and assist with manufacturing and assembly tasks.

Robotics: Robotics is being used to develop robots that can move and manipulate objects in the world. This is essential for robots to be able to perform a wide range of tasks in service robotics. For example, robotics is being used to develop robots that can cook and serve food, assist with personal care, and transport people and goods.

HCI: HCI is being used to design robots that are easy and enjoyable for people to interact with. This is important for robots to be accepted and used by people in a variety of settings. For example, HCI is being used to develop robots that have natural and intuitive interfaces, and that can provide feedback to users in a clear and concise way.

Social Psychology: Social psychology is being used to understand how people will interact with robots and how robots can be designed to interact with people in a socially acceptable way. This is important for ensuring that robots are used in a way that benefits society and does not harm people. For example, social psychology is being used to develop robots that are respectful of human values and that do not discriminate against people.

By building on the foundations of HRI, researchers and engineers are developing service robots that are increasingly capable of interacting with humans in a safe, efficient, and user-friendly way. As these robots become more widespread, they have the potential to revolutionize the way we live and work.

Principles Of Human-Centered Design

Human-centered design (HCD) is a process of designing products and services that are focused on the needs and experiences of the people who will use them. HCD is particularly important for service robots, as they are designed to work closely with humans in a variety of settings.

Exploring Essential Human-Centered Design Principles For Service Robots

Empathy: Designers need to empathize with the users of service robots in order to understand their needs and pain points. This can be done through user research methods such as interviews, surveys, and observation.

Collaboration: Designers should collaborate with users throughout the design process, from ideation to testing. This ensures that the final product meets the needs of the users and is easy to use.

Iteration: Design is an iterative process, and HCD encourages designers to iterate on their designs based on feedback from users. This helps to ensure that the final design is the best possible solution for the users' needs.

Principles of Human-Centered Design

HCD At Work: Real-World Examples In Service Robot Design

Understanding User Needs: Designers can conduct user research to understand the needs of people who might use a service robot. This could involve interviewing people in their homes, workplaces, or other settings where they might interact with a robot. Designers can also observe people performing tasks that a robot might be able to help with.

Designing for Usability: Once designers have a good understanding of user needs, they can start to design a robot that is easy to use and interact with. This involves considering factors such as the robot's size, shape, and movements. Designers should also consider how the robot will communicate with users and how users will be able to control the robot.

Testing and Iterating: Once a prototype of the robot has been built, designers should test it with users to get feedback. This feedback can be used to iterate on the design and make improvements until the robot is as user-friendly as possible.

By following the principles of HCD, designers can create service robots that are safe, efficient, and easy to use. This will help to ensure that service robots are accepted and adopted by humans, and that they can be used to improve the lives of people in a variety of ways.

Enhancing Service Robot Design With HCD: Additional Tips And Insights

Keep it Simple: Service robots should be easy to understand and use. Avoid complex designs that could overwhelm or frustrate users.

Make it Transparent: Service robots should be transparent about their capabilities and intentions. Users should be able to understand what the robot is doing and why.

Make it Adaptable: Service robots should be able to adapt to their environment and the people they are interacting with. This could involve being able to learn from their experiences or adjust their behavior based on user feedback.

Make it Reliable: Service robots should be reliable and consistent in their behavior. Users should be able to trust that the robot will perform as expected.

Cognitive Models For Human-Robot Interaction

Cognitive Models for Human-Robot Interaction (HRI) are computational models that can be used to simulate and predict the cognitive processes involved in human-robot interaction. These models can be used to improve the design and development of service robots by helping engineers and designers to understand how humans will interact with robots and how robots can be designed to interact with humans in a way that is safe, efficient, and natural.

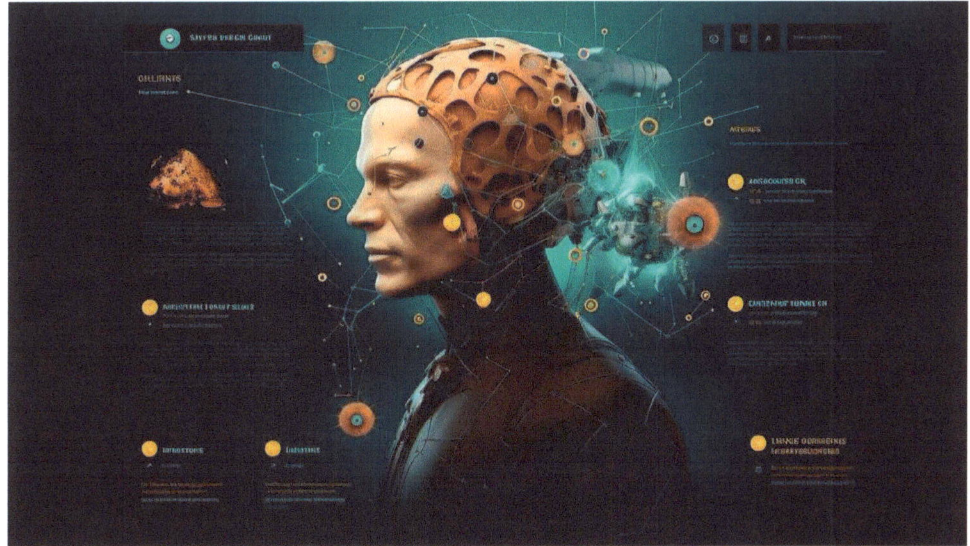

Cognitive Models for Human-Robot Interaction

There Are A Variety Of Different Cognitive Models That Can Be Used For HRI. Some Of The Most Common Models Include:

Mental Models: Mental models are representations of the world that people use to understand and predict how things work. Mental models can be used to explain how people understand and interact with robots. For example, a person's mental model of a robot might include beliefs about the robot's capabilities, limitations, and intentions.

Theory of Mind: Theory of mind is the ability to understand that other people have their own thoughts, beliefs, and desires. Theory of mind is important for HRI because it allows robots to understand and respond to the needs and intentions of their human users.

Joint Action Theory: Joint action theory is a theory of how people coordinate their actions to achieve a common goal. Joint action theory is important for HRI because it allows robots to collaborate with humans and to learn from human users.

Cognitive Models Can Be Used To Improve The Design And Development Of Service Robots In A Number Of Ways. For Example, Cognitive Models Can Be Used To:

Design Robots that are Predictable and Easy to Understand: Cognitive models can be used to understand how humans will interpret the robot's behavior and to design robots that are predictable and easy to understand. This can help to build trust and acceptance between humans and robots.

Design Robots that can Learn and Adapt: Cognitive models can be used to design robots that can learn from their experiences and adapt their behavior to the needs of their users. This can help to improve the efficiency and effectiveness of human-robot interaction.

Design Robots that can Collaborate with Humans: Cognitive models can be used to design robots that can collaborate with humans on tasks and activities. This can help to improve the productivity and effectiveness of both humans and robots.

Overall, cognitive models are a valuable tool for improving the design and development of service robots. By using cognitive models to understand and simulate the cognitive processes involved in HRI, engineers and designers can create robots that are safe, efficient, and natural to interact with.

Enhancing Service Robots With Cognitive Models: Real-World Applications

Personal Service Robots: Cognitive models are being used to design personal service robots that can understand and respond to the needs and intentions of their users. For example, a personal service robot might be able to use a cognitive model to predict what task a user wants it to perform, or to understand why a user is giving it a particular command.

Healthcare Robots: Cognitive models are being used to design healthcare robots that can collaborate with healthcare professionals to provide care to patients. For example, a healthcare robot might be able to use a cognitive model to understand the needs of a patient and to develop a plan of care in collaboration with the patient's healthcare team.

Educational Robots: Cognitive models are being used to design educational robots that can help children learn and develop new skills. For example, an educational robot might be able to use a cognitive model to understand the learning style of a child and to adapt its teaching methods accordingly.

As cognitive models continue to develop, they are likely to play an even greater role in the design and development of service robots. By using cognitive models to create robots that are safe, efficient, and natural to interact with, we can ensure that these robots can be used to improve the lives of people in a variety of ways.

Sensing And Perception In HRI

Sensing and perception are essential for human-robot interaction (HRI) in service robotics. Robots need to be able to perceive their environment and the people in it in order to interact with them safely and effectively. This includes being able to see, hear, and recognize people, as well as understand their gestures and body language.

Sensing and Perception in HRI

Diverse Sensor Technologies For Human-Robot Interaction In Service Robotics

Cameras: Cameras are used to provide visual information about the environment. This information can be used to detect and track people, as well as to identify objects and obstacles.

Microphones: Microphones are used to provide audio information about the environment. This information can be used to detect and recognize speech, as well as to identify sounds such as footsteps or alarms.

Range Sensors: Range sensors, such as lidar and ultrasonic sensors, are used to measure the distance to objects in the environment. This information can be used to avoid collisions and to navigate safely.

Force/torque Sensors: Force/torque sensors are used to measure the forces and torques applied to the robot. This information can be used to prevent the robot from damaging itself or its environment.

Once the robot has collected sensor data, it needs to be able to perceive and interpret the data in order to understand its environment and the people in it. This is done using a variety of perception algorithms, which can be used to:

Detect and Track People: Perception algorithms can be used to detect and track people in the environment. This information can be used to avoid collisions and to interact with people in a natural way.

Recognize Objects and Obstacles: Perception algorithms can be used to recognize objects and obstacles in the environment. This information can be used to navigate safely and to avoid collisions.

Understand Human Gestures and Body Language: Perception algorithms can be used to understand human gestures and body language. This information can be used to infer the intentions of people and to interact with them in a way that is natural and intuitive.

Sensing and perception are essential for safe and effective HRI in service robotics. By using a variety of sensors and perception algorithms, robots can understand their environment and the people in it, which allows them to interact with humans in a safe, efficient, and natural way.

Applications Of Sensing And Perception In Service Robotics: Real-World Instances

Personal Service Robots: Personal service robots use sensing and perception to navigate safely around homes and offices, and to interact with people in a natural way. For example, a personal service robot might use a camera to detect and track people, and then use a microphone to recognize their speech. The robot could then use this information to carry out a task that the person has requested.

Healthcare Robots: Healthcare robots use sensing and perception to assist healthcare professionals with tasks such as patient care and surgery. For example, a healthcare robot might use a camera to track the position of a patient's body during surgery, and then use this information to help the surgeon guide the surgical instruments.

Educational Robots: Educational robots use sensing and perception to interact with children in a natural and engaging way. For example, an educational robot might use a camera to detect and track a child's gaze, and then use this information to focus the child's attention on a particular object or activity.

As sensing and perception technologies continue to develop, they are likely to play an even greater role in service robots. By using sensing and perception to create robots that can understand their environment and the people in it, we can ensure that these robots can be used to improve the lives of people in a variety of ways.

6.3. Modes of Interaction

Modes of interaction in human-robot interaction (HRI) refer to the different ways in which humans and robots can communicate and exchange information. There are a variety of different modes of interaction that can be used in HRI, including:

Verbal Communication: Verbal communication is the use of language to communicate. This can include speech, text, and sign language.

Non-Verbal Communication: Non-verbal communication is the use of body language, facial expressions, and gestures to communicate.

Multimodal Communication: Multimodal communication is the combination of verbal and non- verbal communication.

The best mode of interaction for a particular HRI application will depend on a number of factors, including the capabilities of the robot, the needs of the human user, and the environment in which the interaction is taking place.

Modes of Interaction

Service Robots And Interaction Modes: A Closer Look At Practical Use Cases

Personal Service Robots: Personal service robots often use multimodal communication to interact with humans. For example, a personal service robot might use speech recognition to understand a user's request, and then use gestures to indicate that it is working on the request. The robot might also use facial expressions to communicate its status to the user.

Healthcare Robots: Healthcare robots often use indirect manipulation to interact with humans. For example, a healthcare robot might be programmed to follow a set of instructions to provide care to a patient. The robot might also be able to understand and respond to simple verbal commands.

Educational Robots: Educational robots often use direct manipulation to interact with students. For example, an educational robot might have a joystick or buttons that students can use to control its movements. Educational robots might also use facial expressions and gestures to communicate with students.

The choice of mode of interaction has a significant impact on the usability and safety of HRI systems. It is important to choose a mode of interaction that is appropriate for the capabilities of the robot, the needs of the human user, and the environment in which the interaction is taking place.

Unpacking HRI Interaction Modes: Additional Perspectives

Verbal Communication: Verbal communication is the most natural and intuitive mode of interaction for humans. However, it can be challenging for robots to understand and respond to human language accurately.

Non-Verbal Communication: Non-verbal communication is another natural and intuitive mode of interaction for humans. However, it can be difficult for robots to interpret non-verbal cues accurately.

Multimodal Communication: Multimodal communication is a promising way to improve the naturalness and intuitiveness of HRI. However, it is important to carefully design the multimodal interaction system to ensure that it is easy for both humans and robots to use.

Overall, the best mode of interaction for a particular HRI application will depend on a number of factors. By carefully considering the capabilities of the robot, the needs of the human user, and the environment in which the interaction is taking place, developers can choose a mode of interaction that will lead to a safe, efficient, and user-friendly HRI experience.

Verbal Communication And Natural Language Processing

Verbal communication and natural language processing (NLP) are essential for human-robot interaction (HRI) in service robotics. Verbal communication allows humans to interact with robots in a natural and intuitive way, and NLP allows robots to understand and respond to human language.

Verbal Communication: The Multifaceted Connection Between Humans And Robots

Speech: Robots can use speech to communicate with humans in a natural way. This can be done using text-to-speech (TTS) technology to synthesize speech from text, or using speech recognition technology to convert speech to text.

Text: Robots can also use text to communicate with humans. This can be done using displays or other interfaces to display text messages.

Gestures: Robots can also use gestures to communicate with humans. This can be done using robotic arms or other actuators to generate gestures.

NLP Is Used To Enable Robots To Understand And Respond To Human Language. Nlp Algorithms Can Be Used To:

Recognize Speech: NLP algorithms can be used to recognize speech from humans. This is necessary for robots to be able to understand spoken commands and requests.

Understand Text: NLP algorithms can be used to understand text from humans. This is necessary for robots to be able to understand written commands and requests, as well as to generate responses to text-based questions.

Generate Text: NLP algorithms can be used to generate text in response to human requests. This is necessary for robots to be able to communicate with humans in a natural and informative way.

Verbal communication and NLP are essential for safe and effective HRI in service robotics. By using verbal communication and NLP, robots can interact with humans in a natural and intuitive way, which allows them to carry out tasks and provide assistance in a variety of settings.

Verbal Communication and Natural Language Processing

Harnessing Verbal Communication And NLP In Service Robots: Case Studies

Personal Service Robots: Personal service robots use verbal communication and NLP to interact with people in a natural way. For example, a personal service robot might be able to understand and respond to spoken commands such as "Please make me a cup of coffee" or "Please help me fold my laundry." The robot could also use NLP to generate responses to text-based questions such as "What is the weather like today?"

Healthcare Robots: Healthcare robots use verbal communication and NLP to communicate with patients and healthcare professionals. For example, a healthcare robot might be able to ask a patient questions about their medical history and symptoms, and then use NLP to understand the patient's responses. The robot could then use this information to provide the patient with care or to communicate with the patient's healthcare team.

Educational Robots: Educational robots use verbal communication and NLP to interact with students in a natural and engaging way. For example, an educational robot might be able to ask students questions about a particular subject, and then use NLP to understand the students' responses. The robot could then use this information to provide the students with feedback or to guide them through a learning activity.

As verbal communication and NLP technologies continue to develop, they are likely to play an even greater role in service robots. By using verbal communication and NLP to create robots that can understand and respond to human language in a natural way, we can ensure that these robots can be used to improve the lives of people in a variety of ways.

Non-Verbal Communication: Gestures, Facial Expressions, And Body Language

Non-verbal communication, such as gestures, facial expressions, and body language, is an important aspect of human-robot interaction (HRI) in service robotics. Non-verbal communication can be used by humans to communicate their intentions, emotions, and needs to robots, and robots can use non-verbal communication to communicate with humans in a natural and intuitive way.

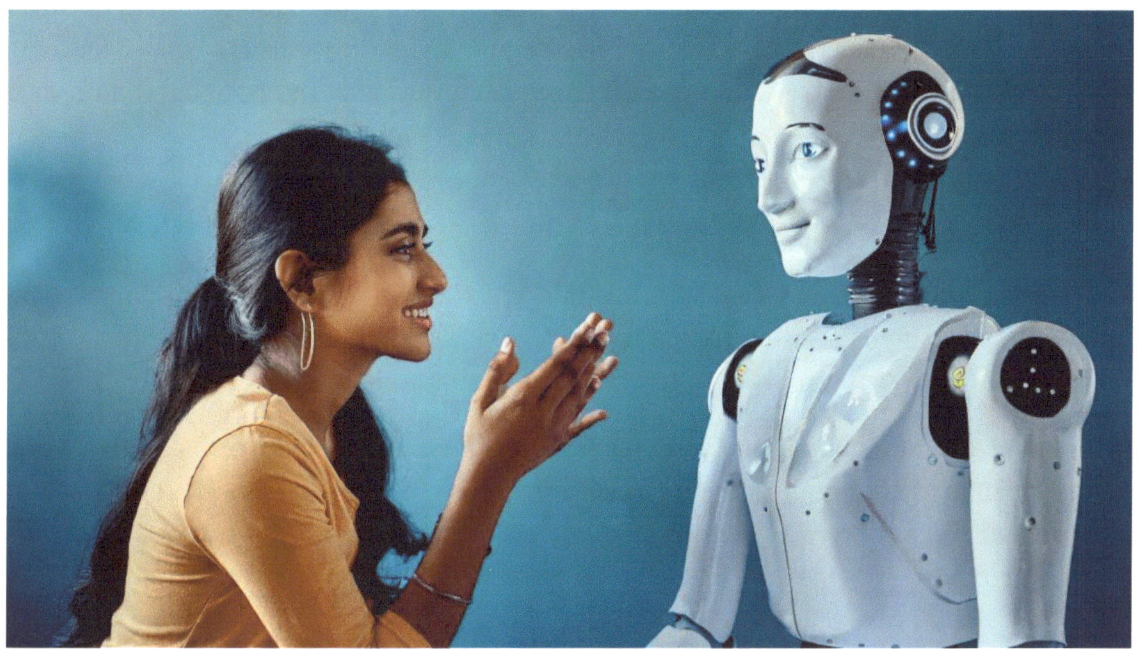

Non-Verbal Communication

Expressing Without Speech: How Service Robots Utilize Non-Verbal Communication

Personal Service Robots: Personal service robots can use non-verbal communication to indicate their availability and readiness to interact with humans. For example, a personal service robot might use a gesture such as nodding its head to indicate that it is ready to receive a command. Personal service robots can also use non-verbal communication to express their emotions and intentions. For example, a personal service robot might use a facial expression such as a smile to indicate that it is happy to help.

Healthcare Robots: Healthcare robots can use non-verbal communication to reassure patients and to communicate with healthcare professionals. For example, a healthcare robot might use a gesture such as patting a patient on the shoulder to reassure them. Healthcare robots can also use non-verbal communication to communicate with healthcare professionals about a patient's condition. For example, a healthcare robot might use a facial expression such as a frown to indicate that a patient is in pain.

Educational Robots: Educational robots can use non-verbal communication to engage students and to provide feedback. For example, an educational robot might use a gesture such as pointing to an object to engage a student's attention. Educational robots can also use non-verbal communication to provide feedback to students. For example, an educational robot might use a facial expression such as a smile to indicate that a student has answered a question correctly.

Non-verbal communication is an important aspect of HRI because it can help to create a more natural and intuitive interaction between humans and robots. By using non-verbal communication, robots can better understand the needs and intentions of humans, and humans can better understand the capabilities and limitations of robots.

As non-verbal communication technologies continue to develop, they are likely to play an even greater role in service robots. By using non-verbal communication to create robots that can interact with humans in a natural and intuitive way, we can ensure that these robots can be used to improve the lives of people in a variety of ways.

HRI And Non-Verbal Interaction: A Deeper Exploration Of Its Importance

Non-verbal communication can help to build trust and rapport between humans and robots. When humans see that robots are able to understand and respond to their non-verbal cues, they are more likely to trust and accept the robots.

Non-verbal communication can help to reduce the cognitive load on humans. When robots are able to communicate using non-verbal cues, humans do not need to spend as much cognitive resources on understanding the robots' communications. This can be especially important in tasks where humans are already under cognitive load.

Non-verbal communication can make HRI more efficient and effective. When robots are able to communicate using non-verbal cues, humans can interact with the robots more quickly and easily.

This can lead to increased efficiency and effectiveness in tasks such as customer service, education, and healthcare.

Overall, non-verbal communication is an essential element of HRI in service robotics. By using non-verbal communication, robots can better understand the needs and intentions of humans, and humans can better understand the capabilities and limitations of robots. This can lead to more natural, intuitive, efficient, and effective interactions between humans and robots.

Multimodal Interaction: Combining Verbal And Non-Verbal Communication

Multimodal interaction is the combination of verbal and non-verbal communication. It is a powerful way for humans to interact with each other, and it is also a promising way for humans to interact with robots.

Multimodal Interaction: Combining Verbal and Non-Verbal Communication

Multimodal Interaction Can Be Used To Improve Hri In A Number Of Ways. For Example, Multimodal Interaction Can Be Used To:

Make HRI More Natural and Intuitive: Humans are accustomed to interacting with each other using both verbal and non-verbal communication. By using multimodal interaction, robots can interact with humans in a way that is more natural and intuitive.

Improve the Efficiency and Effectiveness of HRI: Multimodal interaction can help to reduce the cognitive load on humans and to make HRI more efficient and effective. For example, a robot could use multimodal interaction to communicate with a human in a busy environment, where the human may not have time to read text or listen to speech.

Make HRI more Accessible: Multimodal interaction can make HRI more accessible to people with disabilities. For example, a robot could use multimodal interaction to communicate with a person who is blind or deaf.

Service Robots And Multimodal Communication: Demonstrated Applications

Personal Service Robots: Personal service robots can use multimodal interaction to provide a more natural and intuitive experience for users. For example, a personal service robot could use speech recognition to understand a user's request, and then use gestures to indicate that it is working on the request. The robot could also use facial expressions to communicate its status to the user.

Healthcare Robots: Healthcare robots can use multimodal interaction to improve communication with patients and healthcare professionals. For example, a healthcare robot could use speech recognition to understand a patient's symptoms, and then use gestures to help the patient perform a physical therapy exercise. The robot could also use facial expressions to reassure the patient and to communicate with the patient's healthcare team.

Educational Robots: Educational robots can use multimodal interaction to engage students and to provide feedback in a more natural and intuitive way. For example, an educational robot could use speech recognition to understand a student's answer to a question, and then use gestures to indicate whether the answer is correct or incorrect. The robot could also use facial expressions to motivate the student.

Multimodal interaction is a rapidly developing field, and there are a number of exciting new technologies on the horizon. For example, some researchers are developing robots that can understand and respond to human emotions. Other researchers are developing robots that can communicate with humans using natural language and gestures.

As multimodal interaction technologies continue to develop, they are likely to play an even greater role in service robots. By using multimodal interaction to create robots that can interact with humans in a more natural, intuitive, efficient, and effective way, we can ensure that these robots can be used to improve the lives of people in a variety of ways.

6.4. Challenges and Ethical Considerations

Challenges In Hri For Service Robots:

There are a number of challenges that need to be addressed in order to develop and deploy safe, efficient, and user-friendly service robots. Some of the key challenges include

Technical Challenges: Developing service robots that are capable of interacting with humans in a safe and natural way is a complex technical challenge. Robots need to be able to perceive their environment, understand human language and gestures, and respond in a way that is appropriate and timely.

Social Challenges: The introduction of service robots into society raises a number of social challenges. For example, there is a need to address concerns about safety, privacy, and job displacement. Additionally, it is important to ensure that service robots are designed and deployed in a way that is ethical and socially responsible.

Ethical Considerations In HRI For Service Robots:

In addition to the technical challenges, there are also a number of ethical considerations that need to be taken into account when developing and deploying service robots. It is important to address these challenges and ethical considerations in order to ensure that service robots are developed and deployed in a responsible and beneficial way.

HRI For Service Robots: Addressing Challenges And Ethical Dilemmas:

Technical Challenges: As service robots become more complex and capable, the technical challenges of developing and deploying them will also increase. One of the key challenges will be developing robots that are safe and reliable. Robots need to be able to operate in a variety of environments and to interact with a variety of people without causing harm.

Social Challenges: As service robots become more common in society, it is important to address the social challenges that they raise. For example, it is important to ensure that service robots are designed and deployed in a way that is inclusive and accessible to all people. Additionally, it is important to educate the public about service robots and to ensure that people understand the benefits and risks of interacting with them.

Ethical Considerations: The ethical considerations in HRI for service robots are complex and challenging. As robots become more autonomous and capable, it is important to have a clear understanding of who is responsible for their actions. Additionally, it is important to ensure that service robots are designed and deployed in a way that is ethical and socially responsible.

By carefully considering the challenges and ethical considerations in HRI for service robots, we can ensure that these robots are developed and deployed in a way that benefits society.

Privacy And Data Security In HRI

Privacy and data security are important considerations in human-robot interaction (HRI) for service robots. Service robots often collect a significant amount of data about their environment and the people they interact with. This data can be used to improve the performance of the robot, but it is important to ensure that it is collected and used in a responsible and ethical way.

Privacy and Data Security in HRI

Service Robots And Data Security: Navigating Privacy Challenges In HRI

Data Collection: Service robots may collect a variety of data about their environment and the people they interact with. This data may include video and audio recordings, sensor data, and personal information such as names and addresses. It is important to be transparent about what data is being collected and how it will be used.

Data Storage: Service robots typically store the data they collect on their own internal storage or on a cloud server. It is important to ensure that this data is stored securely and that it is only accessed by authorized personnel.

Data Sharing: Service robots may share the data they collect with their manufacturers, third-party service providers, or other robots. It is important to be transparent about who the data is being shared with and how it will be used.

Data Misuse: Service robots may be used to collect data about people without their knowledge or consent. This data could then be misused for malicious purposes, such as identity theft or surveillance. It is important to have safeguards in place to prevent data misuse.

There Are A Number Of Things That Can Be Done To Address The Privacy And Data Security Concerns In HRI For Service Robots. These Include:

Privacy-Enhancing Technologies: There are a number of privacy-enhancing technologies that can be used to protect the privacy of people who interact with service robots. For example, differential privacy can be used to add noise to data so that it cannot be used to identify individuals.

Transparent Data Collection And Use: It is important to be transparent about what data is being collected and how it will be used. This information should be provided to people before they interact with a service robot.

Data Security: Service robots should use appropriate security measures to protect the data they collect. This includes using strong encryption and access control measures.

Data Governance: It is important to have policies and procedures in place to govern the collection, use, storage, and sharing of data collected by service robots. These policies and procedures should be developed with input from stakeholders such as privacy experts, ethicists, and the public.

By addressing the privacy and data security concerns in HRI for service robots, we can ensure that these robots are developed and deployed in a way that protects the privacy and security of the people who interact with them.

Ethical Considerations In Service Robotics

Ethical considerations in service robotics are important because service robots are increasingly being deployed in settings where they interact with humans in close proximity. This raises a number of ethical concerns, such as:

Autonomy: Service robots should be designed to operate in a way that is safe and beneficial to humans. This means that robots should not be given too much autonomy, as this could lead to unintended consequences. For example, a robot that is too autonomous might not be able to anticipate and respond to unexpected events, which could lead to harm to humans or damage to property.

Privacy: Service robots may collect a significant amount of data about their environment and the people they interact with. It is important to ensure that this data is collected and used in a responsible and ethical way. For example, data collected by service robots should not be used to track people or to discriminate against them.

Accountability: It is important to be able to hold someone accountable for the actions of service robots. This is particularly important if a robot causes harm to a human. For example, if a robot malfunctions and injures someone, it should be clear who is responsible for compensating the victim.

Bias: Service robots should be designed to be fair and unbiased. However, it is important to be aware that robots may be biased due to the data they are trained on or the algorithms they use. For example, a robot that is trained on data that is biased against a particular group of people may be more likely to make decisions that are unfair to that group of people.

Ethical Considerations in Service Robotics

Job Displacement: As service robots become more capable, they may displace human workers from some jobs. It is important to consider the economic and social implications of this job displacement. For example, governments may need to provide retraining programs for workers who are displaced by service robots.

In addition to these general ethical considerations, there are also a number of specific ethical considerations that need to be taken into account when developing and deploying service robots in particular settings. For example, there are different ethical considerations for service robots that are used in healthcare, education, and the workplace.It is important to address the ethical considerations in service robotics in order to ensure that these robots are developed and deployed in a responsible and beneficial way. There are a number of things that can be done to address these ethical considerations, such as:

> **Developing Ethical Guidelines:** Ethical guidelines can be developed to help engineers and designers develop and deploy service robots in a responsible and ethical way. These guidelines should be developed with input from a variety of stakeholders, including ethicists, lawyers, and the public.

> **Conducting Ethical Impact Assessments:** Ethical impact assessments can be conducted to identify and assess the potential ethical impacts of service robots before they are deployed. This can help to identify and mitigate potential ethical risks.

> **Engaging In Public Dialogue:** It is important to engage in public dialogue about the ethical implications of service robots. This can help to raise awareness of these issues and to develop a consensus on how to address them.

By addressing the ethical considerations in service robotics, we can ensure that these robots are developed and deployed in a way that benefits society while minimizing potential harms.

Psychological And Social Impact Of Human-Robot Interaction

The psychological and social impact of human-robot interaction (HRI) is a complex and evolving area of research. On the one hand, robots have the potential to improve our lives in many ways, such as by providing companionship, assistance, and education. On the other hand, there are also some potential negative impacts of HRI, such as job displacement, social isolation, and the fear of being replaced by machines.

Psychological and Social Impact of Human-Robot Interaction

The Psychological And Social Consequences Of Human-Robot Interaction (HRI): Positive Impacts:

Companionship: Robots can provide companionship and social support to people who are lonely or isolated. For example, robots can be used to keep people company in their homes or to help them stay connected with friends and family members.

Assistance: Robots can provide assistance to people with disabilities or other needs. For example, robots can be used to help people with mobility issues get around or to help people with Alzheimer's disease remember their daily tasks.

Education: Robots can be used to provide education and training to people of all ages. For example, robots can be used to help children learn new skills or to help adults learn new languages.

Negative Impacts:

Job Displacement: As robots become more capable, they may displace human workers from some jobs. This could lead to economic hardship and social unrest.

Social Isolation: Robots could lead to social isolation if people are more likely to interact with robots than with other humans. This could have a negative impact on people's mental and emotional health.

Fear Of Being Replaced By Machines: Some people may be afraid of being replaced by machines as robots become more capable. This could lead to anxiety, depression, and other mental health problems.

It is important to note that the psychological and social impacts of HRI will vary depending on the specific context in which robots are used. For example, the impact of robots on social isolation may be different in a setting where robots are used to provide companionship to the elderly than in a setting where robots are used to replace workers in a factory. It is also important to note that the psychological and social impacts of HRI are not inevitable. There are a number of things that can be done to mitigate the potential negative impacts of HRI, such as:

Developing Ethical Guidelines: Ethical guidelines can be developed to help ensure that robots are used in a responsible and ethical way. These guidelines should be developed with input from a variety of stakeholders, including ethicists, lawyers, and the public.

Educating The Public: It is important to educate the public about the potential benefits and risks of HRI. This can help to reduce people's fears and concerns about robots.

Investing In Social Programs: Governments can invest in social programs to help people who are displaced by robots or who are otherwise negatively impacted by HRI.

By taking these steps, we can help to ensure that the psychological and social impacts of HRI are positive and beneficial to everyone.

6.5. Applications And Case Studies

HRI In Healthcare Robotics: Assistive Robots And Patient Care

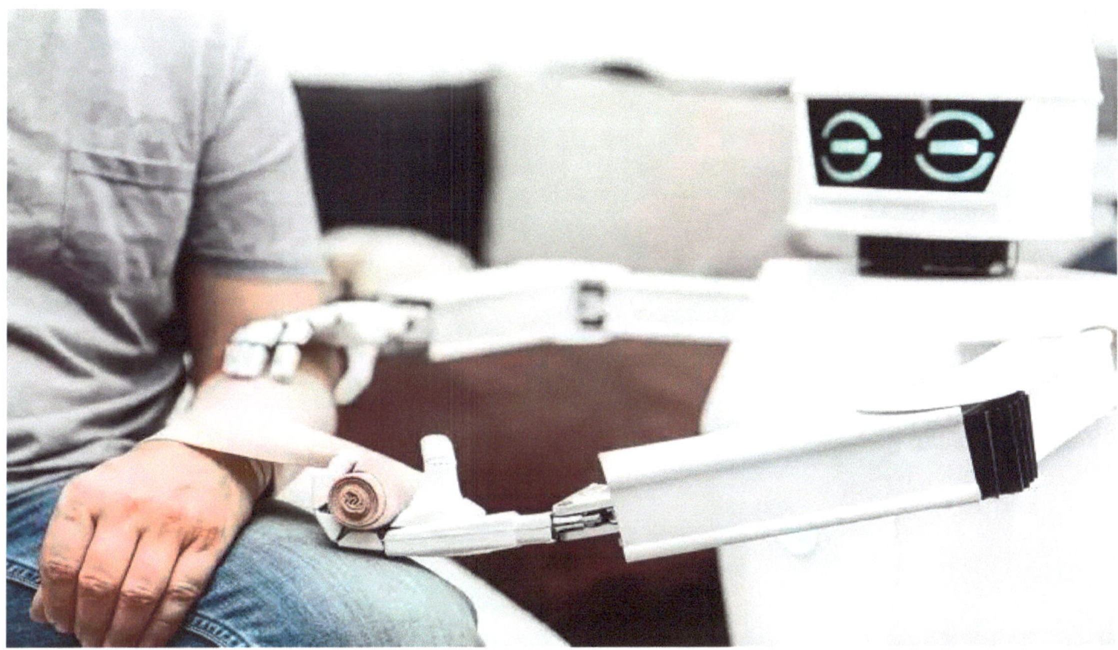

HRI in Healthcare Robotics: Assistive Robots and Patient Care

Assistive robots are being used in a variety of healthcare settings to provide assistance to patients and healthcare professionals. Here are some specific examples:

Patient Care: Assistive robots can be used to provide direct care to patients, such as helping them with mobility, bathing, and dressing. Robots can also be used to provide companionship and social support to patients.

Rehabilitation: Assistive robots can be used to help patients with rehabilitation exercises. For example, robots can be used to help patients with stroke learn to walk again.

Surgery: Assistive robots can be used to assist surgeons with surgery. For example, robotic surgery systems can be used to perform minimally invasive surgery.

HRI In Customer Service: Robots In Retail And Hospitality

Robots are being used in a variety of customer service roles in retail and hospitality settings. Here are some specific examples:

Retail: Robots can be used to greet customers, provide information about products, and help customers find what they are looking for. Robots can also be used to stock shelves and perform other tasks.

Hospitality: Robots can be used to check in guests, deliver food and drinks, and provide information about the hotel or restaurant. Robots can also be used to clean rooms and perform other tasks.

HRI in Customer Service: Robots in Retail and Hospitality

HRI In Industry: Collaborative Robots On The Factory Floor

Collaborative robots, also known as cobots, are being used in a variety of industrial settings to work alongside human workers. Cobots can be used to perform tasks that are dangerous, repetitive, or difficult for humans to do. Here are some specific examples:

Assembly: Cobots can be used to assemble products, such as cars and electronics.
Welding: Cobots can be used to weld parts together.
Painting: Cobots can be used to paint products.
Inspection: Cobots can be used to inspect products for defects.

Exploring HRI Case Studies In Healthcare, Customer Service, And Industry

Healthcare Robotics: The Paro therapeutic robot is a robotic seal that is used to provide companionship and social support to patients in hospitals and nursing homes. Paro has been shown to reduce stress and anxiety in patients, and to improve their mood and quality of life.

Customer Service: The Pepper robot is a humanoid robot that is used in a variety of customer service roles in retail and hospitality settings. Pepper can greet customers, provide information about products, and help customers find what they are looking for. Pepper has been shown to improve the customer experience and to increase sales.

Industry: The UR10 cobot is a collaborative robot that is used in a variety of industrial settings, such as automotive and electronics manufacturing. The UR10 can be used to perform a variety of tasks, such as assembly, welding, and painting. The UR10 has been shown to improve productivity and safety in the workplace.

These are just a few examples of how HRI is being used in healthcare, customer service, and industry. As HRI technology continues to develop, we can expect to see even more innovative and groundbreaking applications of HRI in the future.

6.6. Future Trends and Research Directions

Human-Robot Teaming: The Future Of Service Robotics

Human-robot teaming (HRT) is the collaboration between humans and robots to achieve a common goal. HRT is becoming increasingly important in service robotics, as robots are being deployed in a wider range of settings and are being asked to perform more complex tasks.

Human-Robot Teaming: The Future of Service Robotics

One of the key trends in HRT is the development of robots that are more aware of and responsive to the needs of human users. This includes robots that can understand and respond to human language, gestures, and facial expressions. It also includes robots that can learn and adapt to their environment and to the people they interact with.

Another key trend in HRT is the development of robots that can work more autonomously. This means that robots will be able to perform tasks without the need for constant human supervision. However, even as robots become more autonomous, it will still be important for humans and robots to be able to work together effectively.

Unlocking The Power Of HRT: Service Robotics Case Studies

Healthcare Robotics: HRT is being used in healthcare settings to provide care to patients and to assist healthcare professionals.
For example, HRT is being used to develop robots that can help patients with rehabilitation exercises, to perform surgery, and to provide companionship and social

support.

Customer Service: HRT is being used in customer service settings to provide information and assistance to customers.

For example, HRT is being used to develop robots that can greet customers, answer questions, and help customers find what they are looking for.

Industry: HRT is being used in industrial settings to improve productivity and safety.

For example, HRT is being used to develop robots that can work alongside human workers to assemble products, weld parts, and paint products.

As HRT Technology Continues To Develop, We Can Expect To See Even More Innovative And Groundbreaking Applications Of Hrt In The Future. Here Are Some Specific Research Directions In HRT:

Developing Robots That Are More Aware Of And Responsive To The Needs Of Human Users: This includes research on developing robots that can understand and respond to human language, gestures, and facial expressions. It also includes research on developing robots that can learn and adapt to their environment and to the people they interact with.

Developing Robots That Can Work More Autonomously: This includes research on developing robots that can plan and execute tasks on their own, without the need for constant human supervision.

Developing Methods For Improving Human-Robot Communication And Collaboration: This includes research on developing new interfaces for human-robot communication, and on developing methods for training humans and robots to work together effectively.

HRT has the potential to revolutionize the way we live and work. By developing robots that are more aware of and responsive to the needs of human users, and by developing methods for improving human-robot communication and collaboration, we can ensure that HRT is used to benefit society as a whole.

Chapter 7: Ethical And Legal Considerations In Service Robotics

7.1. The Ethical Framework For Service Robotics

An ethical framework for service robotics is a set of principles and guidelines that can be used to assess the ethical implications of the design, development, deployment, and use of service robots. It is important to have an ethical framework in place to ensure that service robots are used in a way that benefits society and does not harm individuals or groups.

One Ethical Framework For Service Robotics Is Based On The Following Principles:

Autonomy: Service robots should be designed to respect the autonomy of humans. This means that humans should have the right to choose whether or not to interact with a service robot and to control how the robot interacts with them.

Beneficence: Service robots should be designed to benefit humans. This means that the robot should be designed to perform tasks that are safe and helpful to humans.

Justice: Service robots should be designed to be fair and just. This means that the robot should not discriminate against any group of people or cause harm to individuals or groups.

Non-Maleficence: Service robots should be designed to avoid harming humans. This means that the robot should be designed to be safe and to avoid causing any physical or emotional harm to humans.

Transparency: Service robots should be designed to be transparent in their operation. This means that humans should be able to understand how the robot works and what it is doing.

These principles can be used to assess the ethical implications of service robots at all stages of their development and use. For example, when designing a service robot, developers should consider how the robot will respect the autonomy of users, how it will benefit users, how it will be fair and just, how it will avoid harming users, and how it will be transparent in its operation.

In addition to the principles listed above, there are a number of other ethical issues that need to be considered in service robotics, such as privacy, security, responsibility, accountability, and bias. These issues are discussed in more detail in the subtopics of the book chapter.

It is important to note that there is no single ethical framework for service robotics that is universally accepted. The ethical framework outlined above is just one example. As service robots become more widespread and sophisticated, it is likely that new ethical issues will emerge and that the existing ethical framework will need to be adapted.

Ethical Principles In Human-Robot Interaction

Human-robot interaction (HRI) is the field of study that examines the interactions between humans and robots. As service robots become more widespread and sophisticated, it is important to consider the ethical implications of HRI.

One of the most important ethical principles in HRI is autonomy. This principle states that humans should have the right to choose whether or not to interact with a service robot and to control how the robot interacts with them. For example, a service robot should not be able to force a human to interact with it or to perform a task that the human does not want to perform.

Another important ethical principle in HRI is beneficence. This principle states that service robots should be designed to benefit humans. This means that the robot should be designed to perform tasks that are safe and helpful to humans. For example, a service robot should not be designed to perform tasks that are dangerous or that could harm humans.

Ethical Principles in Human-Robot Interaction

Other Important Ethical Principles In HRI Include:

Justice: Service robots should be designed to be fair and just. This means that the robot should not discriminate against any group of people or cause harm to individuals or groups.

Non-Maleficence: Service robots should be designed to avoid harming humans. This means that the robot should be designed to be safe and to avoid causing any physical or emotional harm to humans.

Transparency: Service robots should be designed to be transparent in their operation. This means that humans should be able to understand how the robot works and what it is doing.

Privacy: Service robots may collect a large amount of personal data about their users. It is important to ensure that this data is collected and used in a responsible and ethical manner.

Security: Service robots may be vulnerable to hacking and other cyberattacks. It is important to ensure that service robots are designed and deployed with security in mind.

Responsibility: Who is responsible for the actions of a service robot? Is it the manufacturer, the operator, or the user? This is a complex issue that needs to be addressed in law and regulation.

Accountability: How can we ensure that service robots are accountable for their actions? This is important for ensuring that service robots are used in a safe and responsible manner.

Bias: Service robots may be programmed in a way that biases them against certain groups of people. It is important to identify and mitigate these biases to ensure that service robots are fair and unbiased.

It is important to note that these ethical principles are not mutually exclusive. For example, the principle of beneficence may sometimes conflict with the principle of autonomy. In these cases, it is important to weigh the competing principles carefully and to make a decision that is in the best interests of all stakeholders.

The ethical principles of HRI can be used to assess the ethical implications of service robots at all stages of their development and use. For example, when designing a service robot, developers should consider how the robot will respect the autonomy of users, how it will benefit users, how it will be fair and just, how it will avoid harming users, and how it will be transparent in its operation.

By following the ethical principles of HRI, we can ensure that service robots are used in a way that benefits society and does not harm individuals or groups.

Moral Dilemmas In Service Robotics

A moral dilemma is a situation in which a person is faced with two or more choices, each of which has negative consequences. Moral dilemmas can be particularly challenging in the context of service robotics, as they often involve the need to balance the rights and needs of different individuals and groups.

The Ethics Of Automation: Moral Dilemmas In Service Robotics

Privacy Vs. Security: Service robots may collect a large amount of personal data about their users. This data could be used to improve the robot's performance, but it could also be used to track or monitor users without their consent. Who has the right to access this data? How can we ensure that it is used in a responsible and ethical manner?

Autonomy Vs. Safety: Service robots should be designed to respect the autonomy of users. However, there may be situations where it is necessary to override the user's autonomy in order to prevent harm. For example, a service robot might need to override the user's autonomy to prevent them from driving a car into a wall. How can we balance the right of users to autonomy with the need to protect them from harm?

Responsibility Vs. Accountability: Who is responsible for the actions of a service robot? Is it the manufacturer, the operator, or the user? This is a complex issue that needs to be addressed in law and regulation. However, it is also a moral issue. Who should be held accountable if a service robot causes harm?

Bias Vs. Fairness: Service robots may be programmed in a way that biases them against certain groups of people. For example, a facial recognition robot might be more likely to misidentify people of color. How can we identify and mitigate these biases to ensure that service robots are fair and unbiased?

These are just a few examples of moral dilemmas in service robotics. As service robots become more widespread and sophisticated, it is likely that new moral dilemmas will emerge. It is important to have a framework in place for addressing these moral dilemmas in a thoughtful and ethical manner.

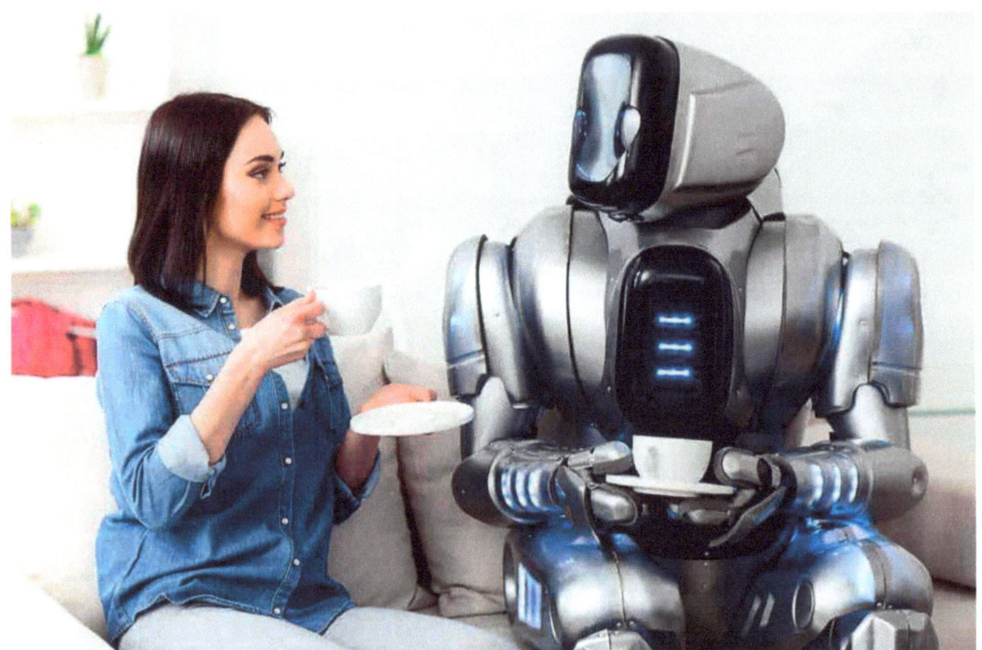
Moral Dilemmas in Service Robotics

The ethical framework for service robotics outlined in the previous section can be used to address moral dilemmas in a number of ways. First, the ethical principles can be used to identify the different moral issues that are at stake in a particular dilemma. Second, the ethical principles can be used to weigh the different moral issues and to develop a solution that is consistent with the ethical framework. Third, the ethical framework can be used to communicate the solution to stakeholders and to explain why the solution is the most ethical one.

It is important to note that there is no single "right" answer to any moral dilemma. The best solution will depend on the specific circumstances of the case. However, the ethical framework for service robotics can help us to think about moral dilemmas in a structured and ethical way.

Ensuring Robot Decision-Making Aligns With Human Values

One of the most important challenges in service robotics is ensuring that robot decision-making aligns with human values. This is a complex challenge, as it requires us to understand and formalize human values, and to develop algorithms and systems that can implement these values in a robust and reliable manner.

There are a number of different approaches to ensuring that robot decision-making aligns with human values. One approach is to involve humans in the design and development of robot algorithms. This can help to ensure that the algorithms reflect human values and that they are able to handle complex ethical situations.

Another approach is to develop algorithms that are transparent and accountable. This means that humans should be able to understand how the algorithms work and to hold them accountable for their decisions.

Finally, it is important to test and validate robot algorithms in real-world situations. This can help to identify any potential problems with the algorithms and to make sure that they are able to handle the ethical challenges that they will face in the real world.

Some Specific Steps That Can Be Taken To Ensure That Robot Decision-Making Aligns With Human Values:

Identify The Relevant Human Values. The first step is to identify the human values that are relevant to the robot's decision-making. This may include values such as autonomy, beneficence, justice, non-maleficence, transparency, privacy, security, responsibility, accountability, and fairness.

Formalize The Human Values. Once the relevant human values have been identified, they need to be formalized in a way that can be understood by robots. This can be done using a variety of methods, such as developing ethical guidelines, creating value-based reward functions, or using machine learning to learn human values from data.

Develop Algorithms That Implement Human Values. Once the human values have been formalized, algorithms need to be developed that can implement these values in a robust and reliable manner. This is a challenging task, as it requires the development of algorithms that can handle complex ethical situations and that are able to generalize to new situations.

Test And Validate The Algorithms. Once the algorithms have been developed, they need to be tested and validated in real-world situations. This can be done through simulations, pilot studies, and field trials.

It is important to note that there is no single solution to the problem of ensuring that robot decision-making aligns with human values. The best approach will depend on the specific robot and the task that it is performing. However, the steps outlined above can provide a starting point for developing robots that are aligned with human values.

In addition to the steps outlined above, there are a number of other things that can be done to ensure that robot decision-making aligns with human values. For example, it is important to have a clear and transparent process for developing and deploying robot algorithms. This process should involve stakeholders from a variety of backgrounds, including ethicists, lawyers, and members of the public. It is also important to have a system in place for monitoring and auditing robot behavior. This system should be able to identify any potential problems with robot decision-making and to take corrective action if necessary.By taking these steps, we can help to ensure that service robots are used in a way that benefits society and does not harm individuals or groups.

7.2. Legal Regulations Governing Service Robots

The legal and regulatory landscape for service robots is still evolving. However, there are a number of existing laws and regulations that apply to service robots, including:

Product Liability Laws: Product liability laws hold manufacturers and sellers responsible for injuries caused by defective products. This includes service robots.

Intellectual Property Laws: Intellectual property laws protect the intellectual property of service robot developers, such as their patents, copyrights, and trademarks.

Data Protection Laws: Data protection laws regulate the collection, use, and disclosure of personal data. Service robots may collect a large amount of personal data about their users, so it is important to comply with data protection laws.

Competition Law: Competition law prohibits businesses from engaging in anti-competitive practices. Service robot developers and vendors need to be aware of competition laws when developing and marketing their products.

Employment Law: Employment law regulates the relationship between employers and employees. Service robots may be used to automate tasks that were previously performed by humans. Employers need to be aware of employment laws when deploying service robots.

In addition to these existing laws and regulations, there are a number of new laws and regulations that are being developed specifically for service robots. For example, the European Union is developing a new regulation on artificial intelligence (AI). This regulation is expected to apply to service robots, as well as other types of AI systems.

It is important to note that the legal and regulatory landscape for service robots is complex and constantly changing. Service robot developers and vendors need to stay up-to-date on the latest laws and regulations to ensure that they are in compliance.

From Legislation To Practice: How Laws And Regulations Impact Service Robots

Product Liability Laws: If a service robot causes injury to a person or property, the manufacturer or seller of the robot could be held liable. This means that they could be sued for damages.

Intellectual Property Laws: Service robot developers may be able to protect their intellectual property using patents, copyrights, and trademarks. This means that they can prevent others from copying or using their intellectual property without permission.

Data Protection Laws: Service robot developers and vendors need to comply with data protection laws when collecting, using, and disclosing personal data about their users. This means that they need to obtain consent from users before collecting their personal data and that they need to use the data in a responsible and ethical manner.

Competition Law: Service robot developers and vendors need to be aware of competition laws when developing and marketing their products. For example, they cannot agree to fix prices or to divide up markets.

Employment Law: Employers who use service robots to automate tasks that were previously performed by humans need to comply with employment laws. For example, they may need to provide retraining to employees who are displaced by service robots.

The legal and regulatory landscape for service robots is still evolving. However, the laws and regulations listed above provide a starting point for understanding how service robots are regulated.

In addition to the laws and regulations listed above, there are a number of other legal and regulatory issues that need to be considered in service robotics. For example, it is important to consider the legal implications of using service robots in public spaces, such as sidewalks and parks. It is also important to consider the legal implications of using service robots in sensitive areas, such as hospitals and schools.

By carefully considering the legal and regulatory landscape for service robots, service robot developers and vendors can help to ensure that their products are used in a safe and responsible manner.

Overview Of Existing Legal Frameworks

There is currently no single legal framework that governs service robots in all jurisdictions. Instead, service robots are subject to a variety of existing laws and regulations, depending on their specific design, function, and intended use.

Some Of The Most Relevant Legal Frameworks For Service Robots Include:

Product Liability Laws: These laws hold manufacturers and sellers responsible for injuries caused by defective products. Service robots are considered products under the law, and their manufacturers and sellers can be held liable if they cause injury or damage.

Intellectual Property Laws: These laws protect the intellectual property of service robot developers, such as their patents, copyrights, and trademarks. This allows developers to prevent others from copying or using their intellectual property without permission.

Data Protection Laws: These laws regulate the collection, use, and disclosure of personal data. Service robots may collect a large amount of personal data about their users, so it is important for developers and operators to comply with data protection laws.

Competition Law: These laws prohibit businesses from engaging in anti-competitive practices. Service robot developers and vendors need to be aware of competition laws when developing and marketing their products.

Employment Law: These laws regulate the relationship between employers and employees. Service robots may be used to automate tasks that were previously performed by humans, so employers need to be aware of employment laws when deploying service robots.

In addition to these general legal frameworks, there are also a number of specific laws and regulations that may apply to service robots in certain jurisdictions. For example, the European Union is currently developing a new regulation on artificial intelligence (AI), which is expected to apply to service robots.

It is important to note that the legal and regulatory landscape for service robots is still evolving. As service robots become more widespread and sophisticated, new legal and regulatory issues are likely to emerge. Service robot developers and operators need to stay up-to-date on the latest laws and regulations to ensure that they are in compliance.

United States: The United States does not have a single federal law that governs service robots. However, there are a number of federal laws and regulations that may apply to service robots, depending on their specific design, function, and intended use. For example, the Food and Drug Administration (FDA) regulates service robots that are used in medical settings.

European Union: The European Union is currently developing a new regulation on AI, which is expected to apply to service robots. The regulation is expected to focus on safety, transparency, and accountability.

China: China has a number of laws and regulations that apply to service robots, including the Law on the Protection of Consumers' Rights and Interests and the Law on the Protection of Personal Information. China is also developing a new AI law, which is expected to be released in the near future.

Service robot developers and operators need to be aware of the legal and regulatory landscape in the jurisdictions where their products are being used. By complying with all applicable laws and regulations,service robot developers and operators can help to ensure that their products are used in a safe and responsible manner.

Liability And Accountability In Robotics

Liability and accountability are two important concepts in the legal regulation of service robots.

Liability refers to the legal obligation to compensate someone for harm that has been caused. In the context of service robots, liability could arise if a robot causes injury or damage. For example, if a service robot malfunctions and causes a car accident, the manufacturer of the robot could be held liable for damages.

Accountability refers to the responsibility to answer for one's actions. In the context of service robots, accountability is important for ensuring that robots are used in a safe and responsible manner. For example, if a service robot causes harm, it is important to be able to identify the person or entity who is accountable for the harm so that corrective action can be taken.

The law of liability and accountability in robotics is still evolving. However, there are a number of existing legal principles that can be applied to service robots. For example, product liability law could be used to hold manufacturers and sellers of service robots liable for injuries or damage caused by defective products. Additionally, negligence law could be used to hold service robot operators liable for injuries or damage caused by their carelessness.

In addition to existing legal principles, there are a number of new legal frameworks that are being developed to address the unique challenges of liability and accountability in robotics. For example, the European Union is currently developing a new regulation on AI, which is expected to include provisions on liability and accountability.

Here Are Some Specific Examples Of How Liability And Accountability Might Be Applied In The Context Of Service Robots:

If a service robot malfunctions and causes a car accident, the manufacturer of the robot could be held liable for damages to the vehicles and any injuries sustained by the occupants.

If a service robot is used in a hospital and causes harm to a patient, the hospital could be held liable for the harm.

If a service robot is used in a factory and causes harm to a worker, the factory owner could be held liable for the harm.

It is important to note that the law of liability and accountability in robotics is complex and evolving. Service robot developers and operators should consult with an attorney to understand their specific legal obligations.

Challenges Of Liability And Accountability In Robotics

There are a number of challenges associated with liability and accountability in robotics. One challenge is that it can be difficult to identify the person or entity who is accountable for harm caused by a service robot. For example, if a service robot is programmed by one company and operated by another company, it may be unclear which company is accountable for harm caused by the robot.

Another challenge is that it can be difficult to prove that a service robot was defective or that the robot operator was negligent. This is especially true for complex autonomous robots that are able to make their own decisions.

Finally, there is a risk that service robot operators could be held liable for harm caused by robots even if the operator could not have reasonably foreseen the harm. This could lead to service robot operators being reluctant to use robots, which could stifle innovation in the robotics industry.

Overcoming The Challenges Of Liability And Accountability In Robotics

There are a number of things that can be done to overcome the challenges of liability and accountability in robotics. One important step is to develop clear and consistent standards for the design, development, and operation of service robots. These standards should help to ensure that service robots are safe and reliable.
Another important step is to develop clear rules on liability and accountability. These rules should be fair and equitable, and they should not discourage innovation in the robotics industry.

Finally, it is important to educate service robot developers, operators, and users about the law of liability and accountability. This will help to ensure that everyone involved in the robotics industry is aware of their legal rights and obligations.

By taking these steps, we can help to ensure that service robots are used in a safe and responsible manner.

Intellectual Property And Patents In The Field Of Service Robotics

Intellectual Property (Ip) is a broad term that refers to creations of the mind, such as inventions, literary and artistic works, designs, and symbols, names and images used in commerce. IP rights give creators the exclusive right to make, use, sell, or license their creations for a set period of time.

Patents are a type of IP protection that grants the inventor the exclusive right to make, use, sell, or offer to sell an invention for a set period of time, typically 20 years from the filing date. Patents can be granted for inventions of any field of technology, including service robotics.

Patents play an important role in the development and commercialization of service robots. By protecting their inventions with patents, service robot developers can gain a competitive advantage and attract investors. Patents can also help service robot developers to license their technology to other companies.

Intellectual Property and Patents in the Field of Service Robotics

The Power Of Patents: Real-World Examples In Service Robotics

A robotic vacuum cleaner that uses artificial intelligence to navigate around obstacles and clean floors more efficiently.

A robotic delivery vehicle that can deliver packages to customers without human intervention. A robotic surgical system that allows surgeons to perform complex surgeries with greater precision and accuracy.

A robotic exoskeleton that can help people with disabilities to walk and perform other tasks.

The process of obtaining a patent for a service robot invention is complex and can take several years. To be eligible for a patent, an invention must be new, useful, and non-obvious. The invention must also be disclosed in a complete and enabling patent application.

Once a patent application is filed, the US Patent and Trademark Office (USPTO) will examine the application to determine whether the invention meets the patentability requirements. If the USPTO finds that the invention is patentable, a patent will be granted.

Service robot developers should consider whether to patent their inventions on a case-by-case basis. There are a number of factors to consider, such as the commercial potential of the invention, the cost of obtaining and maintaining a patent, and the risk of infringement by competitors.

Benefits Of Patenting Service Robot Inventions

There Are A Number Of Benefits To Patenting Service Robot Inventions. These Benefits Include:

Competitive Advantage: Patents can give service robot developers a competitive advantage over their competitors. By preventing competitors from copying their inventions, service robot developers can maintain a market advantage and increase their profits.

Attracting Investors: Investors are more likely to invest in service robot companies that have a strong patent portfolio. Patents demonstrate that a company has innovative technology that is likely to be successful in the marketplace.

Licensing Opportunities: Service robot developers can license their patented technology to other companies. This can generate a new revenue stream for the developer and help to accelerate the adoption of their technology.

Challenges Of Patenting Service Robot Inventions

Patenting Service Robot Inventions: Overcoming Key Challenges

Complexity: The process of obtaining a patent is complex and can be expensive. Service robot developers should consult with a patent attorney to ensure that their patent applications are properly drafted and filed.

Risk Of Infringement: Even if a service robot developer obtains a patent, there is always a risk that competitors will infringe on the patent. If infringement occurs, the patent owner may need to file a lawsuit to protect their rights.

Prior Art: Patent examiners will search for prior art, which is any information that was publicly available before the filing date of the patent application. If the examiner finds prior art that discloses the invention, the patent application may be rejected.

Service robot developers should carefully consider the benefits and challenges of patenting their inventions before making a decision. By consulting with a patent attorney, service robot developers can make informed decisions about protecting their intellectual property.

7.3. Privacy And Data Protection In Service Robotics

Service robots collect a large amount of data about their environment and users. This data can include personal data, such as names, addresses, and images. Service robots may also collect sensitive data, such as health information or financial data.

Privacy and Data Protection in Service Robotics

It is important to protect the privacy of users and to ensure that their data is collected, used, and disclosed in a responsible and ethical manner. This can be done by following a number of privacy and data protection principles, such as:

Transparency: Service robot developers and operators should be transparent about the data that they collect and how they use it. Users should be able to give informed consent to the collection and use of their data.

Data Minimization: Service robots should only collect the data that is necessary for them to perform their intended function.

Data Security: Service robots should implement appropriate security measures to protect user data from unauthorized access, use, disclosure, modification, or destruction.

Data Subject Rights: Users should have the right to access their data, to correct any inaccuracies in their data, and to request that their data be deleted.

In addition to following these principles, service robot developers and operators should also comply with all applicable privacy and data protection laws and regulations.

Navigating Data Privacy Concerns In Service Robotics: Practical Instances

Home Robots: Home robots may collect a large amount of data about the home environment and its occupants. This data could be used to improve the robot's performance, but it could also be used to track or monitor the occupants without their consent. It is important to ensure that home robots are designed with privacy and data protection in mind.

Healthcare Robots: Healthcare robots may collect sensitive data about patients, such as their health history and medical records. This data must be protected from unauthorized access, use, or disclosure. Healthcare robots should also be designed to respect the privacy and dignity of patients.

Financial Robots: Financial robots may collect sensitive data about customers, such as their financial information and account balances. This data must be protected from unauthorized access, use, or disclosure. Financial robots should also be designed to comply with all applicable financial privacy laws and regulations.

Service robot developers and operators should carefully consider the privacy and data protection implications of their products. By following the principles and best practices outlined above, they can help to protect the privacy of users and to ensure that their data is collected, used, and disclosed in a responsible and ethical manner.

Challenges Of Privacy And Data Protection In Service Robotics

There are a number of challenges associated with privacy and data protection in service robotics. One challenge is that service robots may collect a large amount of data without the knowledge or consent of users. This can happen if the robot is not properly designed or if the user is not aware of the data that is being collected.

Another challenge is that service robots may be vulnerable to hacking and other cyberattacks. This could lead to the unauthorized access, use, or disclosure of user data.

Finally, service robots may be used to collect data about people in public places without their consent. This could raise privacy concerns, especially if the data is used to track or monitor people.

Overcoming The Challenges Of Privacy And Data Protection In Service Robotics

There are a number of things that can be done to overcome the challenges of privacy and data protection in service robotics. One important step is to develop clear and consistent standards for the design, development, and operation of service robots. These standards should help to ensure that service robots are designed with privacy and data protection in mind.

Another important step is to develop clear rules on the collection, use, and disclosure of data collected by service robots. These rules should be fair and equitable, and they should not discourage innovation in the robotics industry.

Finally, it is important to educate service robot developers, operators, and users about privacy and data protection. This will help to ensure that everyone involved in the robotics industry is aware of their legal rights and obligations.

Data Collection And Consent In Robot-Assisted Services

Robot-assisted services are services that are provided with the assistance of robots. These services can be provided in a variety of settings, such as homes, hospitals, schools, and factories.

Data Collection and Consent in Robot-Assisted Services

Robot-assisted services may collect a large amount of data about their users. This data can include personal data, such as names, addresses, and images. Robot-assisted services may also collect sensitive data, such as health information or financial data.

It is important to obtain consent from users before collecting their data. This consent should be informed and freely given. Users should be able to understand the type of data that is being collected, how it will be used, and who it will be shared with.

There are a number of ways to obtain consent from users. One way is to have users sign a consent form. Another way is to have users agree to a terms of service agreement that includes a privacy policy. The privacy policy should explain how the data will be collected, used, and shared.

It is important to note that consent should be ongoing. Users should be able to withdraw their consent at any time. If a user withdraws their consent, the robot-assisted service should stop collecting their data and delete any data that has already been collected.

Navigating Data Collection And Consent In Robot-Assisted Services: Real-Life Instances

Home Robots: A home robot may collect data about the home environment and its occupants. This data could be used to improve the robot's performance, but it could also be used to track or monitor the occupants without their consent. It is important to obtain consent from users before collecting any data about them or their home environment.

Healthcare Robots: A healthcare robot may collect data about patients, such as their vital signs or medical history. This data is sensitive and must be protected from unauthorized access, use, or disclosure. It is important to obtain consent from patients before collecting any data about them.

Financial Robots: A financial robot may collect data about customers, such as their account balances or transaction history. This data is sensitive and must be protected from unauthorized access, use, or disclosure. It is important to obtain consent from customers before collecting any data about them.

Robot-assisted service providers should carefully consider the data collection and consent practices that they use. By obtaining informed consent from users and protecting their data, robot-assisted service providers can help to build trust with their users and promote the adoption of robot-assisted services.

Challenges Of Data Collection And Consent In Robot-Assisted Services

There are a number of challenges associated with data collection and consent in robot-assisted services. One challenge is that users may not be aware of the data that is being collected about them or how it will be used. This can happen if the robot-assisted service does not have a clear and transparent privacy policy.

Another challenge is that users may not be able to freely give their consent. This can happen if the robot-assisted service is essential for the user to receive a service. For example, a patient may not be able to refuse to have their data collected by a healthcare robot if the robot is necessary for them to receive treatment.

Finally, it can be difficult to obtain consent from children and other vulnerable populations. This is because these populations may not be able to understand the implications of giving their consent.

Overcoming The Challenges Of Data Collection And Consent In Robot-Assisted Services

There are a number of things that can be done to overcome the challenges of data collection and consent in robot-assisted services. One important step is to develop clear and transparent privacy policies. These privacy policies should explain what data is being collected, how it will be used, and who it will be shared with.

Another important step is to give users control over their data. Users should be able to decide what data is collected about them, how it is used, and who it is shared with. Users should also be able to withdraw their consent at any time.

Finally, it is important to take special care when collecting data from children and other vulnerable populations. These populations may need additional assistance understanding the implications of giving their consent.

Protecting User Privacy In Smart Environments

Smart environments are environments that are equipped with sensors and other devices that collect data about the environment and its occupants. These environments can include homes, offices, schools, and hospitals.

Smart environments can collect a large amount of data about their users. This data can include personal data, such as names, addresses, and images. Smart environments may also collect sensitive data, such as health information or financial data.

Protecting User Privacy in Smart Environments

It Is Important To Protect User Privacy In Smart Environments. This Can Be Done By Following A Number Of Privacy And Data Protection Principles, Such As:

Transparency: Smart environment developers and operators should be transparent about the data that they collect and how they use it. Users should be able to give informed consent to the collection and use of their data.

Data Minimization: Smart environments should only collect the data that is necessary for them to perform their intended function.

> **Data Security:** Smart environments should implement appropriate security measures to protect user data from unauthorized access, use, disclosure, modification, or destruction.

> **Data Subject Rights:** Users should have the right to access their data, to correct any inaccuracies in their data, and to request that their data be deleted.

In addition to following these principles, smart environment developers and operators should also comply with all applicable privacy and data protection laws and regulations.

From Concepts To Safeguards: Protecting User Privacy In Smart Environments

> **Home Automation Systems:** Home automation systems can collect a large amount of data about the home environment and its occupants. This data could be used to improve the performance of the system, but it could also be used to track or monitor the occupants without their consent. It is important to choose a home automation system that has strong privacy and data protection features.

> **Wearable devices:** Wearable devices, such as smartwatches and fitness trackers, can collect a large amount of data about the user's health and activity. This data is sensitive and must be protected from unauthorized access, use, or disclosure. It is important to choose a wearable device that has strong privacy and data protection features.

> **Connected Cars:** Connected cars can collect a large amount of data about the car's location, speed, and occupants. This data is sensitive and must be protected from unauthorized access, use, or disclosure. It is important to choose a connected car that has strong privacy and data protection features.

Smart environment developers and operators should carefully consider the privacy and data protection implications of their products and services. By following the principles and best practices outlined above, they can help to protect the privacy of users and promote the adoption of smart environments.

Challenges Of Protecting User Privacy In Smart Environments

There are a number of challenges associated with protecting user privacy in smart environments. One challenge is that smart environments may collect a large amount of data without the knowledge or consent of users. This can happen if the smart environment is not properly designed or if the user is not aware of the data that is being collected.

Another challenge is that smart environments may be vulnerable to hacking and other cyberattacks. This could lead to the unauthorized access, use, or disclosure of user data.

Finally, smart environments may be used to collect data about people in public places without their consent. This could raise privacy concerns, especially if the data is used to track or monitor people.

Overcoming The Challenges Of Protecting User Privacy In Smart Environments

There are a number of things that can be done to overcome the challenges of protecting user privacy in smart environments. One important step is to develop clear and consistent standards for the design, development, and operation of smart environments. These standards should help to ensure that smart environments are designed with privacy and data protection in mind.

Another important step is to develop clear rules on the collection, use, and disclosure of data collected by smart environments. These rules should be fair and equitable, and they should not discourage innovation in the smart environment industry.

Finally, it is important to educate smart environment developers, operators, and users about privacy and data protection. This will help to ensure that everyone involved in the smart environment industry is aware of their legal rights and obligations.

Handling Sensitive Information In Healthcare And Personal Assistance Robots

Healthcare and personal assistance robots collect a large amount of sensitive information about their users. This information can include health data, financial data, and personal data such as names, addresses, and images.
It is important to handle this sensitive information in a responsible and ethical manner. This means protecting the data from unauthorized access, use, disclosure, modification, or destruction. It also means using the data in a way that is consistent with the user's consent and expectations.

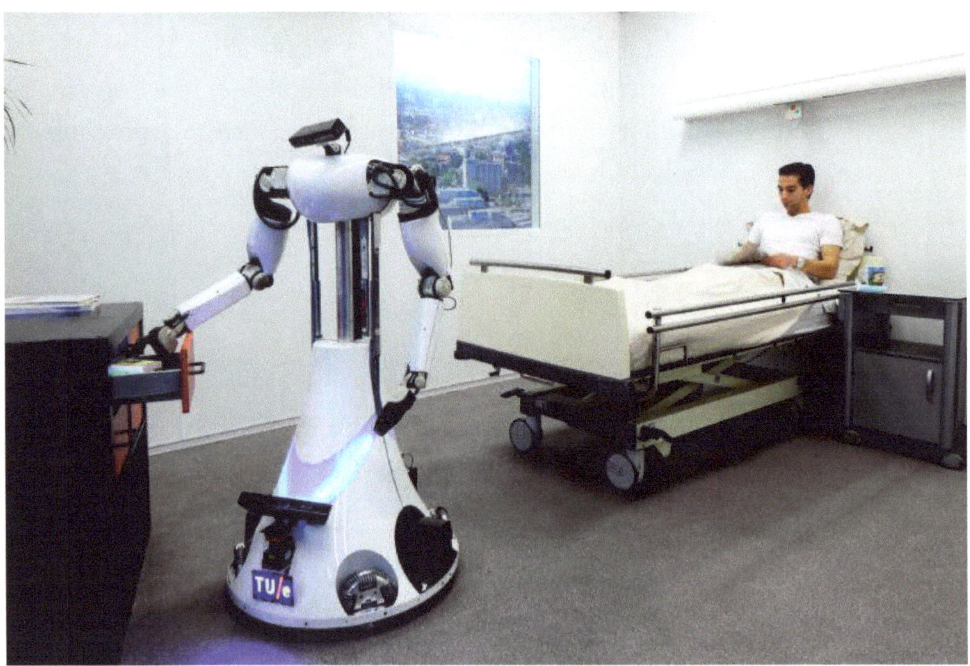

Handling Sensitive Information in Healthcare and Personal Assistance Robots

Safeguarding Sensitive Information:

Implement appropriate security measures. This includes using strong encryption, access controls, and firewalls to protect the data from unauthorized access.

Be transparent about the data that is being collected and how it will be used. This can be done through a privacy policy or consent form.

Give users control over their data. This means allowing users to decide what data is collected about them, how it is used, and who it is shared with. Users should also be able to withdraw their consent at any time.

Use the data in a way that is consistent with the user's consent and expectations. For example, if a user consents to their data being used to improve the robot's performance, the robot should not use the data to track or monitor the user without their consent.

Train employees on privacy and data protection best practices. This will help to ensure that all employees who handle sensitive data are aware of their legal and ethical obligations.

It is also important to note that healthcare and personal assistance robots may be subject to additional privacy and data protection laws and regulations. For example, the Health Insurance Portability and Accountability Act (HIPAA) in the United States protects the privacy of health information. Robot developers and operators should consult with an attorney to ensure that they are in compliance with all applicable privacy and data protection laws and regulations.

Some Specific Examples Of How To Handle Sensitive Information In Healthcare And Personal Assistance Robots:

Healthcare Robots: A healthcare robot may collect data about a patient's vital signs, medical history, or treatment plan. This information is sensitive and must be protected from unauthorized access, use, or disclosure. The robot should use strong encryption to protect the data and should only share the data with authorized personnel.

Personal Assistance Robots: A personal assistance robot may collect data about a user's home environment, schedule, or financial information. This information is also sensitive and must be protected from unauthorized access, use, or disclosure. The robot should use strong encryption to protect the data and should only share the data with authorized personnel.

7.4. Safety Standards And Ethical Design

In the rapidly advancing field of service robotics, the development and deployment of robots in various applications have raised important ethical and legal questions. Ensuring the safety of both robots and humans, as well as adhering to ethical design principles, is of paramount importance. This chapter delves into the critical aspects of safety standards and ethical design in the context of service robotics.

Safety Standards In Service Robotics:

Regulatory Frameworks: Service robots often operate in close proximity to humans, making safety regulations a top priority. Governments and international organizations have started to establish regulatory frameworks to ensure compliance with safety standards. These standards encompass various aspects such as hardware safety, software reliability, and emergency shutdown mechanisms.

Risk Assessment: Service robot developers must conduct thorough risk assessments to identify potential hazards. This process involves evaluating the likelihood and severity of accidents or malfunctions. Mitigation strategies should be implemented based on the findings.

Human-Robot Interaction (Hri): Safety standards extend to the interaction between humans and robots. Guidelines are established for safe human-robot collaboration, taking into account factors like collision avoidance, force and torque limits, and ergonomics to prevent injury during physical interaction.

Testing And Certification: Rigorous testing protocols are essential to verify that a service robot meets safety standards. Certification bodies play a crucial role in ensuring that robots adhere to these standards before they are deployed in real-world scenarios.

Safety Standards and Ethical Design

Ethical Design In Service Robotics:

Transparency And Accountability: Ethical design requires transparency in how robots make decisions and take actions. Developers should ensure that the decision-making processes of robots are understandable and that there is accountability for their actions.

Bias And Fairness: Preventing biases in robot behavior is a significant ethical concern. Developers must strive to eliminate biases in data, algorithms, and design to ensure fair treatment of all individuals, regardless of their characteristics.

Privacy And Data Protection: Robots often collect and process sensitive data. Ethical design principles include robust data protection mechanisms to safeguard individuals' privacy and prevent unauthorized access to personal information.

Human Values And Preferences: Service robots should respect human values and preferences. Ethical design involves customization options and the ability for users to define how robots interact with them within ethical boundaries.

Beneficence And Non-Maleficence: Developers should design robots to maximize benefits while minimizing harm. Ethical considerations extend to ensuring that robots do not engage in harmful actions or behaviors.

Safety standards and ethical design are intertwined in the development of service robots. By adhering to rigorous safety protocols and ethical principles, developers can create robots that not only perform their intended tasks effectively but also do so in a manner that prioritizes the well-being and rights of humans. As the field of service robotics continues to evolve, ongoing collaboration between stakeholders, including governments, industry leaders, and ethicists, will be essential to ensure that ethical and legal considerations are at the forefront of this transformative technology.

Ensuring Robot Safety In Public Spaces

Ensuring robot safety in public spaces is essential for the safe and ethical adoption of service robotics. There are a number of factors to consider when designing and deploying service robots for public use, including the following:

Physical Safety: Service robots should be designed to minimize the risk of physical harm to humans. This includes using safe materials and construction methods, as well as designing robots that are stable and easy to control.

Functional Safety: Service robots should be designed to function safely and reliably in public spaces. This includes being able to detect and avoid obstacles, as well as being able to handle unexpected events.

Behavioral Safety: Service robots should be designed to behave in a way that is safe and predictable. This includes being aware of their surroundings and interacting with humans in a way that is respectful and non-threatening.

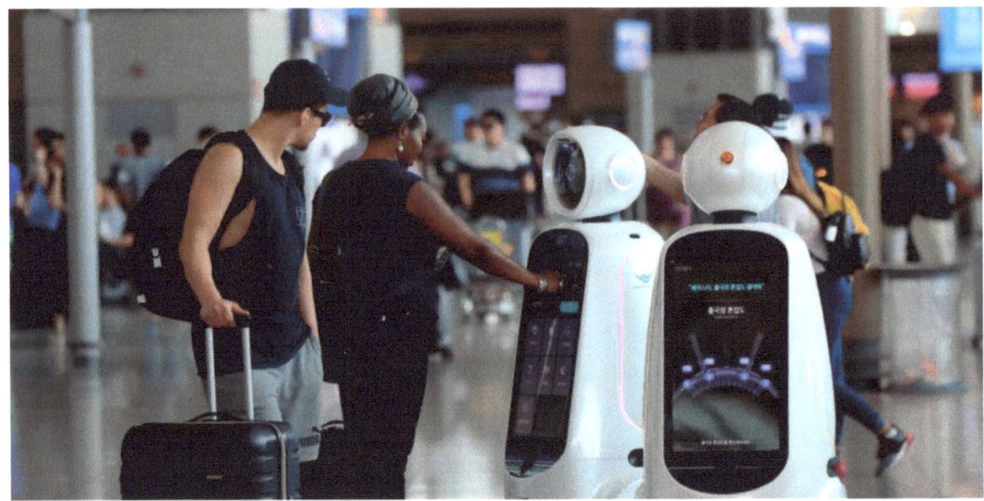

Ensuring Robot Safety in Public Spaces

In addition to these technical considerations, it is also important to consider the ethical implications of deploying service robots in public spaces. For example, it is important to ensure that service robots are used in a way that respects human privacy and autonomy. It is also important to ensure that service robots are not used to discriminate against or disadvantage certain groups of people.

There are a number of safety standards and ethical guidelines that can help to ensure the safe and ethical deployment of service robots in public spaces. For example, the International Organization for Standardization (ISO) has developed a number of standards for the safety of robots, including ISO 13482:2014 "Robots and robotic devices - Safety requirements for personal care robots" and ISO/TS 15066:2016 "Robots and robotic devices - Collaborative robots".

In addition to technical standards, there are also a number of ethical guidelines that have been developed for the development and deployment of service robots. For example, the IEEE Global Initiative on Ethics of Autonomous and Intelligent Systems has developed a set of ethical guidelines for the design and use of autonomous and intelligent systems.

These safety standards and ethical guidelines can help to ensure that service robots are deployed in a safe and ethical way in public spaces.

Here Are Some Additional Specific Considerations For Ensuring Robot Safety In Public Spaces:

Use sensors and algorithms to detect and avoid obstacles and people. Service robots should be equipped with sensors and algorithms that allow them to perceive their surroundings and identify potential hazards. This will help to prevent robots from colliding with people or objects, which could cause injuries.

Limit the speed and force of robots. Service robots should be designed to operate at a safe speed and with a limited force output. This will help to minimize the risk of injuries in the event of a collision. Use soft materials and rounded edges in the robot's design. Service robots should be designed with soft materials and rounded edges to reduce the risk of injuries in the event of a collision.

Provide clear and easy-to-understand instructions for users. Service robots should be accompanied by clear and easy-to-understand instructions for users. This will help to ensure that users are able to operate the robot safely and effectively.

Monitor and maintain robots on a regular basis. Service robots should be monitored and maintained on a regular basis to ensure that they are operating safely. This includes checking the robot's sensors, actuators, and software for any problems.

Designing Ethical Ai Algorithms For Decision-Making

In recent years, there has been a growing awareness of the ethical implications of artificial intelligence (AI). This is particularly true for AI algorithms that are used to make decisions that have a significant impact on people's lives, such as decisions about hiring, lending, and healthcare.

There Are A Number Of Factors To Consider When Designing Ethical Ai Algorithms For Decision- Making. These Include:

Fairness: AI algorithms should be fair and unbiased. This means that they should not discriminate against any particular group of people.

Transparency: AI algorithms should be transparent and explainable. This means that it should be possible to understand how the algorithm works and why it makes the decisions that it does.

Accountability: There should be clear accountability for the decisions that are made by AI algorithms. This means that it should be possible to identify who is responsible for the decisions that the algorithm makes and how those decisions can be challenged or appealed.

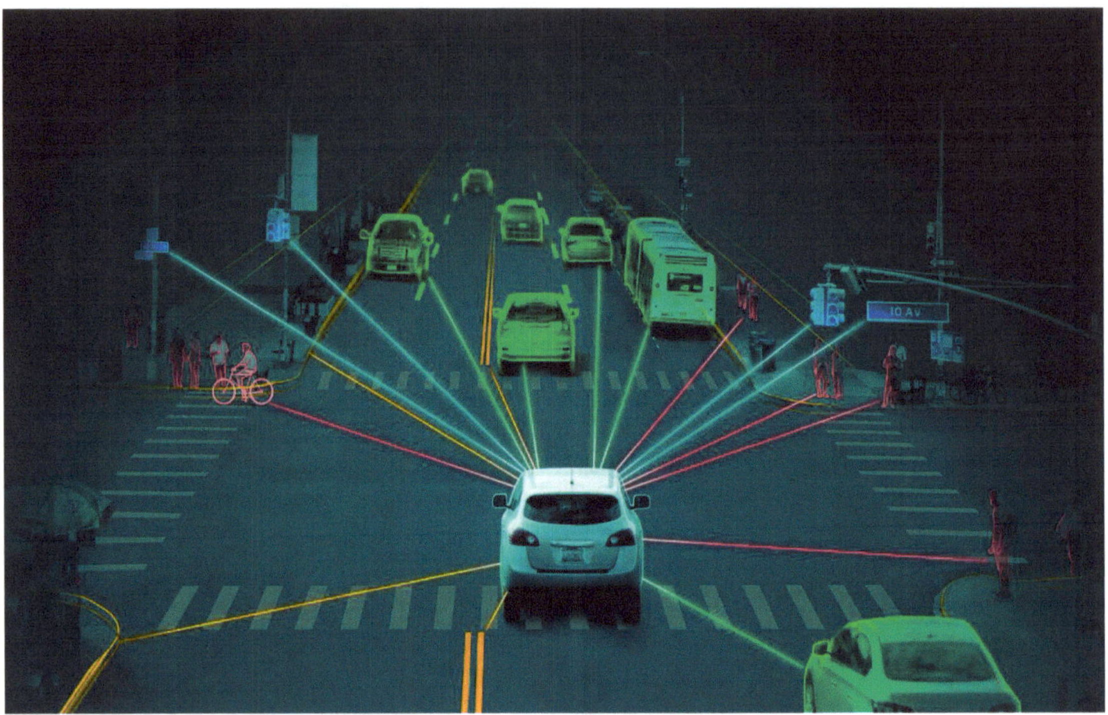

Designing Ethical AI Algorithms for Decision-Making

There Are A Number Of Techniques That Can Be Used To Design Ethical Ai Algorithms For Decision-Making. These Include:

Using Diverse Training Data: One way to reduce bias in AI algorithms is to use training data that is diverse and representative of the population that the algorithm will be used to make decisions about.

Using Fairness Metrics: There are a number of fairness metrics that can be used to evaluate the fairness of AI algorithms. These metrics can help to identify and mitigate bias in the algorithm.

Using Human-In-The-Loop Systems: Human-in-the-loop systems combine the strengths of humans and AI algorithms. In these systems, humans are involved in the decision-making process to ensure that the decisions are fair and ethical.

It is important to note that there is no one-size-fits-all solution to designing ethical AI algorithms for decision-making. The specific techniques that are used will depend on the specific application and the specific ethical concerns that need to be addressed.

Here Are Some Additional Specific Considerations For Designing Ethical Ai Algorithms For Decision-Making In Service Robotics:

Identify And Mitigate Potential Biases In The Training Data. Service robots are often trained on data that is collected from the real world. This data may contain biases that could lead to the robot making unfair decisions. It is important to identify and mitigate these biases in the training data.

Design The Robot To Be Transparent And Explainable. It should be possible to understand how the robot's AI algorithm works and why it makes the decisions that it does. This will help to ensure that humans can trust the robot and that they can challenge or appeal the robot's decisions if necessary.

Design The Robot To Be Accountable. There should be clear accountability for the decisions that are made by the robot. This means that it should be possible to identify who is responsible for the decisions that the robot makes and how those decisions can be challenged or appealed.

Balancing Safety And Autonomy In Service Robots

Balancing safety and autonomy in service robots is a key challenge in the design and deployment of these systems. Service robots are often designed to operate in dynamic and unpredictable environments, which makes it difficult to ensure safety without sacrificing autonomy.

There Are A Number Of Factors To Consider When Balancing Safety And Autonomy In Service Robots. These Include:

The Robot's Environment: The robot's environment is a key factor in determining how much autonomy it can be granted. Robots that operate in dynamic and unpredictable environments, such as public spaces, will need to be more restricted in their autonomy than robots that operate in controlled environments, such as factories.

The Robot's Tasks: The robot's tasks also play a role in determining how much autonomy it can be granted. Robots that perform tasks that are dangerous or complex will need to be more restricted in their autonomy than robots that perform simple and repetitive tasks.

The Robot's Capabilities: The robot's capabilities are also a factor to consider. Robots that are equipped with advanced sensors and actuators will be able to operate more safely and autonomously than robots that are not.

Balancing Safety and Autonomy in Service Robots

There Are A Number Of Techniques That Can Be Used To Balance Safety And Autonomy In Service Robots. These Include:

Using Safety Sensors: Safety sensors can be used to detect potential hazards and trigger the robot to take evasive action. For example, a robot could be equipped with a bumper sensor that causes the robot to stop if it collides with an object.

Using Fail-Safe Mechanisms: Fail-safe mechanisms can be used to ensure that the robot is safe even if something goes wrong. For example, a robot could be equipped with a dead man's switch that causes the robot to stop if it is not being actively controlled.

Using Human Supervision: Human supervision can be used to ensure that the robot is operating safely. For example, a robot could be programmed to stop and wait for human approval before performing certain tasks.

The specific techniques that are used to balance safety and autonomy in service robots will depend on the specific application and the specific safety concerns that need to be addressed. **Some Additional Specific Considerations For Balancing Safety And Autonomy In Service Robots:**

Design The Robot To Be Able To Detect And Avoid Obstacles. Service robots should be equipped with sensors and algorithms that allow them to perceive their surroundings and identify potential hazards. This will help to prevent robots from colliding with people or objects, which could cause injuries.

Design The Robot To Be Able To Handle Unexpected Events. Service robots should be programmed to handle unexpected events in a safe and controlled manner. For example, if a robot encounters an obstacle unexpectedly, it should be able to stop or change direction safely.

Limit The Robot's Speed And Force Output. Service robots should be designed to operate at a safe speed and with a limited force output. This will help to minimize the risk of injuries in the event of a collision.

Provide Clear And Easy-To-Understand Instructions For Users. Service robots should be accompanied by clear and easy-to-understand instructions for users. This will help to ensure that users are able to operate the robot safely and effectively.

Monitor And Maintain Robots On A Regular Basis. Service robots should be monitored and maintained on a regular basis to ensure that they are operating safely. This includes checking the robot's sensors, actuators, and software for any problems

7.5. Emerging Ethical Challenges And Future Considerations

Service robotics is a rapidly evolving field, and with it comes a number of emerging ethical challenges and future considerations. Some of these challenges and considerations include:

The Potential For Job Displacement: As service robots become more sophisticated and capable, they are likely to displace some human workers. This could lead to unemployment and social unrest. It is important to consider how to mitigate the negative impacts of job displacement, such as by providing training and support for displaced workers.

The Potential For Bias And Discrimination: Service robots are often trained on data that is collected from the real world. This data may contain biases that could lead to the robot making unfair decisions. For example, a robot that is used to screen loan applications may be biased against certain groups of people, such as minorities or women. It is important to develop methods to identify and mitigate bias in service robots.

The Potential For Privacy And Security Violations: Service robots may collect a lot of data about their surroundings and the people they interact with. This data could be used to track people's movements, monitor their activities, and even predict their behavior. It is important to develop strong privacy and security protections for data collected by service robots.

The Potential For Misuse And Abuse: Service robots could be misused or abused for malicious purposes. For example, a robot that is designed to assist people with disabilities could be used to spy on them or even harm them. It is important to develop safeguards to prevent service robots from being misused or abused.

In addition to these challenges, there are also a number of future considerations for ethical and legal considerations in service robotics. For example, it is important to consider how to develop ethical guidelines for the design, development, and use of service robots. It is also important to consider how to develop legal frameworks to regulate the use of service robots.

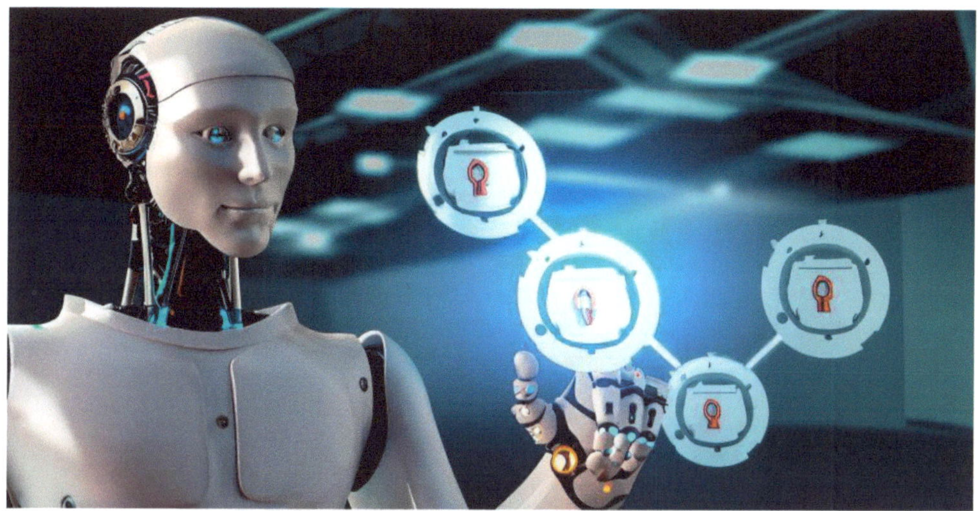

Some Additional Specific Considerations For Emerging Ethical Challenges And Future Considerations In Service Robotics:

The Development Of Ethical Guidelines For Service Robots: There is a need to develop clear and comprehensive ethical guidelines for the design, development, and use of service robots. These guidelines should address issues such as safety, privacy, bias, and accountability.

The Development Of Legal Frameworks For Service Robots: There is also a need to develop legal frameworks to regulate the use of service robots. These frameworks should address issues such as liability, safety, and ethical compliance.

The Impact Of Service Robots On Society: It is important to consider the broader social impact of service robots. For example, how will service robots affect the way we work, live, and interact with each other? It is important to have a public conversation about these issues so that we can develop policies and regulations that promote the responsible and ethical use of service robots.

By addressing these emerging ethical challenges and future considerations, we can ensure that service robotics is used in a way that benefits all of society.

Ethical Implications Of Ai Advancements In Service Robotics

As AI continues to advance, new ethical challenges will emerge for service robots. Some of these challenges include:

Bias: AI systems are trained on data, and if that data is biased, the AI system will also be biased. This could lead to service robots that discriminate against certain groups of people.

Autonomy: As service robots become more autonomous, they will be able to make their own decisions and take their own actions. This raises the question of who is responsible if a service robot makes a mistake or causes harm.

Privacy And Security: Service robots will collect a lot of data about their users and their environment. This data needs to be protected from unauthorized access and use.

Ethical Implications of AI Advancements in Service Robotics

Ethical Considerations In Human-Robot Relationships

As service robots become more sophisticated and commonplace, people will begin to form closer relationships with them. This raises a number of ethical considerations, such as:

Consent: If a service robot is going to provide care or companionship to a person, it is important to obtain the person's consent first.

Emotional Manipulation: Service robots could be used to manipulate people's emotions. This is especially concerning if service robots are used to care for vulnerable populations, such as the elderly or the disabled.

Humanizing Robots: Some people may begin to view service robots as more than just machines. This could lead to people forming inappropriate or unhealthy relationships with service robots.

Ethical Considerations in Human-Robot Relationships

The Role Of Industry And Research In Addressing Ethical Concerns

The industry and research communities have a role to play in addressing the ethical concerns associated with service robots. Some of the things that industry and research can do include:

Developing Ethical Guidelines: Industry and research can work together to develop ethical guidelines for the development and deployment of service robots. These guidelines should address issues such as bias, autonomy, privacy, and security.

Educating The Public: Industry and research can work to educate the public about the ethical implications of service robots. This will help people to make informed decisions about their interactions with service robots.

Investing In Research: Industry and research can invest in research to address the ethical challenges associated with service robots. This research should focus on developing technologies and practices that mitigate the risks associated with service robots.

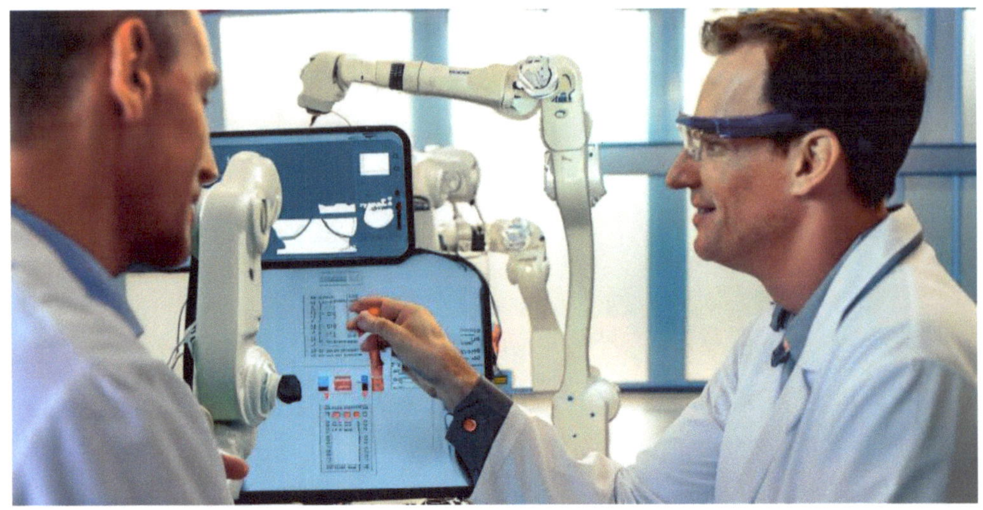

The Role of Industry and Research in Addressing Ethical Concerns

Future Considerations

As service robots become more sophisticated and commonplace, we will need to think carefully about the ethical implications of their use. We need to develop ethical guidelines, educate the public, and invest in research to ensure that service robots are used in a responsible and beneficial way.

In Addition To The Ethical Challenges Discussed Above, There Are A Number Of Other Future Considerations That We Need To Think About, Such As:

The Impact Of Service Robots On The Economy: Service robots could have a significant impact on the economy, both positive and negative. For example, service robots could create new jobs and boost productivity. However, service robots could also displace human workers and lead to increased unemployment.

The Impact Of Service Robots On Society: Service robots could have a significant impact on society, both positive and negative. For example, service robots could improve the quality of life for people with disabilities and the elderly. However, service robots could also lead to social isolation and increased inequality.

Chapter 8: Service Robots of Tomorrow: Emerging Trends and Future Visions

8.1. The Importance Of Anticipating Future Trends

The Importance of Anticipating Future Trends

Introduction To Emerging Trends And Future Directions In Service Robotics

Service robotics is a rapidly evolving field that holds the promise of revolutionizing various aspects of our daily lives. These intelligent machines, designed to assist and interact with humans, are increasingly finding applications in diverse sectors such as healthcare, logistics, hospitality, and more. As we look to the future, several exciting emerging trends and directions are shaping the landscape of service robotics.

This introduction will provide an overview of some of these trends and offer a glimpse into the promising future of service robotics.

Human-Robot Interaction And User Experience: As service robots become more prevalent, ensuring a positive user experience and natural human-robot interaction is crucial. Future developments will focus on making robots more intuitive to interact with and understanding human emotions and intentions better.

Human-Robot Collaboration: One of the most notable trends in service robotics is the emphasis on collaborative robots, often referred to as cobots. These robots are designed to work alongside humans, enhancing productivity and safety in settings like manufacturing, healthcare, and even household chores. The future of service robotics will likely see a proliferation of cobots that seamlessly integrate into our daily routines.

Autonomous Navigation: Advances in artificial intelligence and sensor technologies are driving the development of service robots capable of autonomous navigation in complex environments. This trend is particularly evident in the logistics and delivery sectors, where robots are being deployed for efficient last-mile delivery.

Robotic Assistants In Healthcare: The healthcare industry is witnessing a surge in the use of service robots for tasks ranging from patient care to disinfection and drug delivery. These robots have the potential to alleviate the burden on healthcare professionals and improve patient outcomes, making them a critical part of the future healthcare ecosystem.

Personal Service Robots: As the technology matures, we can expect to see more personal service robots that cater to individual needs. These robots could assist with household chores, provide companionship to the elderly, or even act as educational aids for children.

Ai And Machine Learning Integration: Service robots are becoming smarter and more adaptable thanks to advancements in artificial intelligence and machine learning. These technologies enable robots to learn from their interactions, adapt to changing environments, and provide more personalized services.

Ethical And Regulatory Considerations: With the increasing integration of service robots into society, ethical and regulatory challenges will become more pronounced. Future directions in service robotics will likely involve the development of ethical guidelines and regulatory frameworks to ensure responsible and safe robot deployment.

Robotic Swarms And Swarming Behavior: In some applications, such as search and rescue missions or environmental monitoring, swarms of small robots working together can be highly effective. Research into swarming behavior and its application in service robotics is a burgeoning field with great potential.

Environmental And Sustainability Concerns: The future of service robotics will also consider sustainability. Researchers are working on developing eco-friendly robot designs, efficient power sources, and recyclable materials to minimize the environmental footprint of these machines.

In conclusion, service robotics is poised for remarkable growth and innovation in the coming years. Emerging trends and future directions in this field hold the promise of making our lives more convenient, efficient, and safe. As we explore the exciting developments in service robotics, it is essential to keep a close eye on ethical considerations and regulatory frameworks to ensure responsible and beneficial integration into society.

The Evolving Landscape Of Service Robotics

The landscape of service robotics is evolving rapidly, driven by advances in artificial intelligence, machine learning, and robotics hardware. Service robots are becoming more intelligent, capable, and autonomous, allowing them to perform a wider range of tasks in a variety of settings.

The Evolving Landscape of Service Robotics

One of the most significant trends in service robotics is the increasing use of artificial intelligence (AI). AI enables robots to learn and adapt to their environment, making them more versatile and capable. For example, AI-powered robots can now navigate complex environments, recognize objects and people, and interact with humans in a natural way.

Another important trend is the development of new machine-learning algorithms. Machine learning allows robots to learn from data, which enables them to improve their performance over time. For example, machine learning algorithms are used to train robots to perform tasks such as object detection, image classification, and natural language processing.

Finally, advances in robotics hardware are making robots smaller, faster, and more energy-efficient. This is enabling the development of new types of service robots that can operate in a wider range of environments and perform more complex tasks.

These trends are leading to the development of a new generation of service robots that are more intelligent, capable, and autonomous than ever before. These robots are poised to revolutionize a wide range of industries, including healthcare, logistics, retail, and hospitality.

Here Are Some Specific Examples Of How The Service Robotics Landscape Is Evolving:

Healthcare: Service robots are being used in healthcare to provide a variety of services, including patient care, surgery, and rehabilitation. For example, robots are being used to assist nurses with tasks such as lifting and transferring patients and performing delicate surgical procedures.

Logistics: Service robots are being used in logistics to automate tasks such as order picking, packing, and shipping. For example, robots are being used to pick and pack items in warehouses and to load and unload trucks.

Retail: Service robots are being used in retail to provide customer service, stock shelves, and deliver products. For example, robots are being used to greet customers and answer their questions, to restock shelves, and to deliver products to customers' homes.

Hospitality: Service robots are being used in hospitality to provide a variety of services, such as check-in, room service, and housekeeping. For example, robots are being used to check in guests, deliver room service, and clean hotel rooms.

The evolving landscape of service robotics presents a number of opportunities and challenges. On the one hand, service robots have the potential to improve efficiency and productivity in a wide range of industries. On the other hand, there are concerns about the impact of service robots on jobs and the economy.

Overall, the evolving landscape of service robotics is an exciting one. As service robots become more intelligent, capable, and autonomous, they are poised to revolutionize a wide range of industries. It is important to carefully consider the potential opportunities and challenges of service robotics in order to ensure that they are used in a way that benefits society as a whole.

8.2. Human-Robot Interaction Advances In Natural Language Processing (NLP) For Improved Communication

Advances in natural language processing (NLP) are leading to significant improvements in communication between humans and machines. NLP is a field of artificial intelligence that deals with the interaction between computers and human (natural) languages. It's a subfield of computer science that deals with the ability of computers to understand and process human language, including speech and text.

Human-Robot Interaction

Nlp Is Used For A Variety Of Tasks, Including:

Machine Translation: Translating text from one language to another.

Speech Recognition: Converting spoken language into text.

Text Analysis: Extracting meaning from text, such as sentiment analysis, topic extraction, and question answering.

Natural Language Generation: Generating human-like text, such as creative text formats and code.

NLP can be used to improve communication between humans and machines in a number of ways. For example, NLP can be used to:

Enable machines to understand and respond to natural language commands: This allows humans to interact with machines in a more natural and intuitive way.

Generate Natural Language Responses: This allows machines to communicate with humans in a more natural and engaging way.

Translate Languages: This allows machines to communicate with people from different cultures.

Analyze Text: This allows machines to extract information from text, such as the sentiment of a customer review or the steps involved in a recipe.

Generate Text: This allows machines to create content, such as news articles, blog posts, and poems.

Leveraging Advances In Nlp For Enhanced Human-Machine Communication: Practical Examples

Machine Translation: Machine translation systems have become much more accurate in recent years, thanks to advances in NLP. This is making it easier for people to communicate with each other across language barriers.

Virtual Assistants: Virtual assistants, such as Siri and Alexa, are able to understand and respond to natural language commands more effectively than ever before. This is making it easier for people to control their devices and access information.

Chatbots: Chatbots are being used by businesses to provide customer service and support. NLP is used to enable chatbots to understand and respond to customer inquiries in a natural and helpful way.

Educational Software: NLP is being used to develop educational software that can provide personalized instruction to students. This software can adapt to the student's needs and provide feedback in a way that is more effective than traditional educational methods.

Advances in NLP are having a significant impact on the way we communicate with machines. NLP is making it possible for machines to understand and respond to human language in a more natural and intuitive way. This is leading to the development of new and innovative applications that are improving communication between humans and machines in a variety of ways.

Overall, advances in NLP are a positive development. NLP is making it possible for humans and machines to communicate more effectively, which is leading to the development of new and innovative products and services.

Emotional Intelligence In Robots For Enhanced User Engagement

Emotional intelligence (EI) is the ability to understand, manage, and express one's own emotions, as well as the ability to understand and respond to the emotions of others. EI is important for humans because it allows us to build strong relationships, communicate effectively, and resolve conflict.

While robots do not have emotions in the same way that humans do, they can still be programmed to exhibit EI-like behaviors. This can be done by teaching robots to recognize and respond to human emotions, and by programming them to generate their own emotional responses.

Emotional Intelligence in Robots for Enhanced User Engagement

Ei In Robots Can Be Used To Enhance User Engagement In A Number Of Ways. For Example, Robots With Ei Can:

Provide More Personalized And Engaging Experiences: Robots with EI can tailor their interactions to the individual needs and preferences of users. For example, a robot with EI could use a user's facial expressions to determine if they are happy, sad, or angry, and then adjust its tone of voice and body language accordingly.

Build Stronger Relationships With Users: Robots with EI can build stronger relationships with users by developing an understanding of their emotions and responding to them in a supportive and empathetic way. For example, a robot with EI could comfort a user who is feeling sad or offer encouragement to a user who is feeling frustrated.

Resolve Conflict More Effectively: Robots with EI can resolve conflict more effectively by understanding the emotions of the people involved and by communicating with them in a way that is respectful and mindful of their feelings. For example, a robot with EI could help two people who are arguing to understand each other's perspectives and to find a common solution.

From Emotions To Engagement: Demonstrating Ei Applications In Robot Interaction

Social Robots: Social robots are designed to interact with humans in a natural and engaging way. AI-powered social robots can use EI to recognize and respond to human emotions, such as happiness, sadness, and anger. This allows them to provide more personalized and engaging experiences for users.

Assistive Robots: Assistive robots are designed to help people with disabilities or mobility issues. AI-powered assistive robots can use EI to understand and respond to the needs of users, such as providing comfort and encouragement. This can help to improve the quality of life for users.

Educational Robots: Educational robots are designed to help students learn. AI-powered educational robots can use EI to adapt their teaching style to the individual needs of each student. This can help to improve student engagement and learning outcomes.

Overall, EI is a promising area of research for robots. By developing robots with EI, researchers are enabling them to interact with humans in a more natural and engaging way. This is leading to the development of new and innovative applications that are improving user engagement in a variety of ways.

It is important to note that EI in robots is still in its early stages of development. There are a number of challenges that need to be addressed before EI can be widely implemented in robots. One challenge is that it is difficult to program robots to understand and respond to human emotions in a way that is both natural and effective. Another challenge is that it is important to ensure that EI-powered robots are used in a responsible and ethical way.

Despite these challenges, EI is a promising area of research with the potential to revolutionize the way we interact with robots. By developing EI-powered robots, researchers are enabling them to be more useful and engaging companions for humans.

Gesture And Expression Recognition For Seamless Interactions

Gesture and expression recognition are essential for seamless interactions between humans and robots. Gesture recognition allows robots to understand human body language, while expression recognition allows robots to understand human emotions. Both of these technologies are essential for developing robots that can interact with humans in a natural and effective way.

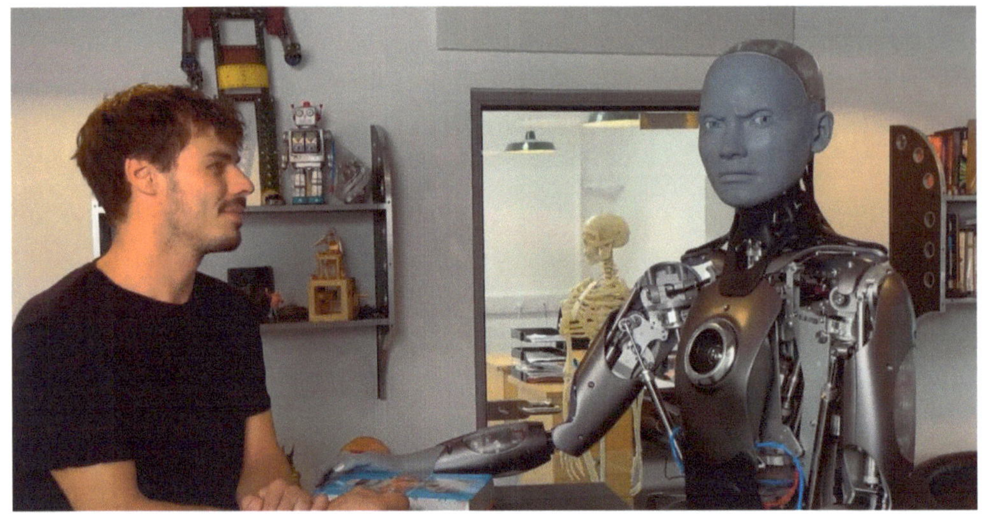

Gesture and Expression Recognition for Seamless Interactions

Gesture Recognition Can Be Used For A Variety Of Tasks, Including:

Controlling Robots: Gestures can be used to control robots, such as commanding a robot to move forward, backward, or turn.

Communicating With Robots: Gestures can be used to communicate with robots, such as giving a thumbs up to indicate approval or a thumbs down to indicate disapproval.

Interpreting Human Intentions: Gestures can be used to interpret human intentions, such as understanding whether someone is asking for help or giving directions.

Providing Feedback: Gestures can be used to provide feedback to robots, such as indicating that they are performing a task correctly or incorrectly.

Expression Recognition Can Be Used For A Variety Of Tasks, Including:

Adapting To Human Emotions: Robots can use expression recognition to adapt their behavior to human emotions, such as speaking more gently to someone who is feeling sad or speaking more enthusiastically to someone who is feeling happy.

Providing Social Support: Robots can use expression recognition to provide social support to humans, such as comforting someone who is feeling sad or encouraging someone who is feeling overwhelmed.

Detecting Dangerous Situations: Robots can use expression recognition to detect dangerous situations, such as when someone is feeling angry or aggressive.

Understanding Human Intentions: Expression recognition can be used to understand human intentions, such as understanding whether someone is interested in a product or service.

Gesture and expression recognition are still in their early stages of development, but they have the potential to revolutionize the way we interact with robots. By enabling robots to understand human body language and emotions, gesture and expression recognition can make robots more accessible and useful to people from all walks of life.

Real-Life Instances: Gesture And Expression Recognition In Service Robots For Enhanced Interactions:

Delivery Robots: Delivery robots can use gesture and expression recognition to understand and respond to customer instructions and reactions. For example, a delivery robot could use gesture recognition to understand when a customer is waving it over or when a customer is giving it a thumbs up or thumbs down. It could also use expression recognition to understand whether a customer is feeling happy, sad, or angry. This information could then be used to adjust the robot's behavior and provide a more personalized and engaging experience for the customer.

Cleaning Robots: Cleaning robots can use gesture and expression recognition to understand and respond to human commands and preferences. For example, a cleaning robot could use gesture recognition to understand when a person wants it to clean a specific area or when a person wants it to stop cleaning. It could also use expression recognition to understand whether a person is feeling comfortable or uncomfortable with its presence. This information could then be used to adjust the robot's cleaning behavior and provide a more satisfying experience for the user.

Security Robots: Security robots can use gesture and expression recognition to detect intruders and assess the safety of a situation. For example, a security robot could use gesture recognition to identify people who are entering a restricted area or who are making threatening gestures. It could also use expression recognition to identify people who are feeling angry or aggressive. This information could then be used to alert security personnel or to take other appropriate action.

Assistive Robots: Assistive robots can use gesture and expression recognition to provide personalized assistance to users. For example, an assistive robot could use gesture recognition to understand when a user wants it to help them with a task or when they want it to move out of the way. It could also use expression recognition to identify when a user is feeling frustrated or overwhelmed. This information could then be used to adjust the robot's assistance behavior and provide a more supportive experience for the user.

Overall, gesture and expression recognition are promising new technologies with the potential to revolutionize the way we interact with robots. By enabling robots to understand human body language and emotions, gesture and expression recognition can make robots more accessible and useful to people from all walks of life.

Ethical Considerations In Human-Robot Interaction

As robots become more commonplace in our lives, it is important to consider the ethical implications of human-robot interaction. There are a number of ethical issues that need to be addressed, such as:

Safety: Robots must be designed and operated in a way that ensures the safety of humans. This means that robots must be able to reliably perceive their environment and avoid collisions with humans. Robots must also be able to respond appropriately to human commands and instructions.

Privacy: Robots may collect personal data about humans, such as images, videos, and voice recordings. It is important to ensure that this data is collected and used in a responsible and ethical way.

Autonomy: Robots are becoming increasingly autonomous, meaning that they are able to make decisions and take actions without human intervention. This raises the question of who is responsible for the actions of robots.

Bias: Robots are trained on data, and this data can be biased. This means that robots may make decisions that are biased against certain groups of people.

Transparency: It is important to be transparent about the capabilities and limitations of robots. This means that humans should be aware of what robots can and cannot do, and how robots make decisions.

These are just some of the ethical considerations that need to be addressed in human-robot interaction. It is important to have a public conversation about these issues so that we can develop ethical guidelines for the development and use of robots.

Here Are Some Specific Examples Of Ethical Considerations In Human-Robot Interaction:

Self-Driving Cars: Self-driving cars need to be designed and operated in a way that ensures the safety of all road users, including pedestrians, cyclists, and other vehicles.

Ai-Powered Facial Recognition Systems: AI-powered facial recognition systems can be used for a variety of purposes, such as law enforcement and security. However, it is important to ensure that these systems are used in a way that respects people's privacy and civil liberties.

Ethical Considerations in Human-Robot Interaction

Social Robots: Social robots are designed to interact with humans in a natural and engaging way. However, it is important to ensure that these robots are not used to manipulate or exploit people.

It is important to note that these are just a few examples, and there are many other ethical considerations in human-robot interaction. It is important to have a public conversation about these issues so that we can develop ethical guidelines for the development and use of robots.

8.3. Autonomous Navigation AndMobility Navigation In Complex And Dynamic Environments

Navigation in complex and dynamic environments is a challenging task for both humans and robots. Complex environments are characterized by a large number of obstacles, such as walls, furniture, and people. Dynamic environments are characterized by obstacles that can move, such as people and vehicles.

There are a number of challenges associated with navigating in complex and dynamic environments. One challenge is that it can be difficult to perceive all of the obstacles in the environment. Another challenge is that it can be difficult to plan a path to the goal that avoids all of the obstacles. Finally, it can be difficult to execute the plan accurately in a dynamic environment where obstacles can move.

Autonomous Navigation and Mobility

There are a number of approaches to navigation in complex and dynamic environments. One approach is to use a map of the environment. The map can be used to plan a path to the goal that avoids all of the obstacles. However, maps can be outdated and inaccurate, and they may not include all of the obstacles in the environment.

Another approach to navigation in complex and dynamic environments is to use sensors to perceive the environment. Sensors such as cameras, lidar, and radar can be used to detect obstacles in the environment. However, sensors can be noisy and unreliable, and they may not be able to detect all of the obstacles in the environment.

Finally, a combination of maps and sensors can be used for navigation in complex and dynamic environments. The map can be used to plan a path to the goal, and the sensors can be used to detect obstacles in the environment and to update the plan as needed.

From Challenges To Solutions: Robot Navigation In Dynamic Environments

Self-Driving Cars: Self-driving cars use a combination of maps and sensors to navigate in complex and dynamic environments. The maps are used to plan a path to the goal, and the sensors are used to detect obstacles in the environment and to update the plan as needed.

Delivery Robots: Delivery robots use a combination of maps and sensors to navigate in complex and dynamic environments, such as sidewalks and streets. The maps are used to plan a path to the goal, and the sensors are used to detect obstacles in the environment and to update the plan as needed.

Cleaning Robots: Cleaning robots use a combination of maps and sensors to navigate in complex and dynamic environments, such as homes and offices. The maps are used to plan a path to the goal, and the sensors are used to detect obstacles in the environment and to update the plan as needed.

Navigation in complex and dynamic environments is a challenging task, but it is essential for the development of autonomous robots. By developing robots that can navigate in complex and dynamic environments, researchers are enabling them to perform a wider range of tasks and to operate more safely in the real world.

Incorporating Ai For Better Path Planning And Obstacle Avoidance

Artificial intelligence (AI) can be used to improve path planning and obstacle avoidance in robots in a number of ways.

For Example, Ai Can Be Used To:

Generate More Efficient Paths: AI can generate more efficient paths by taking into account the robot's dynamics and the constraints of the environment.

Adapt To Changes In The Environment: AI can adapt to changes in the environment, such as moving obstacles, by updating the path plan in real-time.

Handle Complex Environments: AI can handle complex environments with a large number of obstacles and dynamic objects.

Incorporating AI for Better Path Planning and Obstacle Avoidance

From Algorithms To Solutions: Demonstrating Ai's Role In Robot Navigation

Self-Driving Cars: Self-driving cars use AI to generate paths that avoid obstacles and other vehicles, while also optimizing for efficiency and comfort.

Delivery Robots: Delivery robots use AI to navigate complex environments such as sidewalks and streets, and to avoid obstacles such as pedestrians and parked cars.

Cleaning Robots: Cleaning robots use AI to navigate around furniture and other obstacles in homes and offices.

Assistive Robots: Assistive robots use AI to navigate around people and furniture in order to provide assistance to users.

One of the most common AI techniques used for path planning and obstacle avoidance is reinforcement learning. Reinforcement learning is a type of machine learning that allows agents to learn how to behave in an environment by trial and error. Agents are rewarded for taking actions that lead to desired outcomes and penalized for taking actions that lead to undesired outcomes. Over time, the agent learns to choose actions that maximize its expected reward.

Another AI technique used for path planning and obstacle avoidance is deep learning. Deep learning is a type of machine learning that uses artificial neural networks to learn from data. Neural networks can be trained to learn the relationship between the robot's state (e.g., position, velocity, and orientation) and the actions it should take (e.g., move forward, turn left, or turn right) in order to reach its goal while avoiding obstacles.

AI is a powerful tool that can be used to improve path planning and obstacle avoidance in robots. By enabling robots to generate more efficient paths, adapt to changes in the environment, and handle complex environments, AI is enabling them to perform a wider range of tasks and to operate more safely in the real world.

From Innovation To Impact: How Ai Optimizes Robot Navigation

Improved Efficiency: AI can help robots to generate more efficient paths, which can lead to reduced travel time and energy consumption.

Increased Safety: AI can help robots to avoid obstacles, which can lead to reduced collisions and accidents.

Expanded Capabilities: AI can enable robots to operate in more complex and challenging environments, which can expand their range of potential applications.

Reduced Costs: AI can help to reduce the cost of developing and operating robots by making them more efficient and capable.

Overall, AI is a promising technology for improving path planning and obstacle avoidance in robots. By enabling robots to navigate more efficiently and safely, AI is enabling them to perform a wider range of tasks and to operate more safely in the real world.

Multi-Modal Sensing For Improved Localization

Multi-modal sensing is the use of multiple types of sensors to collect data about the environment. This data can then be used to improve the accuracy and reliability of localization systems.

Some Of The Most Common Types Of Sensors Used For Multi-Modal Sensing Include:

Cameras: Cameras can be used to extract visual features from the environment, such as landmarks and objects.

Lidar: Lidar sensors can be used to create a 3D map of the environment.

Radar: Radar sensors can be used to detect objects in the environment, even in low-light or no-light conditions.

Inertial Measurement Units (Imus): IMUs can be used to track the robot's movement and orientation.

Multi-modal sensing can be used to improve localization in a number of ways. For example, multi- modal sensing can be used to:

Improve Accuracy In Challenging Environments: Multi-modal sensing can help to improve the accuracy of localization systems in challenging environments, such as indoor environments with low light or no light conditions.

Reduce The Impact Of Sensor Noise: Multi-modal sensing can help to reduce the impact of sensor noise on localization accuracy. For example, if a camera sensor is noisy, the lidar sensor can be used to provide additional information about the environment.

Improve Robustness To Sensor Failures: Multi-modal sensing can help to improve the robustness of localization systems to sensor failures. For example, if a camera sensor fails, the lidar sensor can be used to continue providing localization information.

Multi-Modal Sensing In Action: Real-World Localization Improvements In Robots

Self-Driving Cars: Self-driving cars use a variety of sensors, including cameras, lidar, and radar, to localize themselves on the road.

Delivery Robots: Delivery robots use a variety of sensors, including cameras and lidar, to localize themselves in indoor and outdoor environments.

Cleaning Robots: Cleaning robots use a variety of sensors, including cameras and IMUs, to localize themselves in homes and offices.

Assistive Robots: Assistive robots use a variety of sensors, including cameras and IMUs, to localize themselves in indoor environments.

Multi-modal sensing is a powerful tool that can be used to improve localization in robots. By combining data from multiple types of sensors, multi-modal sensing can help to improve the accuracy, reliability, and robustness of localization systems.

Maximizing Robot Localization: The Advantages Of Multi-Modal Sensing

More Accurate Localization: Multi-modal sensing can help to improve the accuracy of localization systems, even in challenging environments.

More Robust Localization: Multi-modal sensing can help to improve the robustness of localization systems to sensor noise and failures.

Expanded Capabilities: Multi-modal sensing can enable robots to operate in more complex and challenging environments, which can expand their range of potential applications.

Reduced Costs: Multi-modal sensing can help to reduce the cost of developing and operating robots by making them more accurate and reliable.

Overall, multi-modal sensing is a promising technology for improving localization in robots. By enabling robots to localize themselves more accurately and reliably, multi-modal sensing is enabling them to perform a wider range of tasks and to operate more safely in the real world.

Urban And Outdoor Autonomous Mobility Challenges

Urban and outdoor autonomous mobility (UAM) presents a number of challenges, including:

Safety: UAM vehicles must be able to operate safely in complex and dynamic environments, alongside human-driven vehicles and pedestrians. This requires the development of reliable perception, planning, and control systems.

Infrastructure: UAM vehicles will require new infrastructure, such as landing pads and charging stations. This infrastructure will need to be deployed in a way that is safe and efficient.

Regulation: UAM vehicles will need to be regulated in a way that ensures their safety and security. This regulation will need to be coordinated across different levels of government and with other stakeholders.

Public Acceptance: UAM vehicles are a new technology, and it is important to ensure that the public accepts them before they can be widely deployed. This will require public education and outreach.

Challenges In Urban And Outdoor Autonomous Mobility: Real-Life Examples

Self-Driving Cars: Self-driving cars need to be able to safely navigate complex and dynamic urban environments, such as intersections and construction zones. They also need to be able to interact with human-driven vehicles and pedestrians in a safe and predictable manner.

Delivery Robots: Delivery robots need to be able to safely navigate sidewalks and streets, avoiding obstacles such as pedestrians and parked cars. They also need to be able to interact with humans in a safe and friendly manner.

Flying Cars: Flying cars need to be able to safely navigate complex and dynamic urban airspace, avoiding obstacles such as other aircraft and buildings. They also need to be able to take off and land safely in urban environments.

Despite the challenges, UAM has the potential to revolutionize transportation in urban and outdoor areas. UAM vehicles could help to reduce traffic congestion, improve air quality, and make transportation more accessible to people with disabilities.

Here Are Some Of The Potential Benefits Of Urban And Outdoor Autonomous Mobility:

Reduced Traffic Congestion: UAM vehicles could help to reduce traffic congestion by providing a new mode of transportation that is not subject to the same constraints as traditional vehicles.

Improved Air Quality: UAM vehicles could help to improve air quality by reducing the number of cars on the road.

Increased Accessibility: UAM vehicles could make transportation more accessible to people with disabilities by providing a transportation option that is not limited by physical constraints.

New Economic Opportunities: UAM could create new economic opportunities by enabling new businesses and services.

Overall, urban and outdoor autonomous mobility is a promising technology with the potential to revolutionize transportation. However, there are a number of challenges that need to be addressed before UAM can be widely deployed.

8.4. Cognitive Abilities And Machine Learning

Cognitive Robotics: Robots With Memory And Learning Capabilities

Cognitive robotics is a field of robotics that aims to develop robots that can learn and adapt to their environment. Cognitive robots are typically equipped with memory and learning capabilities, which allow them to store and process information from their experiences. This information can then be used to make decisions and take actions in the real world.

Cognitive Abilities and Machine Learning

Cognitive Robots Can Be Used For A Variety Of Tasks, Including:

Navigation: Cognitive robots can use their memory and learning capabilities to navigate complex and dynamic environments. For example, a cognitive robot could learn a map of its environment over time and use that map to navigate to a specific location.

Object Manipulation: Cognitive robots can use their memory and learning capabilities to manipulate objects in a skillful way. For example, a cognitive robot could learn how to pick up and place objects of different shapes and sizes.

Human-Robot Interaction: Cognitive robots can use their memory and learning capabilities to interact with humans in a natural and engaging way. For example, a cognitive robot could learn to recognize and respond to human emotions.

Cognitive robotics is a rapidly developing field with the potential to revolutionize many industries, including healthcare, manufacturing, and logistics. For example, cognitive robots could be used to provide personalized care to patients in hospitals, to automate tasks in factories, and deliver packages to customers.

Innovations In Progress: Demonstrating Cognitive Robots Under Development

Assistive Robots: Assistive robots are being developed to help people with disabilities or mobility issues. Cognitive assistive robots could learn the needs and preferences of their users over time and adapt their behavior accordingly.

Surgical Robots: Surgical robots are being developed to assist surgeons with complex procedures. Cognitive surgical robots could learn the anatomy of the patient and the surgical plan over time and make adjustments to the surgery as needed.

Delivery Robots: Delivery robots are being developed to deliver packages to customers. Cognitive delivery robots could learn the routes they need to take and avoid obstacles in their path.

Overall, cognitive robotics is a promising field with the potential to develop robots that are more intelligent, capable, and adaptable than ever before. Cognitive robots have the potential to revolutionize many industries and improve the lives of people in many ways.

Reinforcement Learning In Service Robotics

Reinforcement learning (RL) is a type of machine learning that allows agents to learn how to behave in an environment by trial and error. Agents are rewarded for taking actions that lead to desired outcomes and penalized for taking actions that lead to undesired outcomes. Over time, the agent learns to choose actions that maximize its expected reward.

Rl Is A Powerful Tool That Can Be Used To Train Robots To Perform A Variety Of Tasks, Including:

Navigation: RL can be used to train robots to navigate complex and dynamic environments. For example, an RL-trained robot could learn to navigate a warehouse environment, avoiding obstacles and people.

Object manipulation: RL can be used to train robots to manipulate objects in a skillful way. For example, an RL-trained robot could learn to pick up and place objects of different shapes and sizes.

Human-robot interaction: RL can be used to train robots to interact with humans in a natural and engaging way. For example, an RL-trained robot could learn to recognize and respond to human emotions.

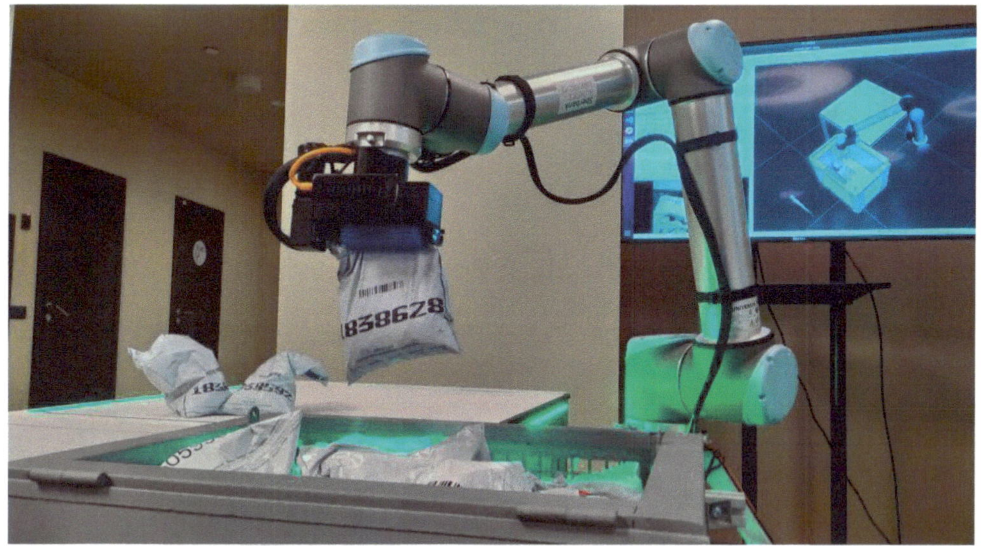

Reinforcement Learning in Service Robotics

Rl-Trained Robot Navigating A Warehouse Environment Opens In A New Window

RL-trained robot navigating a warehouse environment

RL is particularly well-suited for service robotics applications because it allows robots to learn from their experiences in the real world. This is important because service robots often operate in complex and dynamic environments where it is difficult to pre-program all of the possible scenarios that the robot may encounter.

For example, an RL-trained robot could be used to deliver packages to customers in an urban environment. The robot would learn how to navigate the streets, avoid obstacles, and interact with people. It could also learn how to handle unexpected situations, such as traffic jams or pedestrians crossing the street.

RL is still a relatively new technology, but it is rapidly developing and has the potential to revolutionize service robotics. RL-trained robots could be used to perform a wide variety of tasks in a wide variety of environments.

Here Are Some Specific Examples Of Rl In Service Robotics:

Delivery Robots: RL-trained delivery robots could be used to deliver packages to customers in urban and suburban environments.

Cleaning Robots: RL-trained cleaning robots could be used to clean homes, offices, and other public spaces.

Customer Service Robots: RL-trained customer service robots could be used to provide customer assistance in stores, banks, and other businesses.

Assistive Robots: RL-trained assistive robots could be used to help people with disabilities or mobility issues.

Overall, RL is a promising technology for service robotics applications. RL-trained robots have the potential to be more intelligent, adaptable, and capable than traditional robots. RL-trained robots could be used to perform a wide variety of tasks and improve the lives of people in many ways.

Transfer Learning For Rapid Adaptation To New Tasks

Transfer learning is a machine learning technique that allows models to learn from data that was collected for a different task. This can be useful for training robots to perform new tasks quickly and efficiently.

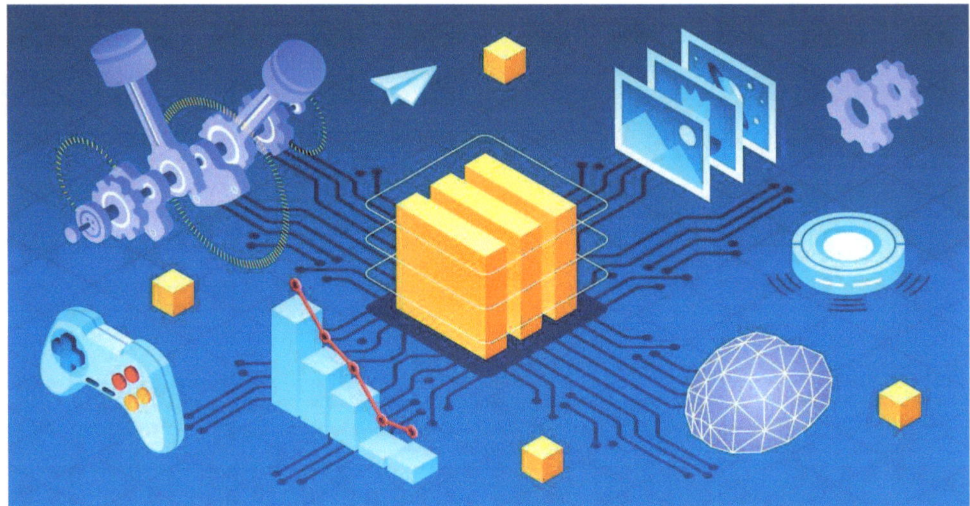

Transfer Learning for Rapid Adaptation to New Tasks

For example, a robot that has been trained to navigate a warehouse using RL could use transfer learning to learn how to navigate a new warehouse environment. This would allow the robot to adapt to the new environment quickly without having to start from scratch.

For example, a transfer learning-enabled robot could be used to perform different tasks in a warehouse, such as picking and packing orders, transporting goods, and loading and unloading trucks. The robot could also be used to perform different tasks in a hospital, such as delivering medications, transporting patients, and assisting with medical procedures.

Transfer learning is still a relatively new technology, but it is rapidly developing and has the potential to revolutionize service robotics. Transfer learning-enabled robots could be used to perform a wide variety of tasks in a wide variety of environments. Transfer learning is particularly well-suited for service robotics applications because it can help robots adapt to new tasks quickly and efficiently. This is important because service robots often need to be able to perform a variety of tasks in different environments.

Some Specific Examples Of Transfer Learning In Service Robotics:

Delivery Robots: Transfer learning-enabled delivery robots could be used to deliver packages to customers in a variety of environments, such as urban, suburban, and rural areas.

Cleaning Robots: Transfer learning-enabled cleaning robots could be used to clean homes, offices, hospitals, and other public spaces with different types of flooring and obstacles.

Customer Service Robots: Transfer learning-enabled customer service robots could be used to provide customer assistance in stores, banks, and other businesses with different types of products and services.

Assistive Robots: Transfer learning-enabled assistive robots could be used to help people with disabilities or mobility issues with different types of tasks.

Overall, transfer learning is a promising technology for service robotics applications. Transfer learning- enabled robots have the potential to be more adaptable and capable than traditional robots. Transfer learning-enabled robots could be used to perform a wide variety of tasks and improve the lives of people in many ways.

Rapid Task Mastery: Real-Life Gains From Transfer Learning In Service Robotics

Faster Training: Transfer learning can help robots learn new tasks more quickly than traditional training methods. This is because robots can use their existing knowledge from previous tasks to learn new tasks.

Improved Performance: Transfer learning can also help robots improve their performance on new tasks. This is because robots can use their existing knowledge from previous tasks to avoid making common mistakes.

Reduced Cost: Transfer learning can help reduce the cost of training robots for new tasks. This is because robots do not need to be trained from scratch for each new task.

Overall, transfer learning is a promising technology for service robotics applications. Transfer learning- enabled robots have the potential to be more adaptable, capable, and efficient than traditional robots. Transfer learning-enabled robots could be used to perform a wide variety of tasks and improve the lives of people in many ways.

Explainable Ai For Trust And Transparency

Explainable AI (XAI) is a field of artificial intelligence that aims to develop AI systems that are understandable to humans. XAI systems can explain their decisions and recommendations in a way that humans can understand.

XAI is important for service robotics applications because it can help to build trust and transparency between humans and robots. For example, an XAI-powered robot could explain to a human why it is taking a particular action or why it is making a particular recommendation. This would help the human to understand and trust the robot.

There Are A Number Of Different Xai Techniques That Can Be Used To Explain The Decisions Of Ai Systems. Some Of The Most Common Xai Techniques Include:

Rule-Based Explanations: Rule-based explanations explain the decisions of AI systems by providing a set of rules that were used to make the decision.

Counterfactual Explanations: Counterfactual explanations explain the decisions of AI systems by showing how the decision would change if one or more inputs were changed.

Example-Based Explanations: Example-based explanations explain the decisions of AI systems by providing examples of similar situations where the AI system made the same decision.

XAI is still a relatively new field of research, but it is rapidly developing. There are a number of different XAI techniques that can be used to explain the decisions of AI systems, and new techniques are being developed all the time.

Here Are Some Specific Examples Of Xai In Service Robotics:

Delivery Robots: XAI-powered delivery robots could explain to humans why they are taking a particular route or why they are avoiding a particular obstacle.

Explainable AI for Trust and Transparency

Cleaning Robots: XAI-powered cleaning robots could explain to humans why they are focusing on a particular area of the room or why they are using a particular cleaning method.

Customer Service Robots: XAI-powered customer service robots could explain to humans why they are recommending a particular product or service.

Assistive Robots: XAI-powered assistive robots could explain to humans why they are taking a particular action or why they are making a particular recommendation.

Overall, XAI is a promising technology for service robotics applications. XAI-powered robots have the potential to be more trustworthy and transparent than traditional robots. XAI-powered robots could be used to perform a wide variety of tasks and improve the lives of people in many ways.

Trust And Transparency In Service Robots: Unveiling The Advantages Of XAI

Increased Trust: XAI can help to increase trust between humans and robots by explaining the decisions and actions of the robots.

Improved Transparency: XAI can help to improve transparency by providing humans with more information about how robots work and why they make the decisions they do.

Reduced Bias: XAI can help to reduce bias in robots by identifying and eliminating biases in the data that robots are trained on.

Increased safety: XAI can help to increase safety by explaining the decisions and actions of robots so that humans can understand and respond to potential problems.

Overall, XAI is a promising technology for service robotics applications. XAI-powered robots have the potential to be more trustworthy, transparent, safe, and unbiased than traditional robots. XAI-powered robots could be used to perform a wide variety of tasks and improve the lives of people in many ways.

8.5. Service Robotics In Home And Daily Life

Personal Assistants And Household Chores

Personal assistants can help with a variety of household chores, including:

Cleaning And Organizing: Personal assistants can help with cleaning tasks such as vacuuming, dusting, mopping, and cleaning bathrooms and kitchens. They can also help with organizing tasks such as putting away laundry, decluttering closets, and arranging furniture.

Meal Planning And Preparation: Personal assistants can help with meal planning and preparation by creating shopping lists, grocery shopping, cooking meals, and cleaning up afterwards.

Pet Care: Personal assistants can help with pet care tasks such as feeding, walking, and grooming pets.

Child Care: Personal assistants can help with childcare tasks such as playing with children, helping them with homework, and transporting them to and from activities.

Errands: Personal assistants can help with errands such as running to the store, picking up prescriptions, and dropping off and picking up packages.

In addition to these specific tasks, personal assistants can also provide general support and assistance around the house. For example, they can answer the phone, respond to emails, and manage schedules.

Personal assistants can be a valuable asset for busy people who need help with household chores. By outsourcing these tasks, personal assistants can free up people's time so that they can focus on other things, such as work, family, and hobbies.

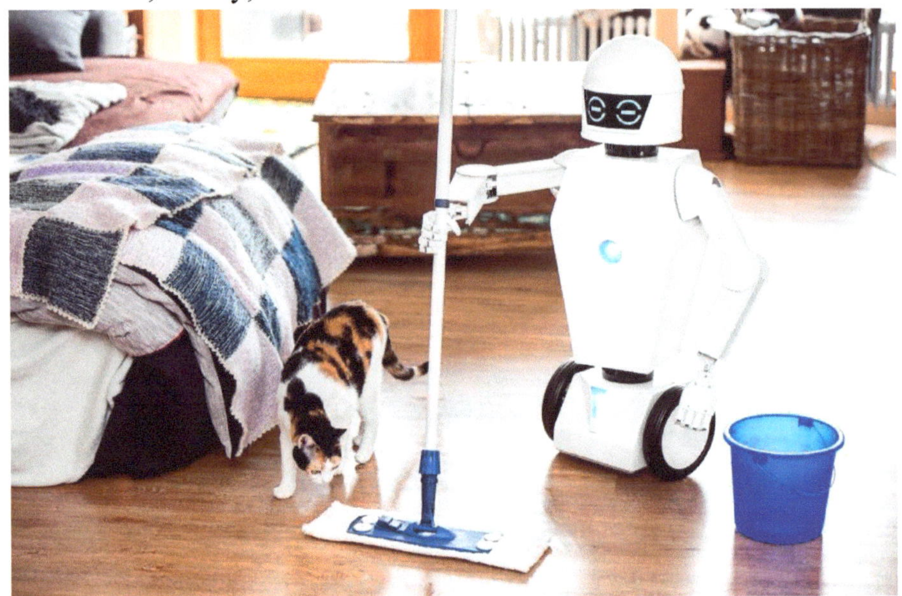

Service Robotics in Home and Daily Life

Household Chores Made Easy: Real-World Examples of Personal Assistant Support

- A personal assistant could help a busy parent with young children by preparing meals, cleaning the house, and helping with childcare tasks.
- A personal assistant could help an elderly person with mobility issues by running errands, cooking meals, and helping with household chores.
- A personal assistant could help a person with a disability by managing their schedule, assisting with personal tasks, and providing transportation.

Overall, personal assistants can be a valuable asset for people of all ages and abilities. By providing help and support with household chores, personal assistants can make people's lives easier and more enjoyable.

Home Security And Surveillance Applications

Personal Assistants Can Be Used For A Variety Of Home Security And Surveillance Applications, Including

Monitoring For Intruders: Personal assistants can be equipped with cameras and sensors to monitor the home for intruders. If an intruder is detected, the personal assistant can alert the homeowner or the authorities.

Locking And Unlocking Doors And Windows: Personal assistants can be used to lock and unlock doors and windows remotely. This can be useful for homeowners who are away from home or who want to allow guests into their homes without having to be present.

Controlling Lights And Appliances: Personal assistants can be used to control lights and appliances remotely. This can be useful for deterring burglars, simulating occupancy when the homeowner is away, or simply making it more convenient to control the home environment.

Monitoring For Smoke, Fire, And Carbon Monoxide: Personal assistants can be equipped with smoke, fire, and carbon monoxide detectors to alert the homeowner to potential hazards.

Providing Emergency Assistance: Personal assistants can be used to call for emergency assistance in the event of a break-in, fire, or medical emergency.

Home Security Reinvented: Real-World Examples Of Personal Assistants In Action

- A personal assistant could be used to monitor a home while the homeowner is on vacation. If the personal assistant detects an intruder, it could send an alert to the homeowner's smartphone and call the police.
- A personal assistant could be used to lock the doors and windows of a home at night or when the homeowner leaves the house.
- A personal assistant could be used to turn on the lights and appliances in a home when the homeowner arrives home, creating the illusion that someone is home.
- A personal assistant could be used to monitor the air quality in a home and alert the homeowner to dangerous levels of smoke, fire, or carbon monoxide.
- A personal assistant could be used to call for emergency assistance in the event of a break-in, fire, or medical emergency.

Home Security and Surveillance Applications

Overall, personal assistants can be a valuable tool for home security and surveillance. By providing a variety of features and capabilities, personal assistants can help to protect homes and families from harm.

Home Surveillance Reimagined: Case Studies Of Benefits From Personal Assistants

 Personal assistants can be controlled remotely using a smartphone or other device, making them convenient to use even when the homeowner is away from home.

Peace Of Mind: Personal assistants can provide peace of mind by monitoring the home for intruders and other hazards.

Protection: Personal assistants can help to protect homes and families from harm by deterring burglars, alerting the authorities to intruders, and providing emergency assistance.

Overall, personal assistants are a promising technology for home security and surveillance applications. Personal assistants have the potential to make homes more secure and provide homeowners with peace of mind.

Delivery And Logistics Robots For Everyday Needs

Delivery and logistics robots have the potential to revolutionize the way we receive our everyday needs. These robots can be used to deliver a wide variety of items, including food, groceries, packages, and even medicine.

Delivery and logistics robots offer a number of advantages over traditional delivery methods. First, they are more efficient. Robots can travel faster and more directly than human delivery drivers, and they can operate 24/7. This means that customers can receive their deliveries more quickly, even during off-peak hours.

Delivery and Logistics Robots for Everyday Needs

Second, delivery and logistics robots are more cost-effective. Robots do not require salaries or benefits, and they do not need to pay for fuel. This means that businesses can save money on delivery costs, which can translate into lower prices for consumers.

Third, delivery and logistics robots are more environmentally friendly. Robots do not produce emissions, so they can help to reduce air pollution and greenhouse gas emissions.

Delivery and logistics robots are still under development, but they have the potential to revolutionize the way we live and work. Here are some specific examples of how delivery and logistics robots could be used to meet our everyday needs:

Food Delivery robots could be used to deliver food from restaurants to customers' homes or offices. This would make it more convenient for people to get their meals without having to leave their homes or offices.

Grocery delivery robots could be used to deliver groceries from grocery stores to customers' homes. This would make it easier for people to get their groceries, especially for those who have mobility issues or who do not have access to a car.

Package delivery robots could be used to deliver packages from retailers to customers' homes. This would make it more convenient for people to receive their packages, especially for those who live in urban areas where parking can be difficult.

Medicine delivery robots could be used to deliver medicine from pharmacies to customers' homes. This would make it easier for people to get their medications, especially for those who have mobility issues or who do not have access to a car.

In addition to these specific examples, delivery and logistics robots could be used to deliver a wide variety of other items, such as books, clothes, and electronics. As delivery and logistics robots become more sophisticated and widespread, they have the potential to make our lives much easier and more convenient.

Robotic Delivery Solutions: Practical Scenarios For Everyday Needs

Convenience: Delivery robots can make it more convenient for people to get the items they need without having to leave their homes or offices.

Affordability: Delivery robots can help to reduce the cost of delivery, which can translate into lower prices for consumers.

Sustainability: Delivery robots are more environmentally friendly than traditional delivery methods, as they do not produce emissions.

Accessibility: Delivery robots can help to improve access to goods and services for people with mobility issues or who do not have access to a car.

Overall, delivery and logistics robots have the potential to revolutionize the way we live and work. These robots can make our lives more convenient, affordable, sustainable, and accessible.

Aging Population And The Role Of Robots In Elderly Care

As the global population ages, there is a growing need for elderly care. Robots can play an important role in elderly care by providing assistance with tasks such as:

- **Activities Of Daily Living (Adls):** Robots can help with ADLs such as bathing, dressing, and eating. This can help elderly people maintain their independence and quality of life.
- **Instrumental Activities Of Daily Living (Iadls):** Robots can help with IADLs such as cooking, cleaning, and shopping. This can help elderly people stay in their homes longer and avoid institutionalization.
- **Medication Management:** Robots can help elderly people manage their medications by reminding them to take their medications and dispensing the correct dosage. This can help to improve medication adherence and reduce the risk of adverse drug events.
- **Companionship:** Robots can provide companionship to elderly people who live alone or who have limited social interaction. This can help to reduce loneliness and social isolation.

Robots are still under development, but they have the potential to revolutionize elderly care.

Robots In Action: Practical Scenarios For Elderly Care Advancements

Assistive Robots: Assistive robots are designed to help elderly people with ADLs and IADLs. For example, the Paro robot is a therapeutic robot that looks like a baby seal. Paro has been shown to reduce stress and anxiety in elderly people.

Medication Management Robots: Medication management robots are designed to help elderly people manage their medications. For example, the MedMinder robot reminds elderly people to take their medications and dispenses the correct dosage.

Companion Robots: Companion robots are designed to provide companionship to elderly people. For example, the ElliQ robot is a social robot that can engage in conversation and provide companionship to elderly people.

As robots become more sophisticated and affordable, they are likely to play an increasingly important role in elderly care. Robots can help to improve the quality of life for elderly people, reduce the burden on caregivers, and save healthcare costs.

Aging Population and the Role of Robots in Elderly Care

Aging Gracefully With Robotics: Real-Life Gains In Senior Care

Improved Quality Of Life: Robots can help elderly people maintain their independence and quality of life by assisting them with ADLs and IADLs.

Reduced Burden On Caregivers: Robots can help to reduce the burden on caregivers by providing assistance with ADLs and IADLs. This can free up caregivers to spend more time on other tasks, such as providing emotional support.

Reduced Healthcare Costs: Robots can help to reduce healthcare costs by preventing institutionalization and reducing the need for hospitalization.

Overall, robots have the potential to revolutionize elderly care. Robots can help to improve the quality of life for elderly people, reduce the burden on caregivers, and save healthcare costs.

8.6. Sustainability And Eco-Friendly Robotics

Green Robotics: Designing Environmentally Conscious Robots

Green robotics is the design and development of robots that are environmentally friendly and sustainable. Green robots are designed to minimize their environmental impact throughout their life cycle, from the extraction of the raw materials used to build them to their disposal at the end of their useful life.

There are a number of ways to design green robots. One way is to use recycled materials in the construction of the robots. Another way is to design robots that are energy-efficient. Green robots can also be designed to be biodegradable or recyclable at the end of their useful life.

Sustainability and Eco-Friendly Robotics

Robots For A Greener World: Practical Scenarios Of Green Robotics

Solar-Powered Robots: Solar-powered robots use solar energy to power their operations. This reduces the need for fossil fuels and helps to reduce greenhouse gas emissions.

Robots Made From Recycled Materials: Robots made from recycled materials reduce the need to extract new raw materials, which can have a negative impact on the environment.

Biodegradable Robots: Biodegradable robots are made from materials that can be broken down by microorganisms. This helps to reduce the amount of waste that goes to landfills.

Robots that can be repaired and reused: Robots that can be repaired and reused extend their useful life and reduce the need to produce new robots.

Green robots are still under development, but they have the potential to revolutionize the way we design and build robots. Green robots can help to reduce the environmental impact of robotics and make robotics more sustainable.

A Greener Tomorrow: Real-Life Gains From Embracing Green Robotics

Reduced Environmental Impact: Green robots can help to reduce the environmental impact of robotics by minimizing their use of energy and resources, and by reducing the amount of waste they produce.

Increased Sustainability: Green robots can help to make robotics more sustainable by using recycled materials, designing robots that can be repaired and reused, and making robots biodegradable.

Reduced Costs: Green robots can help to reduce the costs of robotics by reducing the need for fossil fuels, raw materials, and waste disposal.

Improved Public Image: Green robots can help to improve the public image of robotics by showing that robots can be designed and operated in an environmentally friendly and sustainable manner.

Overall, green robotics is a promising field with the potential to revolutionize the way we design and build robots. Green robots can help to reduce the environmental impact of robotics, and make robotics more

Energy-Efficient Robotics For Extended Operation

Energy-efficient robotics is the design and development of robots that consume less energy. Energy- efficient robots are important for a number of reasons. First, they can help to reduce costs. Second, they can help to reduce greenhouse gas emissions. Third, they can extend the operation time of robots, which can be important for applications such as disaster response and space exploration.

There are a number of ways to design energy-efficient robots. One way is to use lightweight materials and efficient motors. Another way is to design robots that can reuse energy. Energy-efficient robots can also be designed to operate in low-power modes.

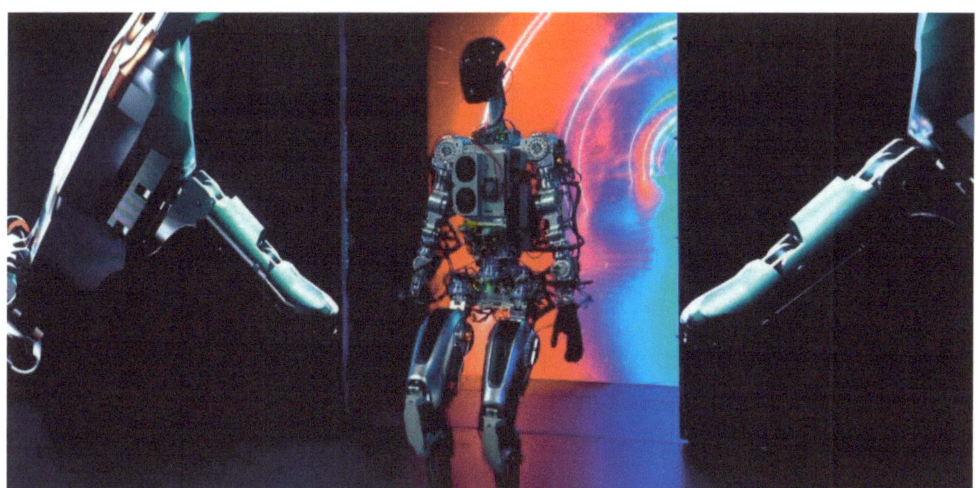

Energy-Efficient Robotics for Extended Operation

Efficiency In Motion: Demonstrating Energy-Efficient Robot Innovations

Solar-Powered Robots: Solar-powered robots use solar energy to power their operations. This reduces the need for batteries and helps to extend the operation time of the robots.

Robots With Regenerative Braking: Robots with regenerative braking can capture energy when they slow down or stop. This energy can then be reused to power the robot.

      Robots With Multiple Power Modes: Robots with multiple power modes can switch to low-power modes when they are not performing tasks. This helps to conserve energy.

Energy-efficient robots are still under development, but they have the potential to revolutionize the way we design and build robots. Energy-efficient robots can help to reduce costs, and greenhouse gas emissions, and extend the operation time of robots.

Here Are Some Of The Benefits Of Energy-Efficient Robotics:

      Reduced Costs: Energy-efficient robots consume less energy, which can lead to significant cost savings over the lifetime of the robot.

      Reduced Greenhouse Gas Emissions: Energy-efficient robots help to reduce greenhouse gas emissions by consuming less energy.

      Extended Operation Time: Energy-efficient robots can operate for longer periods of time on a single battery charge, which is important for applications where access to power is limited.

      Improved Safety: Energy-efficient robots can help to improve safety by reducing the risk of fires and explosions caused by overheating batteries.

      Increased Reliability: Energy-efficient robots are less likely to fail due to overheating or power loss, which can improve their reliability.

Overall, energy-efficient robotics is a promising field with the potential to revolutionize the way we design and build robots. Energy-efficient robots can help to reduce costs, greenhouse gas emissions, extend operation time, improve safety, and increase reliability.

Robotics In Agriculture And Sustainable Food Production

Robotics is playing an increasingly important role in agriculture and sustainable food production. Robots can be used to automate a wide range of tasks, from planting and harvesting crops to weeding and pest control. This can help to improve efficiency, reduce costs, and reduce the use of pesticides and herbicides.

Robots On The Farm: Practical Scenarios For Sustainable Food Production

      Planting And Harvesting Robots: Planting and harvesting robots can automate the tasks of planting and harvesting crops. This can help to improve efficiency and reduce costs.

      Weeding Robots: Weeding robots can automate the task of weeding. This can help to reduce the use of herbicides and improve the sustainability of agriculture.

      Pest Control Robots: Pest control robots can automate the task of pest control. This can help to reduce the use of pesticides and improve the safety of food.

Precision Agriculture Robots: Precision agriculture robots can be used to monitor crops and apply inputs, such as water and fertilizer, with greater precision. This can help to improve yields and reduce waste.

In addition to these specific examples, robots can be used to automate a wide range of other tasks in agriculture, such as milking cows, feeding chickens, and cleaning barns.

The use of robots in agriculture is still in its early stages, but it has the potential to revolutionize the industry. Robots can help to make agriculture more efficient, sustainable, and productive.

Robotics in Agriculture and Sustainable Food Production

Robotic Solutions For Sustainable Agriculture: Benefits Of Modern Farming Technology

Improved Efficiency: Robots can automate many tasks that are currently performed manually by humans. This can lead to significant improvements in efficiency.

Reduced Costs: Robots can help to reduce costs by automating tasks and reducing the need for human labor.

Reduced Use Of Pesticides And Herbicides: Robots can be used to weed and control pests without the use of pesticides and herbicides. This can improve the sustainability of agriculture and produce safer food.

Improved Yields: Precision agriculture robots can help to improve yields by applying inputs, such as water and fertilizer, with greater precision.

Reduced Waste: Precision agriculture robots can help to reduce waste by applying inputs only where and when they are needed.

Overall, robotics is a promising technology for agriculture and sustainable food production. Robots have the potential to make agriculture more efficient, sustainable, productive, and profitable.

Cleaning And Maintenance Robots For Greener Facilities

Cleaning and maintenance robots can play an important role in making facilities greener. These robots can automate a wide range of tasks, from cleaning floors and windows to maintaining equipment. This can help to reduce the use of chemicals, energy, and water, and can also help to improve the overall sustainability of facilities.

Facility Sustainability With Robots: Practical Scenarios For Eco-Friendly Solutions

Floor Cleaning Robots: Floor cleaning robots can be used to clean floors without the use of chemicals. This can help to improve indoor air quality and reduce the environmental impact of cleaning.

Window Cleaning Robots: Window cleaning robots can be used to clean windows without the use of harsh chemicals. This can help to protect the environment and improve the safety of workers.

Equipment Maintenance Robots: Equipment maintenance robots can be used to maintain equipment without the need for human intervention. This can help to reduce energy consumption and extend the life of equipment.

Leak Detection Robots: Leak detection robots can be used to detect leaks in pipes and other infrastructure. This can help to prevent water waste and environmental damage.

In addition to these specific examples, cleaning and maintenance robots can be used to automate a wide range of other tasks in facilities, such as emptying trash cans, delivering supplies, and transporting people.

The use of cleaning and maintenance robots in facilities is still in its early stages, but it has the potential to revolutionize the way that facilities are managed. Robots can help to make facilities greener, more efficient, and more productive.

A Greener Approach To Facility Maintenance: Real-Life Gains From Robotic Solutions

Reduced Use Of Chemicals: Cleaning and maintenance robots can automate many tasks that currently require the use of chemicals. This can help to improve indoor air quality and reduce the environmental impact of cleaning.

Reduced Energy Consumption: Cleaning and maintenance robots can help to reduce energy consumption by automating tasks and operating more efficiently than humans.

Reduced Water Consumption: Cleaning and maintenance robots can help to reduce water consumption by using less water to clean and maintain facilities.

Improved Indoor Air Quality: Cleaning and maintenance robots can help to improve indoor air quality by reducing the use of chemicals and by removing dust and other allergens from the air.

Improved Worker Safety: Cleaning and maintenance robots can help to improve worker safety by automating tasks that are dangerous or hazardous.

Overall, cleaning and maintenance robots are a promising technology for making facilities greener. Robots have the potential to reduce the use of chemicals, energy, and water, improve indoor air quality, and improve worker safety.

Cleaning and Maintenance Robots for Greener Facilities

Robotic Solutions For Greener Facilities: Real-World Scenarios And Innovations

- Use robots to clean and maintain renewable energy systems, such as solar panels and wind turbines. This can help to reduce the environmental impact of energy production.
- Use robots to clean and maintain green spaces, such as parks and gardens. This can help to improve the quality of life for residents and reduce the environmental impact of human activity.
- Use robots to clean and maintain public transportation systems. This can help to make transportation more sustainable and reduce air pollution.

Overall, cleaning and maintenance robots have the potential to play a major role in making facilities and communities greener.

Conclusion
The Ongoing Evolution Of Service Robotics

The ongoing evolution of service robotics is a rapidly developing field with the potential to revolutionize many aspects of our lives. Service robots are designed to assist humans with tasks in a variety of settings, including homes, businesses, hospitals, and public spaces.

Service robots are becoming increasingly sophisticated and capable, thanks to advances in artificial intelligence, machine learning, and robotics hardware. These robots are able to perform a wide range of tasks, such as:

- ❖ Cleaning and maintenance

- ❖ Food preparation and delivery

- ❖ Customer service and assistance

- ❖ Healthcare and eldercare

- ❖ Logistics and transportation

- ❖ Education and entertainment

Service robots are already being used in a variety of ways around the world. For example, cleaning robots are used to clean floors, windows, and other surfaces in homes, businesses, and public spaces. Food delivery robots are used to deliver food to customers' homes and offices. Customer service robots are used to provide information and assistance to customers in stores and other businesses. Healthcare robots are used to assist with patient care, rehabilitation, and surgery. Logistics robots are used to pick and pack orders in warehouses and deliver packages to customers. Educational robots are used to teach children about a variety of subjects.

The ongoing evolution of service robotics is likely to have a significant impact on many aspects of our lives. As service robots become more sophisticated and capable, they will be able to take on an even wider range of tasks, freeing up humans to focus on other activities. Service robots are also likely to play a major role in addressing some of the challenges facing society, such as the aging population and the need for more sustainable transportation.

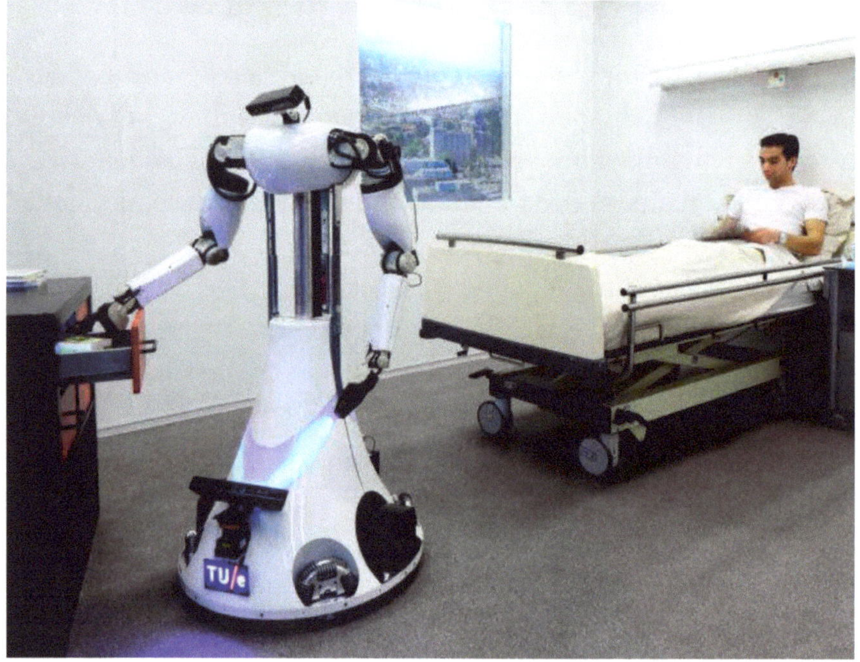

source Link: https://www.medicaldevice-network.com/comment/what-are-the-main-types-of-robots-used-in-healthcare/

The Future Unveiled: Examples Of How Service Robotics Is Reshaping Our Lives

In Healthcare: Service robots can help to address the shortage of healthcare workers and provide high-quality care to patients. For example, service robots can be used to assist with patient care, rehabilitation, and surgery. Service robots can also be used to provide companionship and support to patients.

In Transportation: Service robots can help to make transportation more efficient and sustainable. For example, service robots can be used to deliver packages and transport passengers. Service robots can also be used to develop self-driving cars and trucks.

In The Workplace: Service robots can help to improve productivity and efficiency in the workplace. For example, service robots can be used to automate tasks such as cleaning, maintenance, and customer service. Service robots can also be used to assist with tasks such as assembly and manufacturing.

In Homes: Service robots can help to make our homes more convenient and comfortable. For example, service robots can be used to clean, cook, and provide companionship. Service robots can also be used to assist with tasks such as childcare and eldercare.

Overall, the ongoing evolution of service robotics is a promising development with the potential to improve our lives in many ways. As service robots become more sophisticated and capable, they will be able to play an even greater role in our society.

Preparing For The Future: Challenges And Opportunities

The ongoing evolution of service robotics presents both challenges and opportunities.

Challenges:

Ethics: Service robots raise a number of ethical questions, such as how to ensure that they are used in a responsible and ethical manner, and how to protect the privacy and safety of humans.

Job Displacement: Some people worry that service robots could displace human workers. However, it is important to note that service robots are likely to create new jobs as well, as they will need to be developed, maintained, and operated.

Safety: Service robots need to be designed to be safe and reliable. This is especially important for robots that will be used in close contact with humans, such as healthcare and eldercare robots.

Opportunities:

Improved productivity and efficiency: Service robots can help to improve productivity and efficiency in a variety of industries. This can lead to lower costs and higher profits for businesses.

Enhanced quality of life: Service robots can help to improve the quality of life for people in many ways. For example, service robots can help people with disabilities to live more independently, and they can provide companionship and support to people who are lonely or isolated.

New products and services: Service robots can enable the development of new products and services. For example, service robots can be used to deliver food and packages on demand, and they can be used to provide personalized education and training.

Overall, the ongoing evolution of service robotics is a promising development with the potential to improve our lives in many ways. However, it is important to be aware of the challenges that service robots pose, such as ethical concerns, job displacement, and safety issues. By carefully considering these challenges and opportunities, we can ensure that service robots are used in a responsible and beneficial manner.

Navigating The Future Of Service Robotics: Additional Considerations And Insights

Invest In Education And Training: We need to invest in education and training programs to prepare people for the jobs that will be created by the service robotics industry. We also need to help people who may be displaced by service robots to transition to new jobs.

Develop Ethical Guidelines: We need to develop ethical guidelines for the development and use of service robots. These guidelines should address issues such as privacy, safety, and accountability.

Promote Responsible Innovation: We need to promote responsible innovation in the service robotics industry. This means ensuring that service robots are designed and developed in a way that minimizes the risks and maximizes the benefits.

By taking these steps, we can help to ensure that the future of service robotics is bright.

Chapter 9: Advanced Manufacturing with Industrial Robots

9.1. Introduction

In today's rapidly evolving industrial landscape, the integration of technology has paved the way for a remarkable transformation in the manufacturing sector. At the forefront of this revolution are advanced manufacturing processes, driven by the remarkable capabilities of industrial robots. These specialized machines, often referred to as service robots, have transcended the realm of traditional human labor, taking on tasks that were once exclusively within the human domain.

Industrial robots have become indispensable in redefining manufacturing and industrial applications. These intelligent and versatile machines have brought forth a wave of innovation, optimizing production lines, enhancing precision, and elevating efficiency. As we delve into the realm of "Advanced Manufacturing with Industrial Robots," we will explore the incredible potential and real-world applications of these technological marvels. From automating complex assembly tasks to ensuring consistent quality control, industrial robots are reshaping the manufacturing landscape, proving that the future of industry lies in their capable, mechanical hands.

Service Robots' Multifaceted Roles In Manufacturing And Industry

Material Handling: Service robots can be used to move materials from one location to another, such as from a warehouse to a production line, or from a production line to a packaging area.

Assembly: Service robots can be used to assemble products, such as electronic devices or automotive parts.

Inspection And Maintenance: Service robots can be used to inspect products for defects and to perform maintenance tasks on equipment.

Security And Surveillance: Service robots can be used to patrol industrial facilities and to monitor for security threats.

Increased Productivity And Efficiency: Service robots can work 24/7 and can perform tasks much faster and more efficiently than humans.

Improved Quality And Accuracy: Service robots can perform tasks with a high degree of precision and accuracy. This can help to improve the quality of products and to reduce the number of defects.

Reduced Costs: Service robots can help to reduce labor costs and other operational costs.

Improved Safety: Service robots can be used to perform dangerous or repetitive tasks, which can help to improve safety for workers.

Versatility And Flexibility: Service robots can be programmed to perform a wide variety of tasks, which makes them very versatile and flexible.

Image Source: https://www.genesis-systems.com/blog/robots-automotive-manufacturing-top-6-applications

Hurdles In The Path Of Industrial Service Robot Implementation

High Upfront Costs: Service robots can be expensive to purchase and implement.

Integration With Existing Systems And Processes: Service robots need to be integrated with existing manufacturing and industrial systems and processes. This can be a complex and challenging task.

Training And Support Requirements: Employees need to be trained on how to use and maintain service robots. This can require significant time and resources.

Safety Concerns: Service robots need to be designed and implemented in a way that minimizes safety risks for workers.

Overall, service robots offer a number of potential benefits for manufacturing and industrial applications. However, it is important to carefully consider the challenges associated with implementing service robots before making a decision.

Future Of Service Robots In Manufacturing And Industrial Applications

The future of service robots in manufacturing and industrial applications is very bright. As service robot technology continues to advance, service robots will become more capable, affordable, and easier to use. This will lead to the adoption of service robots in a wider range of manufacturing and industrial applications.

Emerging Trends Fueling The Growth Of Service Robots In Manufacturing

The Increasing Adoption Of Industry 4.0: Industry 4.0 is the fourth industrial revolution, which is characterized by the use of advanced technologies such as cyber-physical systems, the Internet of Things, and big data analytics. Service robots are a key component of Industry 4.0, and they are expected to play a major role in the future of manufacturing and industrial production.

The Growing Demand For Labor-Saving Solutions: The manufacturing and industrial sectors are facing a growing shortage of skilled workers. Service robots can help to address this shortage by automating tasks that are currently performed by humans.

The Declining Cost Of Service Robots: The cost of service robots is declining due to advances in technology and economies of scale. This is making service robots more affordable for a wider range of businesses.

As a result of these trends, the market for service robots in manufacturing and industrial applications is expected to grow significantly in the coming years.

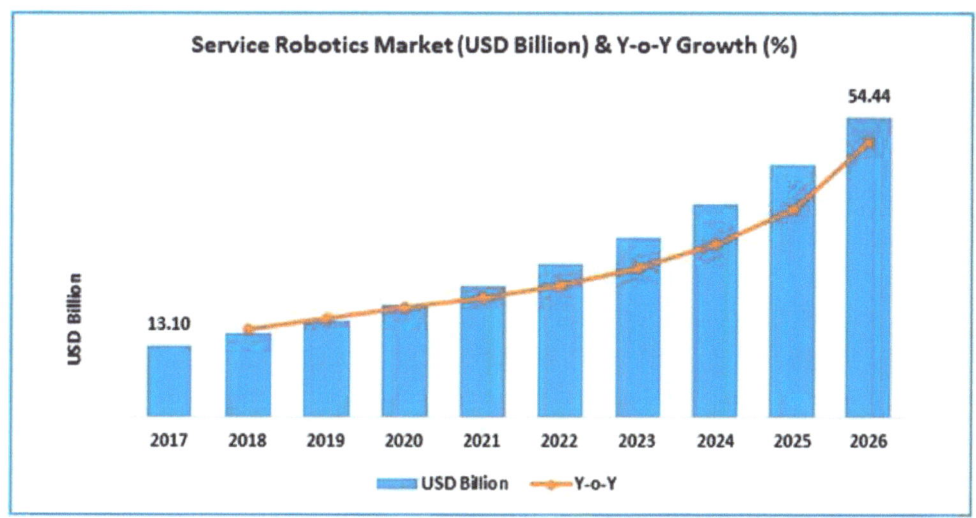

Future of Service Robots in Manufacturing and Industrial Applications

Significance Of Service Robotics In Manufacturing :

Service robots are playing an increasingly significant role in manufacturing. They can be used to perform a wide variety of tasks, from material handling to assembly to inspection and maintenance. Service robots can help to improve productivity, efficiency, quality, and safety in manufacturing operations.

In addition to their well-established roles, service robots are continuously expanding their footprint in modern manufacturing, introducing new dimensions of versatility and productivity. Some of the emerging roles for service robots in manufacturing include

Significance of Service Robotics in Manufacturing

Inventory Management: Service robots are now actively involved in inventory management, where they autonomously track and manage stock levels, ensuring that materials and components are always available when needed.

Collaborative Manufacturing: They work in tandem with human workers on intricate assembly tasks, capitalizing on their precision and consistent performance to achieve higher levels of product quality.

Data Analytics: Service robots are being employed to gather and analyze production data in real- time, aiding in optimizing manufacturing processes and identifying areas for improvement.

Customization And Personalization: In an era of increasing product customization, service robots are stepping in to handle personalized manufacturing, adjusting product specifications on the fly to meet individual customer requirements.

Environmental Impact: Beyond their impact on productivity, service robots are contributing to greener manufacturing by implementing energy-efficient processes and reducing waste.

Predictive Maintenance: Service robots are increasingly used for predictive maintenance, identifying and addressing equipment issues before they cause costly breakdowns. This proactive approach minimizes downtime and optimizes machine lifespan.

Collaborative Material Handling: Service robots work alongside human operators in materials handling tasks, ensuring seamless movement of materials between workstations and storage areas.

Supply Chain Optimization: In manufacturing environments with complex supply chains, service robots play a pivotal role in streamlining the movement and distribution of goods and materials, contributing to just-in-time production.

Waste Management: Service robots are employed to manage waste and recycling processes within manufacturing facilities, promoting sustainability and compliance with environmental regulations.

Robotic Inspections: They are used for in-depth inspections and quality control, using sensors and AI to identify defects, variations, or irregularities in the production process, ensuring consistently high-quality products.

These new roles are elevating manufacturing capabilities to unprecedented heights, with service robots serving as dynamic partners in achieving precision, productivity, and environmental responsibility. Their adaptability and ability to learn and evolve with evolving production requirements make them indispensable assets in modern manufacturing landscapes. As service robot technology continues to advance, service robots are expected to play an even greater role in manufacturing in the future.

Current Manufacturing Applications Of Service Robots

Amazon: Amazon uses service robots to pick and pack items in its warehouses.
Tesla: Tesla uses service robots to assemble electric cars in its factories.
Bmw: BMW uses service robots to inspect car bodies for defects.
Foxconn: Foxconn uses service robots to assemble iPhones and other electronic devices.
Siemens: Siemens uses service robots to maintain and repair industrial equipment.

These are just a few examples of the many ways in which service robots are being used in manufacturing today. As service robot technology continues to advance, service robots are expected to play an even greater role in manufacturing in the future.

9.2. Types Of Service Robots In Manufacturing

Service robots can be classified into a variety of types based on their design, capabilities, and applications. Some of the most common types of service robots used in manufacturing include:

Mobile Manipulation Robots (Mmrs): MMRs are robots equipped with arms and grippers that allow them to manipulate objects. They are highly versatile and can be used for a variety of tasks in manufacturing, such as assembly, packaging, inspection, and machine tending.

Autonomous Guided Vehicles (Agvs): AGVs are robots that can navigate autonomously through a manufacturing facility using sensors and guidance systems. They are typically used to transport materials, products, and people around a facility, freeing up human workers to focus on more complex tasks.

Collaborative Robots (Cobots): Cobots are robots that are designed to work safely alongside human workers. They are typically smaller and less powerful than MMRs, but they are also more affordable and easier to program. Cobots can be used for a variety of tasks in manufacturing, such as assembly, inspection, and quality control.

Aerial Robots (Drones): Aerial robots are becoming increasingly popular in manufacturing for tasks such as inspection, inventory management, and security. They can be used to inspect large areas or difficult-to-reach areas quickly and efficiently.

Inspection And Maintenance Robots: These robots are designed to inspect and maintain equipment and infrastructure. They can help to improve the safety and reliability of manufacturing operations by identifying potential problems before they cause downtime or accidents.

In addition to these general types of service robots, there are also a number of specialized service robots that are used in specific manufacturing industries. For example, in the food and beverage industry, there are service robots that are used to palletize and depalletize products, to package products, and to clean and disinfect equipment. In the pharmaceutical industry, there are service robots that are used to fill vials and syringes, to package products, and to clean and disinfect equipment.

The type of service robot that is best suited for a particular manufacturing application will depend on the specific requirements of the application. Factors to consider include the type of tasks that the robot will be performing, the environment in which the robot will be operating, and the budget available.

Automated Guided Vehicles (Agvs)

Automated guided vehicles (AGVs) are a type of service robot that is commonly used in manufacturing and industrial applications. AGVs are self-propelled vehicles that can navigate autonomously through a manufacturing facility using sensors and guidance systems. They are typically used to transport materials, products, and people around a facility, freeing up human workers to focus on more complex tasks.

Image Source : https://blog.toyota-forklifts.co.uk/what-are-automated-guided-vehicle-agv-robots

AGVs Offer A Number Of Benefits For Manufacturing And Industrial Applications, Including:

Increased Productivity And Efficiency: AGVs can work 24/7 and can transport materials and products much faster and more efficiently than humans. This can help to increase productivity and efficiency in manufacturing operations.

Improved Safety: AGVs can be used to transport heavy or dangerous materials, which can help to improve safety for workers.

Reduced Costs: AGVs can help to reduce labor costs and other operational costs.

Versatility And Flexibility: AGVs can be programmed to transport a variety of different materials and products, and they can be used in a variety of different manufacturing and industrial settings.

AGVs Are Typically Guided By One Of The Following Technologies:

Wire guidance: AGVs follow a wire that is embedded in the floor.

Laser guidance: AGVs use lasers to navigate through a facility.

Vision guidance: AGVs use cameras to navigate through a facility.

AGVs Can Be Used For A Variety Of Tasks In Manufacturing And Industrial Applications, Including:

Transporting Materials And Products: AGVs can be used to transport materials and products from one location to another in a manufacturing facility. For example, AGVs can be used to transport raw materials from a warehouse to a production line, or to transport finished products from a production line to a packaging area.

Loading And Unloading Trucks: AGVs can be used to load and unload trucks from docks and loading bays. This can help to reduce the amount of manual labor required and to improve the safety of loading and unloading operations.

Stocking Shelves In Warehouses: AGVs can be used to stock shelves in warehouses. This can help to improve the efficiency of warehouse operations and to reduce the amount of manual labor required.

Moving People: AGVs can be used to move people around a manufacturing facility. For example, AGVs can be used to transport workers to and from their workstations, or to transport visitors around a facility.

The use of AGVs in manufacturing and industrial applications is expected to grow significantly in the coming years, driven by factors such as the increasing demand for labor-saving solutions, the growing adoption of Industry 4.0, and the declining cost of AGVs. AGVs are having a significant impact on the manufacturing and industrial sectors, helping to improve productivity, efficiency, safety, and versatility.

Autonomous Mobile Robots (AMRs)

Autonomous mobile robots (AMRs) are a type of service robot that can navigate autonomously through a manufacturing or industrial environment without the need for human intervention. AMRs are typically equipped with a variety of sensors, including cameras, lasers, and ultrasonic sensors, which allow them to perceive their surroundings and avoid obstacles.

Image Source: https://www.thefabricator.com

Amrs' Growing Popularity In Industry

Increased Productivity And Efficiency: AMRs can work 24/7 and can transport materials and products much faster and more efficiently than human workers.

Improved Quality And Accuracy: AMRs can perform tasks with a high degree of precision and accuracy. This can help to improve the quality of products and to reduce the number of defects.

Reduced Costs: AMRs can help to reduce labor costs and other operational costs.

Improved Safety: AMRs can be used to perform dangerous or repetitive tasks, which can help to improve safety for workers.

Versatility And Flexibility: AMRs can be programmed to perform a wide variety of tasks, which makes them very versatile and flexible.

Scalability: AMRs can be easily scaled to accommodate changing workloads, making them adaptable to fluctuating demand and reducing the need for significant infrastructure changes.

Non-Invasive Implementation: Unlike traditional conveyor systems, AMRs can be integrated into existing facilities without the need for significant structural modifications, saving time and resources.

Data Collection And Analysis: AMRs can collect data during their operations, providing valuable insights that can be used for process optimization, predictive maintenance, and continuous improvement.

Dynamic Path Planning: AMRs can adapt to dynamic environments and reroute themselves in real-time in response to changing conditions, making them highly efficient in complex, busy spaces. **Reduced Environmental Impact:** AMRs are often designed to be energy-efficient, contributing to reduced power consumption and environmental sustainability in industrial and manufacturing processes.

Minimal Downtime: AMRs can operate continuously, reducing the downtime associated with shift changes and breaks, which is particularly beneficial in 24/7 manufacturing environments.

Remote Monitoring And Control: AMRs can be remotely monitored and controlled, allowing for real-time adjustments, troubleshooting, and maintenance, even from off-site locations.

Customizable End-Effectors: The end-effectors on AMRs can be easily swapped or customized to handle different tasks, providing versatility across various applications.

Integration With Iot And Industry 4.0: AMRs can seamlessly integrate with the Internet of Things (IoT) and Industry 4.0 initiatives, enhancing overall automation and connectivity in smart manufacturing.

Adherence To Regulations: AMRs can be programmed to follow strict safety and regulatory guidelines, ensuring compliance with industry standards and reducing the risk of accidents or errors.

Versatile Applications Of AMRs In Manufacturing And Industry

Material Handling: AMRs can be used to transport materials and products from one location to another. This can free up human workers to focus on more complex tasks.

Assembly: AMRs can be used to assemble products, such as electronic devices or automotive parts. This can help to improve the accuracy and consistency of the assembly process.

Inspection And Maintenance: AMRs can be used to inspect products for defects and to perform maintenance tasks on equipment. This can help to improve the quality of products and to reduce downtime.

Security And Surveillance: AMRs can be used to patrol manufacturing and industrial facilities and to monitor for security threats. This can help to improve the security of these facilities and to protect against theft and vandalism.

AMRs In Industrial Settings Today :

General Motors: General Motors employs AMRs to transport materials and components within their automotive manufacturing facilities.

Procter & Gamble: Procter & Gamble utilizes AMRs for optimizing and streamlining packaging processes in their consumer goods manufacturing plants.

Airbus: Airbus integrates AMRs into their aerospace manufacturing to assist in parts handling and logistics, enhancing overall efficiency.

Samsung: Samsung employs AMRs for quality control and inspection tasks, ensuring precision and consistency in their electronic manufacturing processes.

General Electric: General Electric uses AMRs for predictive maintenance, enabling proactive servicing of industrial machinery in their manufacturing operations.

These are just a few examples of the many ways in which AMRs are being used in manufacturing and industrial applications today. As AMR technology continues to advance, AMRs are expected to play an even greater role in manufacturing and industrial operations in the future.

Issues Encountered With AMRs In Industrial And Manufacturing Uses Challenges:

High Upfront Costs: AMRs can be expensive to purchase and implement.

Integration With Existing Systems And Processes: AMRs need to be integrated with existing manufacturing and industrial systems and processes. This can be a complex and challenging task. **Training And Support Requirements:** Employees need to be trained on how to use and maintain AMRs. This can require significant time and resources.

Safety Concerns: AMRs need to be designed and implemented in a way that minimizes safety risks for workers.

Scalability: Adapting AMR deployments to changing production needs and expanding operations can be challenging.

Navigation In Complex Environments: AMRs may face difficulties in navigating intricate industrial settings with dynamic layouts.

Battery Life And Charging: Ensuring that AMRs have sufficient battery life and efficient charging solutions can be a challenge.

Interoperability: Ensuring compatibility with a variety of equipment and systems can be complex in diverse industrial environments.

Security: Protecting AMRs and the data they collect from cyber threats and unauthorized access poses challenges.

Overall, AMRs offer a number of potential benefits for manufacturing and industrial applications. However, it is important to carefully consider the challenges associated with implementing AMRs before making a decision.

Aerial Robots (Drones)

Aerial robots, also known as drones, are increasingly being used in manufacturing and industrial applications. Drones offer a number of advantages over traditional methods of inspection, maintenance, and other tasks, including:

Access To Difficult-To-Reach Areas: Drones can fly to and inspect areas that are difficult or dangerous for humans to access, such as rooftops, towers, and confined spaces.

Speed And Efficiency: Drones can quickly and efficiently cover large areas, which can save time and money.

Data Collection: Drones can be equipped with a variety of sensors to collect data on things like temperature, humidity, and air quality. This data can be used to improve safety, efficiency, and quality in manufacturing and industrial operations.

Aerial robots (drones)

Current Utilizations Of Drones In Manufacturing And Industry

Inspection: Drones can be used to inspect equipment, infrastructure, and products for defects. For example, drones can be used to inspect bridges for cracks, roofs for damage, and solar panels for hot spots.

Maintenance: Drones can be used to perform maintenance tasks on equipment and infrastructure that is difficult or dangerous for humans to access. For example, drones can be used to clean and inspect wind turbines, inspect power lines, and repair cell towers.

Inventory Management: Drones can be used to track and manage inventory in warehouses and other facilities. This can help to improve accuracy and efficiency in inventory management operations.

Security: Drones can be used to patrol facilities and monitor for security threats. This can help to improve security and reduce the risk of theft and vandalism.

Environmental Monitoring: Drones are employed to monitor air and water quality, track emissions, and assess environmental impact in manufacturing processes.

Agricultural Surveys: Drones assist in agricultural applications by monitoring crop health, soil conditions, and pest infestations, leading to better crop management.

Emergency Response: Drones are used for rapid assessment and response in industrial accidents, natural disasters, and search and rescue operations.

Infrastructure Inspection: Beyond bridges and roofs, drones inspect pipelines, railways, and roads to identify maintenance needs and ensure safety.

Mapping And Surveying: Drones are employed for land surveys, site mapping, and creating 3D models of industrial sites, aiding in planning and development.

Quality Control: Drones support quality control processes by inspecting products on production lines, identifying defects, and ensuring consistency.

Temperature Monitoring: In high-temperature environments like foundries or glass manufacturing, drones measure and monitor temperatures for safety and process optimization.

Waste Management: Drones help in monitoring and managing waste disposal sites, ensuring compliance with regulations and minimizing environmental impact.

The use of drones in manufacturing and industrial applications is expected to grow significantly in the coming years, driven by factors such as the increasing demand for labor-saving solutions, the growing adoption of Industry 4.0, and the declining cost of drones. Drones are having a significant impact on the manufacturing and industrial sectors, helping to improve safety, efficiency, and quality.

Further Upsides Of Aerial Robots In Manufacturing And Industrial Applications

Reduced Costs: Drones can help to reduce labor costs and other operational costs. For example, drones can be used to perform tasks that would otherwise require multiple workers or specialized equipment.

Improved Safety: Drones can be used to perform dangerous or repetitive tasks, which can help to improve safety for workers. For example, drones can be used to inspect hazardous areas or to perform maintenance on equipment that is difficult or dangerous to access.

Increased Productivity: Drones can help to increase productivity by automating tasks and by providing real-time data. For example, drones can be used to track inventory levels or to inspect products for defects, which can help to identify and resolve problems early on.

Rapid Response: Drones can be quickly deployed to assess emergency situations, such as fire outbreaks, chemical spills, or structural damage, providing immediate insights to aid in crisis management.

Data Visualization: Drones capture high-resolution images and videos that offer a visual overview of large facilities or infrastructure, facilitating better decision-making and remote inspections.

Enhanced Precision Agriculture: Drones equipped with specialized sensors and cameras can optimize crop management, monitor soil conditions, and assist in precision agriculture practices, ultimately increasing agricultural yields.

Time Savings: Drones can complete tasks in a fraction of the time it would take human workers, reducing project durations and improving overall efficiency.

Environmental Monitoring: Drones can be used to assess environmental conditions, air quality, and emissions in industrial zones, contributing to improved environmental compliance and sustainability.

Reduced Equipment Wear And Tear: By taking over repetitive inspection tasks, drones can extend the operational life of equipment and infrastructure by reducing physical wear and tear.

Non-Disruptive Inspection: Drones can perform inspections without interrupting regular operations, minimizing downtime and production losses.

Global Connectivity: Drones can transmit data and images from remote or challenging locations, enabling real-time decision-making from anywhere in the world.

Elevated Employee Morale: By automating tedious or dangerous tasks, drones can enhance the job satisfaction and well-being of human workers who are freed from such duties.

Asset Tracking: Drones can help in tracking the movement of assets, equipment, and inventory within large manufacturing and warehouse facilities, reducing the risk of loss or theft.

Overall, aerial robots offer a number of potential benefits for manufacturing and industrial applications. However, it is important to carefully consider the challenges associated with implementing aerial robots, such as safety regulations and the need for training and support.

Cobots

Cobots, or collaborative robots, are robots that are designed to work safely alongside human workers. They are typically smaller and less powerful than industrial robots, but they are also more affordable and easier to program. Cobots can be used for a variety of tasks in manufacturing and industrial applications, such as assembly, inspection, and machine tending.

Cobots Offer A Number Of Benefits For Manufacturing And Industrial Applications, Including:

Increased Productivity And Efficiency: Cobots can work 24/7 and can perform tasks much faster and more efficiently than humans. This can help to increase productivity and efficiency in manufacturing operations.

Improved Quality And Accuracy: Cobots can perform tasks with a high degree of precision and accuracy. This can help to improve the quality of products and to reduce the number of defects.

Reduced Costs: Cobots can help to reduce labor costs and other operational costs.

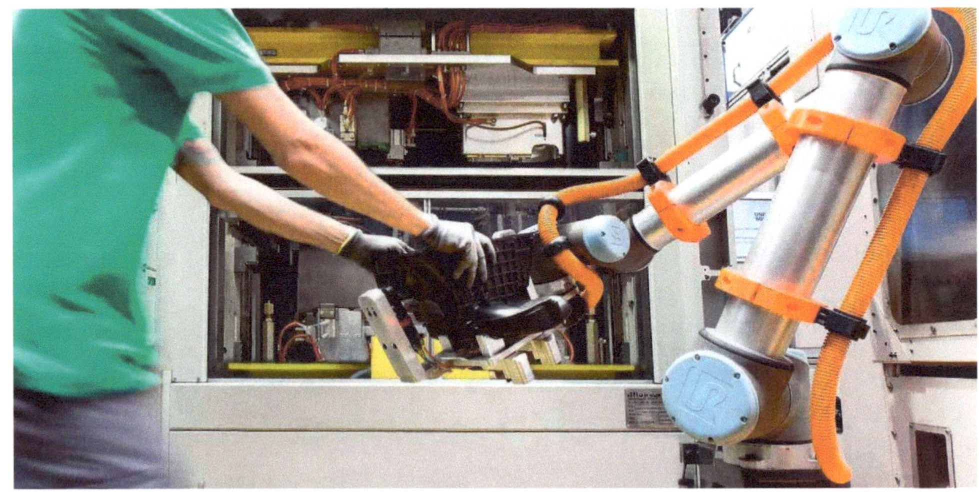

Image Source : https://www.wevolver.com/article/what-is-a-collaborative-robot

Improved Safety: Cobots can be used to perform dangerous or repetitive tasks, which can help to improve safety for workers.

Versatility And Flexibility: Cobots can be programmed to perform a wide variety of tasks, which makes them very versatile and flexible.

Human Collaboration: Cobots are designed to work alongside human employees, fostering collaboration rather than replacement. They can assist workers in tasks that require strength, precision, or repetitive actions.

Rapid Reconfiguration: Cobots can be quickly reprogrammed to adapt to changing production requirements or to take on new tasks, allowing for agile and responsive manufacturing.

Space Optimization: Due to their compact size and ability to work in confined spaces, cobots can help optimize the utilization of available workspace within manufacturing facilities.

Minimized Learning Curve: Cobots are often designed with intuitive interfaces and programming that do not require extensive training, reducing the time it takes to implement them in a manufacturing environment.

Consistent Performance: Cobots consistently perform tasks without fatigue or variations in quality, ensuring uniform product quality across production cycles.

Shorter Setup Times: Cobots are designed for easy integration and setup, reducing the time and resources required for implementation, even in small-scale or medium-sized manufacturing operations.

Real-Time Data Collection: Cobots are equipped with sensors that can collect and transmit data, providing insights for process optimization, predictive maintenance, and quality control.

Energy Efficiency: Cobots are typically energy-efficient, contributing to reduced power consumption and lower operating costs. This aligns with sustainability and cost-saving initiatives in manufacturing.

Cross-Training: Cobots can be cross-trained to perform a range of tasks within a single facility, offering versatility and adaptability across various manufacturing processes.

Improved Employee Engagement: By taking over strenuous or monotonous tasks, cobots can enhance the overall job satisfaction and morale of human workers, allowing them to focus on more creative and value-added activities.

Practical Uses Of Cobots In Industrial And Manufacturing Settings

Assembly: Cobots can be used to assemble products, such as electronic devices, automotive parts, and medical devices.

Inspection: Cobots can be used to inspect products for defects. For example, cobots can be used to inspect car bodies for paint defects or to inspect electronic devices for assembly errors.

Machine Tending: Cobots can be used to tend machines, such as CNC machines and injection molding machines. This can free up human workers to focus on more complex tasks.

The use of cobots in manufacturing and industrial applications is expected to grow significantly in the coming years, driven by factors such as the increasing demand for labor-saving solutions, the growing adoption of Industry 4.0, and the declining cost of cobots. Cobots are having a significant impact on the manufacturing and industrial sectors, helping to improve productivity, efficiency, quality, safety, and versatility.

Advantages Of Implementing Cobots In Manufacturing And Industrial Settings

Reduced Downtime: Cobots can help to reduce downtime by automating tasks that are prone to human error. For example, cobots can be used to load and unload machines or to package products, which can help to ensure that production continues uninterrupted.

Improved Employee Satisfaction: Cobots can help to improve employee satisfaction by taking on dangerous or repetitive tasks. This can free up human workers to focus on more challenging and rewarding work.

Enhanced Competitiveness: Cobots can help businesses to stay competitive by improving productivity and efficiency. This can lead to lower costs and higher profits.

Overall, cobots offer a number of potential benefits for manufacturing and industrial applications. They are a key part of the Industry 4.0 revolution, and they are expected to play an even greater role in manufacturing and industrial operations in the future.

Inspection And Quality Control Robots

Inspection and quality control robots are service robots that are designed to inspect products and components for defects. They are typically used in manufacturing and industrial settings to ensure that products meet quality standards and to reduce the risk of shipping defective products to customers.

Inspection And Quality Control Robots Offer A Number Of Benefits, Including:

Increased Accuracy And Consistency: Inspection and quality control robots can inspect products and components with a high degree of accuracy and consistency. This is because they are not subject to human error or fatigue.

Reduced Costs: Inspection and quality control robots can help to reduce labor costs and other operational costs.

Improved Quality: Inspection and quality control robots can help to improve the quality of products by identifying defects early on in the production process. This can help to reduce the number of defective products that are shipped to customers.

Increased Productivity: Inspection and quality control robots can help to increase productivity by automating tasks that would otherwise be performed by human workers.

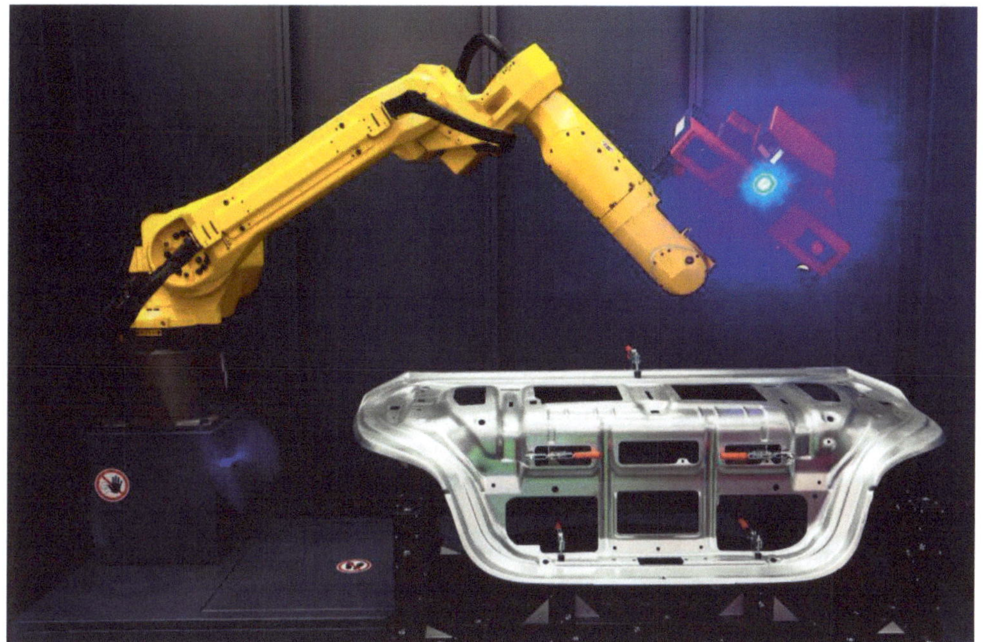

Image Source : https://madox.in/solutions/robot-inspection/

Inspection And Quality Control Robots Can Be Used To Inspect A Wide Variety Of Products And Components, Including:

Electronic Devices: Inspection and quality control robots can be used to inspect electronic devices for defects such as incorrect component placement, solder defects, and assembly errors.

Automotive Parts: Inspection and quality control robots can be used to inspect automotive parts for defects such as cracks, surface defects, and dimensional errors.

Food And Beverage Products: Inspection and quality control robots can be used to inspect food and beverage products for defects such as foreign objects, contamination, and incorrect packaging.

Pharmaceutical Products: Inspection and quality control robots can be used to inspect pharmaceutical products for defects such as incorrect packaging, contamination, and missing or damaged components.

Contemporary Applications Of Inspection And Quality Control Robots

Amazon: Amazon uses inspection and quality control robots to inspect products in its warehouses.

Tesla: Tesla uses inspection and quality control robots to inspect car parts in its factories.

Bmw: BMW uses inspection and quality control robots to inspect car bodies in its factories.

Foxconn: Foxconn uses inspection and quality control robots to inspect electronic components in its factories.

Siemens: Siemens uses inspection and quality control robots to inspect equipment and products in its factories.

The use of inspection and quality control robots in manufacturing and industrial applications is expected to grow significantly in the coming years. This is due to a number of factors, including the increasing demand for high-quality products, the growing adoption of Industry 4.0, and the declining cost of inspection and quality control robots.

Extended Gains Of Quality Control Robots In Modern Industry

Reduced Waste: Inspection and quality control robots can help to reduce waste by identifying and removing defective products from the production process early on. This can save businesses money and reduce their environmental impact.

Improved Brand Reputation: By shipping high-quality products to customers, businesses can improve their brand reputation and attract new customers.

Enhanced Customer Satisfaction: Customers are more likely to be satisfied with products that meet their quality expectations. Inspection and quality control robots can help businesses to ensure that their products meet the expectations of their customers.

Overall, inspection and quality control robots offer a number of potential benefits for manufacturing and industrial applications. They can help businesses to improve product quality, reduce costs, increase productivity, and reduce waste. Inspection and quality control robots are a key part of the Industry 4.0 revolution, and they are expected to play an even greater role in manufacturing and industrial operations in the future.

Material Handling Robots

Material handling robots are service robots that are designed to transport materials and products from one location to another. They are typically used in manufacturing and industrial settings to automate tasks such as loading and unloading machines, palletizing and depalletizing products, and transporting materials between different areas of a facility.

Material Handling Robots Offer A Number Of Benefits, Including:

Increased Productivity And Efficiency: Material handling robots can work 24/7 and can transport materials and products much faster and more efficiently than humans. This can help to increase productivity and efficiency in manufacturing and industrial operations.

Reduced costs: Material handling robots can help to reduce labor costs and other

operational costs.

Improved safety: Material handling robots can be used to perform dangerous or repetitive tasks, which can help to improve safety for workers.

 Material handling robots can be programmed to perform a wide variety of tasks, which makes them very versatile and flexible.

Material Handling Robots

Material handling robots can be used to transport a wide variety of materials and products, including:

Raw materials: Material handling robots can be used to transport raw materials from storage areas to production lines.

Work-in-process: Material handling robots can be used to transport work-in-process between different stages of the production process.

Finished goods: Material handling robots can be used to transport finished goods from production lines to packaging and shipping areas.

Automotive factories: Material handling robots are used to transport car bodies, parts, and other materials around automotive factories.

Food And Beverage Processing Plants: Material handling robots are used to transport food and beverage products around food and beverage processing plants.

Warehouses: Material handling robots are used to pick, pack, and ship products in warehouses.

Distribution Centers: Material handling robots are used to move products around distribution centers.

The use of material handling robots in manufacturing and industrial applications is expected to grow significantly in the coming years. This is due to a number of factors, including the increasing demand for labor-saving solutions, the growing adoption of Industry 4.0, and the declining cost of material handling robots.

Added Advantages Of Employing Material Handling Robots In Industry

Reduced Downtime: Material handling robots can help to reduce downtime by automating tasks that are prone to human error. For example, material handling robots can be used to load and unload machines or to palletize products, which can help to ensure that production continues uninterrupted.

Improved Ergonomics: Material handling robots can help to improve ergonomics by taking on physically demanding tasks. This can help to reduce the risk of injuries and improve the overall well-being of workers.

Enhanced Competitiveness: Material handling robots can help businesses to stay competitive by improving productivity and efficiency. This can lead to lower costs and higher profits.

Overall, material handling robots offer a number of potential benefits for manufacturing and industrial applications. They are a key part of the Industry 4.0 revolution, and they are expected to play an even greater role in manufacturing and industrial operations in the future.

Mobile Manipulation Robots (Mmrs)

Mobile manipulation robots (MMRs) are service robots that are equipped with both mobility and manipulation capabilities. This means that they can move around independently and interact with their environment using their arms and grippers. MMRs are highly versatile and can be used for a wide variety of tasks in manufacturing and industrial applications, such as:

Assembly: MMRs can be used to assemble products, such as electronic devices, automotive parts, and medical devices.

Inspection: MMRs can be used to inspect products and components for defects. For example, MMRs can be used to inspect car bodies for paint defects or to inspect electronic devices for assembly errors.

Machine Tending: MMRs can be used to tend machines, such as CNC machines and injection molding machines. This can free up human workers to focus on more complex tasks.

Packaging And Palletizing: MMRs can be used to package and palletize products. This can help to improve productivity and efficiency in packaging and shipping operations.

Warehousing And Logistics: MMRs can be used to pick, pack, and ship products in warehouses and distribution centers.

Other Tasks: MMRs can also be used for a variety of other tasks, such as cleaning and maintenance, hazardous materials handling, and search and rescue.

Mobile Manipulation Robots (MMRs)

MMRs Offer A Number Of Benefits For Manufacturing And Industrial Applications, Including:

Increased Productivity And Efficiency: MMRs can work 24/7 and can perform tasks much faster and more efficiently than humans. This can help to increase productivity and efficiency in manufacturing and industrial operations.

Improved Quality And Accuracy: MMRs can perform tasks with a high degree of precision and accuracy. This can help to improve the quality of products and to reduce the number of defects.

Reduced costs: MMRs can help to reduce labor costs and other operational costs.

Improved Safety: MMRs can be used to perform dangerous or repetitive tasks, which can help to improve safety for workers.

Versatility And Flexibility: MMRs can be programmed to perform a wide variety of tasks, which makes them very versatile and flexible.

MMRs are still a relatively new technology, but they are rapidly gaining popularity in manufacturing and industrial settings. As MMRs become more sophisticated and affordable, they are expected to play an even greater role in the future of manufacturing and industry.

Current Applications Of Mobile Manipulation Robots In Manufacturing And Industry

Automotive Factories: MMRs are being used to assemble cars, inspect car parts, and tend machines in automotive factories.

Food And Beverage Processing Plants: MMRs are being used to package and palletize food and beverage products in food and beverage processing plants.

Warehouses: MMRs are being used to pick, pack, and ship products in warehouses.

Distribution Centers: MMRs are being used to move products around distribution centers.

The use of MMRs in manufacturing and industrial applications is expected to grow significantly in the coming years. This is due to a number of factors, including the increasing demand for labor-saving solutions, the growing adoption of Industry 4.0, and the declining cost of MMRs.

Further Advantages Of Employing Mmrs In Manufacturing And Industrial Applications

Reduced Downtime: MMRs can help to reduce downtime by automating tasks that are prone to human error. For example, MMRs can be used to load and unload machines or to palletize products, which can help to ensure that production continues uninterrupted.

Improved Ergonomics: MMRs can help to improve ergonomics by taking on physically demanding tasks. This can help to reduce the risk of injuries and improve the overall well-being of workers.

Enhanced Competitiveness: MMRs can help businesses to stay competitive by improving productivity and efficiency. This can lead to lower costs and higher profits.

Overall, MMRs offer a number of potential benefits for manufacturing and industrial applications. They are a key part of the Industry 4.0 revolution, and they are expected to play an even greater role in manufacturing and industrial operations in the future.

9.3. Enhancing Manufacturing With Service Robotics:

Real-World Cases

Service robots are robots that are designed to perform tasks that assist humans in a variety of settings, including manufacturing and industrial environments. Service robots can be used to automate a wide range of tasks, from simple repetitive tasks to more complex tasks that require dexterity and intelligence.

Service Robots Transforming Manufacturing Environments

Material Handling And Transportation: Service robots can be used to move materials and products around a factory or warehouse. This can include tasks such as loading and unloading machines, palletizing and depalletizing goods, and transporting goods to and from different stages of the production process.

Inspection And Quality Control: Service robots can be used to inspect products for defects and to ensure that they meet quality standards. This can include tasks such as visually inspecting products for flaws, checking dimensions and tolerances, and testing products for functionality.

Assembly And Packaging: Service robots can be used to assemble products and to package them for shipping. This can include tasks such as picking and placing components, inserting parts into assemblies, and sealing and labeling packages.

Maintenance And Repair: Service robots can be used to perform maintenance and repair tasks on equipment and machinery. This can include tasks such as greasing and lubricating moving parts, replacing worn components, and identifying and troubleshooting problems.

Customer Service: Service robots can be used to provide customer service in manufacturing and industrial environments. This can include tasks such as greeting customers, answering questions, and providing directions.

Applications of Service Robots

Service Robots Offer A Number Of Benefits For Manufacturing And Industrial Applications, Including:

Increased Productivity And Efficiency: Service robots can automate tasks that are repetitive, time-consuming, or dangerous. This can free up human workers to focus on more complex and value-added tasks.

Improved Quality And Consistency: Service robots can perform tasks with a high degree of accuracy and repeatability. This can help to improve the quality and consistency of manufactured products.

Reduced Costs: Service robots can help to reduce labor costs and other operating costs.

Improved Safety: Service robots can perform tasks that are dangerous or hazardous for human workers. This can help to improve safety in the workplace.

Overall, service robots have the potential to revolutionize manufacturing and industrial applications. By automating tasks and improving productivity, quality, and safety, service robots can help businesses to become more competitive and efficient.

Here Are A Few Specific Examples Of How Service Robots Are Being Used In Manufacturing And Industrial Settings Today:

Automated guided vehicles (AGVs) and autonomous mobile robots (AMRs) are used to transport materials and products around factories and warehouses.

Collaborative robots (cobots) are used to work alongside human workers on assembly lines and other tasks.

Inspection robots are used to inspect products for defects and to ensure that they meet quality standards.

Maintenance robots are used to perform maintenance and repair tasks on equipment

and machinery.

Customer service robots are used to greet customers, answer questions, and provide directions in manufacturing and industrial settings.

These are just a few examples of the many ways that service robots are being used in manufacturing and industrial applications today. As service robot technology continues to develop, we can expect to see even more innovative and groundbreaking applications in the years to come.

Assembly And Disassembly Operations

Assembly and disassembly operations are essential parts of many manufacturing and industrial processes. However, these operations can be time-consuming, repetitive, and error-prone. Service robots can be used to automate assembly and disassembly operations, which can improve productivity, quality, and safety.

Assembly: Service robots can be used to assemble products of all shapes and sizes. This can include tasks such as picking and placing components, inserting parts into assemblies, and fastening components together.

Disassembly: Service robots can be used to disassemble products at the end of their life cycle. This can include tasks such as removing components from assemblies, separating different types of materials, and sorting materials for recycling or reuse.

Assembly and Disassembly Operations

Service robots can be equipped with a variety of tools and sensors to perform assembly and disassembly operations. For example, service robots can be equipped with grippers to pick and place components, cameras to inspect parts for defects, and force sensors to apply the correct amount of force when fastening components together.

Service robots can also be equipped with artificial intelligence (AI) and machine learning (ML) capabilities. This allows service robots to learn and adapt to new assembly and disassembly tasks, and to perform these tasks with a high degree of accuracy and repeatability.

Here Are A Few Specific Examples Of How Service Robots Are Being Used For Assembly And Disassembly Operations In Manufacturing And Industrial Settings Today:

Collaborative robots (cobots) are being used to assemble small electronic devices, such as smartphones and tablets.

Service robots equipped with AI and ML capabilities are being used to disassemble complex products, such as aircraft engines and automobiles.
Service robots are also being used to assemble and disassemble large products, such as furniture and appliances.

Benefits Of Using Service Robots For Assembly And Disassembly Operations

There are a number of benefits to using service robots for assembly and disassembly operations in manufacturing and industrial applications, including:

Increased Productivity And Efficiency: Service robots can automate tasks that are repetitive, time-consuming, or dangerous. This can free up human workers to focus on more complex and value-added tasks.

Improved Quality And Consistency: Service robots can perform tasks with a high degree of accuracy and repeatability. This can help to improve the quality and consistency of manufactured products.

Reduced costs: Service robots can help to reduce labor costs and other operating costs.

Improved Safety: Service robots can perform tasks that are dangerous or hazardous for human workers. This can help to improve safety in the workplace.

Overall, service robots have the potential to revolutionize assembly and disassembly operations in manufacturing and industrial applications. By automating these tasks and improving productivity, quality, and safety, service robots can help businesses to become more competitive and efficient.

Welding And Fabrication

Welding and fabrication are essential processes in many manufacturing and industrial applications. However, these processes can be dangerous and require a high level of skill and experience. Service robots can be used to automate welding and fabrication tasks, which can improve productivity, quality, and safety.

Discover Some Specific Examples Of How Service Robots Can Be Used For Welding And Fabrication Tasks In Manufacturing And Industrial Applications:

Welding: Service robots can be used to perform a variety of welding tasks, including arc welding, resistance welding, and laser welding. Service robots can be equipped with a variety of welding tools and sensors to perform these tasks with a high degree of accuracy and repeatability.

Fabrication: Service robots can be used to perform a variety of fabrication tasks, such as cutting, bending, and forming metal. Service robots can be equipped with a variety of tools and sensors to perform these tasks with a high degree of precision.

Service robots can also be equipped with artificial intelligence (AI) and machine learning (ML) capabilities. This allows service robots to learn and adapt to new welding and fabrication tasks, and to perform these tasks with a high degree of accuracy and repeatability.

Welding and Fabrication

Here Are A Few Specific Examples Of How Service Robots Are Being Used For Welding And Fabrication Tasks In Manufacturing And Industrial Settings Today:

Collaborative robots (cobots) are being used to perform arc welding tasks on small to medium-sized parts.

Service robots equipped with AI and ML capabilities are being used to perform laser welding tasks on complex and/or high-precision parts.

Service robots are also being used to perform a variety of fabrication tasks, such as cutting, bending, and forming metal in the automotive, aerospace, and electronics industries.

Robotic Welding Cells: Automated welding cells with robotic arms are employed in automotive manufacturing for spot welding and seam welding of vehicle frames and components.

3D Printing Robots: Service robots with 3D printing capabilities are used in rapid prototyping and custom part manufacturing for various industries.

CNC Machining Robots: Computer Numerical Control (CNC) robots assist in milling, drilling, and machining tasks in precision manufacturing processes.

Robotic Laser Cutting: Service robots equipped with laser cutting tools are used to precisely cut metal sheets and other materials in manufacturing and metalworking.

Robotic Grinding and Polishing: Robots are employed in grinding and polishing operations to ensure consistent surface finishes in industries like furniture production and metalworking.

Benefits Of Using Service Robots For Welding And Fabrication

Service Robots Offer Manifold Advantages In Manufacturing And Industry, Notably For Welding And Fabrication Tasks.

Increased Productivity And Efficiency: Service robots can automate tasks that are repetitive, time-consuming, or dangerous. This can free up human workers to focus on more complex and value-added tasks.

Improved Quality And Consistency: Service robots can perform tasks with a high degree of accuracy and repeatability. This can help to improve the quality and consistency of manufactured products.

Reduced Costs: Service robots can help to reduce labor costs and other operating costs. **Improved Safety:** Service robots can perform tasks that are dangerous or hazardous for human workers. This can help to improve safety in the workplace.

Overall, service robots have the potential to revolutionize welding and fabrication operations in manufacturing and industrial applications. By automating these tasks and improving productivity, quality, and safety, service robots can help businesses to become more competitive and efficient.

Inventory Management And Warehousing

Inventory management and warehousing are essential functions for many manufacturing and industrial businesses. However, these functions can be complex and time-consuming, and they can be prone to errors. Service robots can be used to automate many inventory management and warehousing tasks, which can improve accuracy, efficiency, and productivity.

These are some specific examples of how service robots can be used for inventory management and warehousing tasks in manufacturing and industrial applications:

Receiving And Inspection: Service robots can be used to receive and inspect incoming shipments. This can include tasks such as scanning barcodes, checking for damage, and verifying product quality.

Picking And Packing: Service robots can be used to pick and pack orders. This can include tasks such as locating products in the warehouse, picking products from shelves, and packing products into boxes.

Storage And Retrieval: Service robots can be used to store and retrieve products in the warehouse. This can include tasks such as placing products on shelves, moving products to and from different areas of the warehouse, and retrieving products for orders.

Inventory Tracking: Service robots can be used to track inventory levels in the warehouse. This can include tasks such as scanning barcodes, counting products, and updating inventory records.

Service robots can also be equipped with artificial intelligence (AI) and machine learning (ML) capabilities. This allows service robots to learn and adapt to new inventory management and warehousing tasks, and to perform these tasks with a high degree of accuracy and repeatability.

Modernizing Inventory Management In Industrial Settings

Autonomous mobile robots (AMRs) are being used to pick and pack orders in e-commerce fulfillment warehouses.

Collaborative robots (cobots) are being used to inspect products for defects and to verify product quality.

Service robots equipped with AI and ML capabilities are being used to optimize inventory storage and retrieval, and to predict future inventory needs.

Inventory Management and Warehousing

Benefits Of Using Service Robots For Inventory Management And Warehousing

Increased Accuracy And Efficiency: Service robots can automate tasks that are repetitive, time- consuming, or error-prone. This can help to improve the accuracy and efficiency of inventory management and warehousing operations.

Reduced Costs: Service robots can help to reduce labor costs and other operating costs.

Improved Customer Service: Service robots can help to improve customer service by ensuring that orders are fulfilled accurately and on time.

Improved Safety: Service robots can perform tasks that are dangerous or hazardous for human workers. This can help to improve safety in the workplace.

Overall, service robots have the potential to revolutionize inventory management and warehousing operations in manufacturing and industrial applications. By automating tasks and improving accuracy, efficiency, and productivity, service robots can help businesses to reduce costs and improve customer service.

Inspection And Defect Detection

Inspection and defect detection are essential tasks in manufacturing and industrial applications. However, these tasks can be time-consuming and repetitive, and they can be prone to human error. Service robots can be used to automate inspection and defect detection tasks, which can improve accuracy, efficiency, and productivity.

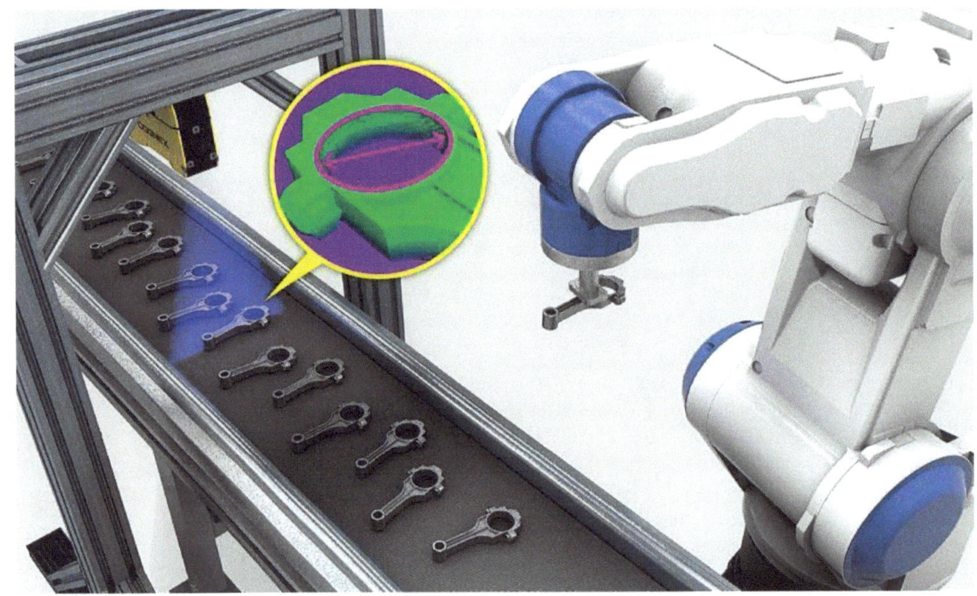

Inspection and Defect Detection

Precision Inspection In Manufacturing: Service Robots At Work

Visual Inspection: Service robots can be equipped with cameras and other sensors to visually inspect products for defects. This can include tasks such as identifying surface defects, checking dimensions and tolerances, and verifying product completeness.

Non-Destructive Testing (Ndt): Service robots can be equipped with NDT tools and sensors to inspect products for internal defects without damaging them. This can include tasks such as performing ultrasonic testing, radiographic testing, and magnetic particle testing.

Leak Testing: Service robots can be equipped with leak testing tools to identify leaks in products and systems. This can include tasks such as testing pipes, valves, and seals for leaks.

Service robots can also be equipped with artificial intelligence (AI) and machine learning (ML) capabilities. This allows service robots to learn and adapt to new inspection and defect detection tasks, and to perform these tasks with a high degree of accuracy and repeatability.

Service Robots Excel In Inspecting And Detecting Defects In Modern Manufacturing And Industrial Environments.

Collaborative robots (cobots) are being used to visually inspect products on assembly lines.

Service robots equipped with AI and ML capabilities are being used to identify defects in complex products, such as aircraft and automobiles.

Service robots are also being used to perform NDT and leak testing tasks on pipelines, pressure vessels, and other industrial equipment.

Benefits Of Using Service Robots For Inspection And Defect Detection

Unlocking The Advantages Of Service Robots In Inspection And Defect Detection For Manufacturing And Industry

Increased Accuracy And Efficiency: Service robots can automate tasks that are repetitive, time- consuming, or error-prone. This can help to improve the accuracy and efficiency of inspection and defect detection operations.

Improved Product Quality: Service robots can help to improve product quality by identifying defects early in the manufacturing process.

Reduced Costs: Service robots can help to reduce labor costs and other operating costs. **Improved Safety:** Service robots can perform tasks that are dangerous or hazardous for human workers. This can help to improve safety in the workplace.

Overall, service robots have the potential to revolutionize inspection and defect detection operations in manufacturing and industrial applications. By automating tasks and improving accuracy, efficiency, and productivity, service robots can help businesses to reduce costs and improve product quality.

Hazardous Environment Exploration

Hazardous environments are those that pose a risk to human health and safety. This could include environments that are exposed to extreme temperatures, toxic chemicals, or radiation. Service robots can be used to explore hazardous environments and collect data, which can help to reduce the risk to human workers.

Utilizing Service Robots For Exploring Hazardous Environments In Manufacturing

Nuclear Power Plants: Service robots can be used to inspect nuclear reactors and other equipment in nuclear power plants for damage and wear and tear. This can help to identify potential problems before they lead to accidents.

Chemical Plants: Service robots can be used to inspect chemical plants for leaks and other hazards. This can help to protect workers from exposure to toxic chemicals.

Hazardous Waste Sites: Service robots can be used to map and characterize hazardous waste sites. This can help to develop plans for safely cleaning up these sites.

Disaster Zones: Service robots can be used to search for survivors and assess damage in disaster zones. This can help to speed up rescue and recovery efforts.

Service robots can be equipped with a variety of sensors and tools to explore hazardous environments. For example, service robots can be equipped with cameras and other sensors to inspect equipment and identify hazards. They can also be equipped with tools to collect samples and perform other tasks.

Service robots can also be equipped with artificial intelligence (AI) and machine learning (ML) capabilities. This allows service robots to learn and adapt to new hazardous environments and to perform exploration tasks with a high degree of autonomy.

Hazardous Environment Exploration

Service Robots In Manufacturing And Industry: Exploring Hazardous Environments

Service robots equipped with AI and ML capabilities are being used to inspect nuclear reactors and other equipment in nuclear power plants.

Service robots are also being used to inspect chemical plants for leaks and other hazards.

Service robots are being used to map and characterize hazardous waste sites in preparation for cleanup.

Service robots are also being used to search for survivors and assess damage in disaster zones.

Benefits Of Using Service Robots For Hazardous Environment Exploration

There are a number of benefits to using service robots for hazardous environment exploration in manufacturing and industrial applications, including:

Reduced Risk To Human Workers: Service robots can perform tasks in hazardous environments that would be too dangerous or unhealthy for human workers.

Improved Safety: Service robots can help to identify hazards and prevent accidents.

Increased Efficiency: Service robots can automate tasks that are time-consuming or repetitive.

Improved Data Collection: Service robots can collect data from hazardous environments that would be difficult or impossible for human workers to collect.

Overall, service robots have the potential to revolutionize hazardous environment exploration in manufacturing and industrial applications. By automating tasks and reducing the risk to human workers, service robots can help businesses to improve safety and efficiency.

9.4. Technological Advancements

Sensing And Perception Technologies:

Sensing and perception technologies are essential for service robots in manufacturing and industrial applications. Robots need to be able to perceive their environment in order to navigate autonomously, interact with objects and humans safely, and perform tasks accurately.

Sensing and Perception Technologies

Recent Advancements In Sensing And Perception Technologies Include:

Vision Sensors: Vision sensors have become more affordable, powerful, and smaller in recent years. This has led to the development of new types of robots that can use vision to navigate, interact with objects, and perform tasks.

Lidar Sensors: LiDAR sensors are becoming increasingly popular for robotic applications. LiDAR sensors can provide 3D maps of the environment, which can be used for navigation, obstacle detection, and object recognition.

Force Sensors: Force sensors are becoming more sensitive and accurate, which allows robots to interact with objects in a more delicate and controlled way.

Other Sensors: Other sensors that are becoming increasingly important for service robots include ultrasonic sensors, infrared sensors, and chemical sensors.

Artificial Intelligence And Machine Learning

Artificial intelligence (AI) and machine learning (ML) are two rapidly developing fields that are having a major impact on robotics. AI and ML can be used to enable robots to learn from data, make decisions autonomously, and adapt to changing environments.

Recent Advancements In Ai And Ml For Service Robots Include:

Deep Learning: Deep learning is a type of AI that allows robots to learn from large amounts of data without being explicitly programmed. This has led to the development of robots that can perform tasks such as object recognition, natural language processing, and decision making at a much higher level than previous generations of robots.

Reinforcement Learning: Reinforcement learning is a type of ML that allows robots to learn by trial and error. This is particularly useful for robots that need to learn to perform complex tasks in dynamic environments.

Ai For Safety: AI can also be used to improve the safety of service robots. For example, AI can be used to develop robots that can predict and avoid collisions, and robots that can recognize and respond to human emotions.

Artificial Intelligence and Machine Learning

Human-Robot Interaction (Hri) Interfaces

HRI interfaces are essential for service robots that need to interact with humans in a safe and effective way. HRI interfaces can include natural language processing, gesture recognition, and speech synthesis.

Recent Advancements In Hri Interfaces For Service Robots Include:

Natural Language Processing: Natural language processing (NLP) allows robots to understand and respond to human language. This is essential for robots that need to interact with humans in a natural way.

Gesture Recognition: Gesture recognition allows robots to interpret human gestures. This can be used for a variety of tasks, such as controlling robots and interacting with objects.

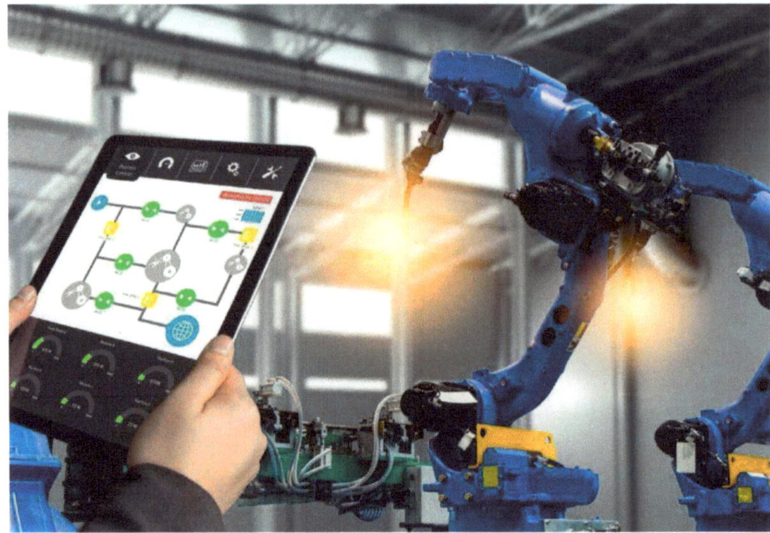

Human-Robot Interaction (HRI) Interfaces

Connectivity And Iot Integration

Connectivity and IoT integration are essential for service robots that need to operate in complex
and dynamic environments. Robots need to be able to communicate with each other, with
humans, and with other devices in order to coordinate their activities and perform tasks
effectively.

Recent Advancements In Connectivity And Iot Integration For Service Robots Include:

5g: 5G is a new wireless communication standard that offers high bandwidth and low
latency. This makes it ideal for robotic applications that require fast and reliable
communication.

Industrial Internet Of Things (Iiot): The IIoT is a network of interconnected devices
that can collect and share data. This data can be used to improve the efficiency and
productivity of manufacturing and industrial operations. Service robots can be integrated
into the IIoT to collect data and perform tasks

Connectivity and IoT Integration

Industrial Ethernet: Industrial Ethernet is a type of Ethernet that is designed for use in industrial environments. It provides reliable and secure communication between devices and systems on the factory floor.

Mqtt: MQTT is a messaging protocol that is designed for use in IoT applications. It is lightweight and efficient, making it ideal for use in service robots.

Safety Systems And Standards

Safety is a top priority for service robots that operate in manufacturing and industrial environments. Robots need to be designed and operated in a way that minimizes the risk of injury to humans and damage to property.

Recent Advancements In Safety Systems And Standards For Service Robots Include:

Collision Avoidance Systems: Collision avoidance systems use sensors to detect obstacles and prevent robots from colliding with them.

Safety Standards: New safety standards are being developed for service robots. These standards will help to ensure that robots are designed and operated in a safe manner.

Iso 10218: ISO 10218 is a set of international standards that define safety requirements for industrial robots. These standards are widely followed by robot manufacturers and users.

Collaborative Robots: Collaborative robots (cobots) are robots that are designed to work safely alongside human workers. Cobots have features such as soft bodies and torque sensing to prevent injuries in the event of a collision with a human worker.

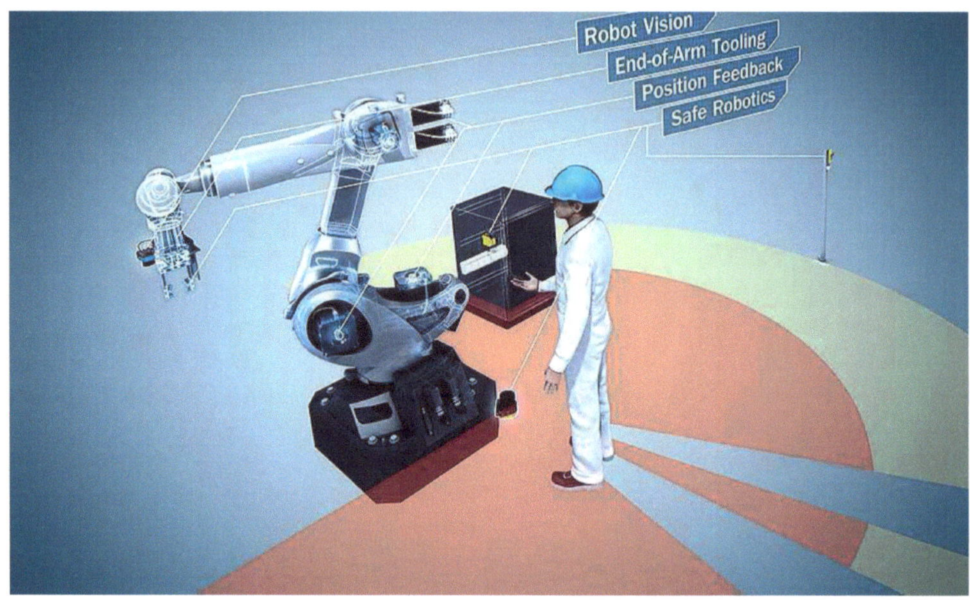

Safety Systems and Standards

The technological advancements in sensing and perception technologies, AI and ML, HRI interfaces, connectivity and IoT integration, and safety systems and standards are enabling the development of new and innovative service robots for manufacturing and industrial applications.

9.5. Challenges And Future Trends

Economic Considerations And Roi

One of the main challenges for service robots in manufacturing and industrial applications is the high initial cost. However, the long-term cost savings can be significant, especially for tasks that are dangerous, repetitive, or time-consuming. For example, a study by the International Federation of Robotics found that industrial robots can reduce labor costs by up to 50%.

Another challenge is the difficulty of calculating the ROI for service robots. This is because there are many factors to consider, such as the cost of the robot, the cost of integration, and the expected productivity gains. However, there are a number of tools and resources available to help manufacturers calculate the ROI for service robots.

Workforce Implications And Training

The deployment of service robots in manufacturing and industrial applications is likely to have a significant impact on the workforce. Some jobs may be lost to automation, while others may be created. However, overall, the impact is expected to be positive, as service robots can free up workers to focus on more complex and skilled tasks.

Workforce Implications and Training

One of the main challenges for the workforce is the need for training on how to use and maintain service robots. This training can be expensive and time-consuming. However, it is essential to ensure that workers are able to use service robots safely and effectively.

Regulatory And Ethical Concerns

There are a number of regulatory and ethical concerns surrounding the use of service robots in manufacturing and industrial applications. For example, there are concerns about the safety of workers and the potential for job displacement. There are also concerns about the privacy and security of data collected by service robots.

It is important to develop clear regulations and ethical guidelines for the use of service robots in manufacturing and industrial applications. These regulations and guidelines should address issues such as safety, privacy, security, and job displacement.

Sustainability And Environmental Impact

Service robots can have a positive impact on sustainability and the environment. For example, service robots can help to reduce energy consumption and waste. Service robots can also help to improve the working environment for workers.

Sustainability and Environmental Impact

However, it is important to design and manufacture service robots in a sustainable way. This includes using recycled materials and renewable energy sources. It is also important to ensure that service robots are disposed of responsibly at the end of their lifespan.

Emerging Innovations And Research Directions

There are a number of emerging innovations and research directions in the field of service robots for manufacturing and industrial applications. For example, researchers are developing service robots that are more intelligent, autonomous, and flexible. Researchers are also developing service robots that are able to collaborate with humans safely and effectively.

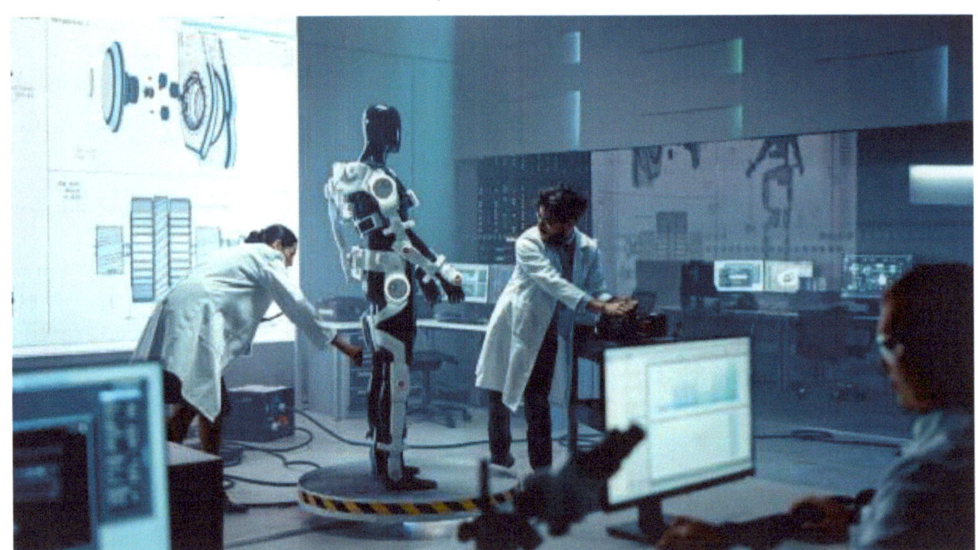
Emerging Innovations and Research Directions

One of the main challenges for emerging innovations in service robots is the need for integration with existing manufacturing and industrial systems. It is important to develop standards and protocols that will facilitate the integration of service robots with existing systems.

Another challenge is the need to ensure that service robots are safe and reliable. Researchers are developing new technologies and methods to improve the safety and reliability of service robots.

Overall, the future of service robots in manufacturing and industrial applications is promising. Service robots have the potential to improve productivity, safety, and sustainability in the manufacturing and industrial sectors. However, there are a number of challenges that need to be addressed before service robots can be widely deployed.

9.6. Case Studies

Case 1: Automotive Manufacturing

Automotive Manufacturing

Real-World Example: Tesla's Fremont, California factory is one of the most automated in the world, with over 1,600 robots performing a variety of tasks, including welding, painting, and assembly. One example is a fleet of mobile robots that transport car bodies between different workstations. These robots are equipped with sensors and cameras that allow them to navigate autonomously and avoid obstacles.

Lessons Learned: Tesla's experience shows that service robots can be used to automate a wide range of tasks in automotive manufacturing, including those that are dangerous or repetitive. This can help to improve productivity, quality, and safety.

Best Practices: When implementing service robots in automotive manufacturing, it is important to carefully consider the specific needs of the application and to choose robots that are designed for the task at hand. It is also important to properly integrate the robots with existing manufacturing systems and to train employees on how to operate and maintain them.

Case 2: Electronics Production

Electronics Production

Real-World Example: Foxconn, a major manufacturer of electronics products, uses a variety of service robots in its factories. One example is a fleet of robots that transport materials between different workstations. These robots are able to navigate autonomously and avoid obstacles, even in crowded and dynamic environments.

Lessons Learned: Foxconn's experience shows that service robots can be used to improve efficiency and reduce costs in electronics production. By automating the transportation of materials, robots can free up workers to focus on more complex tasks.

Best Practices: When implementing service robots in electronics production, it is important to choose robots that are designed for the specific needs of the application. For example, robots that transport delicate electronic components need to be very precise and gentle. It is also important to properly integrate the robots with existing manufacturing systems and to train employees on how to operate and maintain them.

Case 3: Industrial Automation And Assembly Line Robots

Real-World Example: BMW uses a variety of service robots in its assembly lines. One example is a robot that transports car bodies between different workstations. The robot is equipped with a camera and laser scanner that allow it to navigate autonomously and avoid obstacles.

Lessons Learned: BMW's experience shows that service robots can be used to improve efficiency and reduce costs in industrial automation and assembly lines. By automating the transportation of materials, robots can free up workers to focus on more complex tasks. Robots can also be used to perform tasks that are dangerous or repetitive, such as welding and painting.

Industrial Automation and Assembly Line Robots

Best Practices: When implementing service robots in industrial automation and assembly lines, it is important to choose robots that are designed for the specific needs of the application. It is also important to properly integrate the robots with existing manufacturing systems and to train employees on how to operate and maintain them.

Case 4: Warehouse And Logistics Robots

Warehouse and Logistics Robots

Real-World Example: Amazon uses a variety of service robots in its warehouses. One example is a fleet of robots that pick and pack items for shipment. These robots are able to navigate autonomously and identify and pick items from shelves.

Lessons Learned: Amazon's experience shows that service robots can be used to improve efficiency and reduce costs in warehouse and logistics operations. By automating the picking and packing of items, robots can free up workers to focus on other tasks, such as customer service and order fulfillment.

Best Practices: When implementing service robots in warehouse and logistics operations, it is important to choose robots that are designed for the specific needs of the application.

For example, robots that need to pick and pack a wide variety of items need to be very flexible and adaptable. It is also important to properly integrate the robots with existing warehouse management systems and to train employees on how to operate and maintain them.

Case 5: Inspection And Maintenance Robots In Hazardous Environments

Real-World Example: BP uses service robots to inspect and maintain oil and gas platforms. These robots are able to navigate autonomously in hazardous environments and perform tasks such as inspecting welds for cracks and leaks.

Lessons Learned: BP's experience shows that service robots can be used to improve safety and efficiency in hazardous environments. By automating inspection and maintenance tasks, robots can reduce the need for human workers to be exposed to dangerous conditions.

Best Practices: When implementing service robots in hazardous environments, it is important to choose robots that are designed for the specific needs of the application. For example, robots that need to operate in explosive environments need to be certified as intrinsically safe. It is also important to properly train employees on how to operate and maintain the robots in a safe manner.

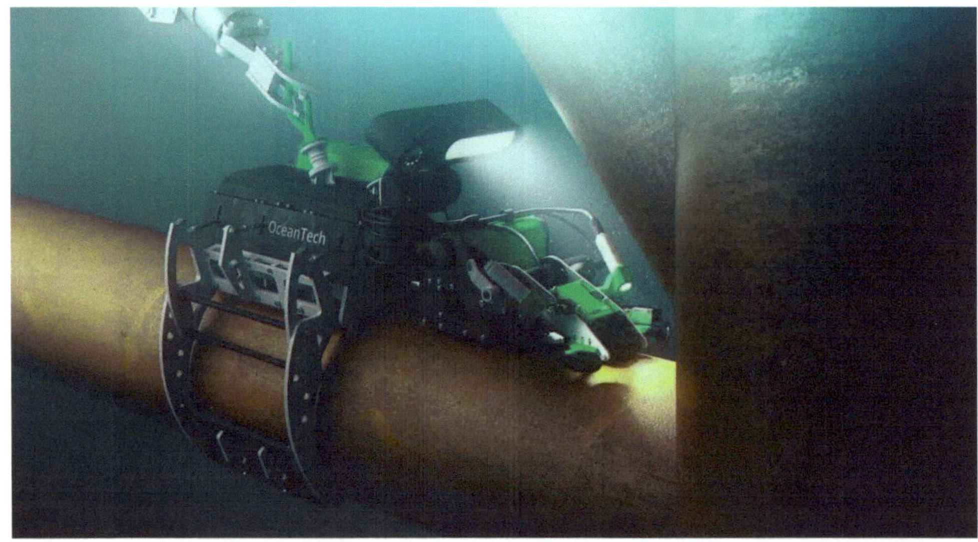

Inspection and Maintenance Robots in Hazardous Environments

Chapter 10: Service Robots In Public Spaces And Hospitality

10.1. Introduction To Service Robots In Hospitality Scope Of Service Robotics

The scope of service robotics is broad and encompasses a wide range of applications. Service robots can be used to perform a variety of tasks, including:

Customer service: Service robots can be used to provide customer service in a variety of settings, such as hotels, restaurants, and retail stores. These robots can answer customer questions, provide recommendations, and even process payments.

Manufacturing: Service robots can be used to automate tasks in manufacturing facilities. These robots can work alongside human workers to perform tasks that are dangerous, repetitive, or difficult for humans to perform.

Logistics: Service robots can be used to perform tasks in the logistics industry, such as picking and packing orders. This can help logistics companies to improve efficiency and reduce costs.

Healthcare: Service robots can be used in a variety of healthcare applications, such as surgery, rehabilitation, and patient care. For example, robots are being used to perform minimally invasive surgery and to help patients with physical therapy.

Education: Service robots can be used in education to provide students with personalized learning experiences and to help teachers with tasks such as grading papers and providing feedback.

Home care: Service robots can be used to provide home care for elderly and disabled people. For example, robots can be used to help people with tasks such as cooking, cleaning, and taking medication.

Scope of Service robotics

These are just a few examples of the many different ways that service robots can be used. As service robots become more sophisticated and more widely available, we can expect to see even more innovative and groundbreaking applications of these robots in the years to come.

The scope of service robotics is also expanding to include new and emerging applications. For example, service robots are being developed to help with disaster relief, environmental monitoring, and space exploration. As these new applications develop, the scope of service robotics will continue to grow.

The future of service robotics is very bright. These robots have the potential to revolutionize a wide range of industries and to make our lives easier and more efficient.

Evolution Of Service Robotics In Hospitality

The evolution of service robotics in hospitality can be traced back to the early 1970s when robots were first used in hotels to deliver food and drinks to guests. Since then, there has been a steady increase in the use of service robots in hospitality, as robots have become more sophisticated and capable.

Early Examples Of Service Robots In Hospitality

Early examples of service robots in hospitality

1974: The first robot to be used in a hotel was a delivery robot called "Hubot". Hubot was used at the Sherry-Netherland Hotel in New York

City to deliver food and drinks to guests.

1980s: Robots began to be used in more hotels, including robots that could clean rooms and vacuum floors.

1990s: Robots began to be used in restaurants, including robots that could take orders from customers and deliver food to their tables.Recent developments in service robotics in hospitality

2000s: The development of artificial intelligence (AI) has led to the development of more intelligent and capable service robots. For example, AI-powered robots can now understand and respond to human language, which allows them to provide more personalized service to guests.

2010s: There has been a growing trend towards the use of service robots in hotels and restaurants. This is due to a number of factors, including the desire to improve efficiency, reduce costs, and provide a more personalized experience for guests.

Examples Of Current Service Robots In Hospitality

Porter robots: These robots can transport luggage and other items around hotels

Cleaning robots: These robots can clean rooms and vacuum floors.

Receptionist robots: These robots can greet guests, answer questions, and provide directions.

Delivery robots: These robots can deliver food and drinks to guests' rooms.

Customer service robots: These robots can answer customer questions, provide recommendations, and even process payments.

Service Robotics In Hospitality: New Launches And Trends For 2023

Bear Robotics Servi Plus: The Servi Plus is a self-driving service robot that can deliver food and drinks, clear tables, and bus dishes. It is designed to help restaurants and hotels improve efficiency and reduce costs.

Pudu Robotics BellaBot Pro: The BellaBot Pro is a multifunctional service robot that can perform a variety of tasks, including delivering food and drinks, welcoming guests, and providing information. It is designed to be interactive and engaging, and it can be customized to match the branding of any business.

Robotario BellaBot: The BellaBot is a humanoid service robot that can deliver food and drinks, welcome guests, and provide information. It is designed to be cute and friendly, and it can be used to create a more unique and memorable experience for guests.

Ubtech Cruzr: The Cruzr is a mobile service robot that can deliver food and drinks, welcome guests, and provide information. It is designed to be easy to use and navigate, and it can be used in a variety of settings, including hotels, restaurants, and hospitals.

Relay Robotics Briggo: The Briggo is a fully autonomous coffee robot that can brew barista-quality coffee in under a minute. It is designed to be placed in high-traffic areas, such as airports, train stations, and office buildings.

These are just a few examples of the many new service robots that are being launched for the hospitality industry in 2023. As the technology continues to develop, we can expect to see even more innovative and useful service robots in hotels and restaurants around the world.

In addition to these new launches, there is also a growing trend of existing service robots being adopted by more and more hospitality businesses. For example, the **Yobot robot,** which was first launched in 2015, is now being used in over **200 hotels** around the world.

The adoption of service robots in the hospitality industry is being driven by a number of factors, including the rising cost of labor, the need to improve efficiency, and the desire to provide guests with a more personalized and engaging experience.

Service robots can help hospitality businesses to save money by automating repetitive tasks, such as delivering food and drinks and cleaning floors. They can also help to improve efficiency by freeing up staff to focus on more complex and customer-facing tasks. Additionally, service robots can help to create a more personalized and engaging experience for guests by providing them with information and assistance in a fun and interactive way.

Overall, the launch of new service robots and the adoption of existing robots by more hospitality businesses is a sign of the growing importance of robotics in this industry. As the technology continues to develop and become more affordable, we can expect to see service robots play an increasingly important role in providing guests with a memorable and enjoyable experience.

Future Of Service Robotics In Hospitality

The future of service robotics in hospitality looks very promising. As robots become more sophisticated and more affordable, we can expect to see even more innovative and groundbreaking applications of these robots in the years to come.

For example, robots could be used to provide personalized concierge services to guests or to help guests with tasks such as packing and unpacking their luggage. Robots could also be used to create more immersive and interactive dining experiences for guests.

Overall, the future of service robotics in hospitality is very bright. These robots have the potential to revolutionize the way that we experience hospitality.

Benefits Of Service Robots In Hospitality:

Improved efficiency: Service robots can help hotels and restaurants to improve efficiency by automating tasks such as cleaning, delivery, and customer service. This can free up staff to focus on other tasks, such as providing personalized service to guests.

Reduced costs: Service robots can help hotels and restaurants to reduce costs by automating tasks that are currently performed by human workers. This can lead to savings on labor costs, as well as other costs such as training and benefits.

Improved guest experience: Service robots can help to improve the guest experience by providing a more personalized and efficient service. For example, robots can be used to greet guests, answer questions, and provide directions. Robots can also be used to deliver food and drinks to guests' rooms, which can save guests time and hassle.

New opportunities for innovation: Service robots can open up new opportunities for innovation in the hospitality industry. For example, robots could be used to create more immersive and interactive dining experiences for guests. Robots could also be used to provide personalized concierge services to guests.

Enhanced cleanliness and hygiene: Service robots can help maintain high cleanliness and hygiene standards in hotels and restaurants, contributing to the well-being and safety of guests.

Increased operational reliability: Robots are less prone to fatigue, ensuring consistent and reliable service 24/7 without the risk of human error.

Reduced labor turnover: The use of service robots can help reduce issues related to staff turnover and absenteeism, as robots do not experience burnout or fatigue.

Language diversity: Service robots can overcome language barriers by offering multilingual communication, catering to diverse guest needs.

Enhanced branding and customer engagement: The deployment of service robots can be a unique selling point, attracting tech-savvy customers and increasing brand recognition in the competitive hospitality sector.

Challenges Of Service Robots In Hospitality:

Cost: Service robots can be expensive to purchase and maintain. This can be a barrier for some hotels and restaurants.

Acceptance: Some guests may be hesitant to interact with service robots. This is because they may feel that robots are impersonal or untrustworthy.

Safety: It is important to ensure that service robots are safe for guests and staff. This means that robots need to be properly designed and tested to avoid accidents.

Regulation: There is currently a lack of clear regulation governing the use of service robots in hospitality. This can make it difficult for hotels and restaurants to deploy robots safely and responsibly.

Overall, the benefits of service robots in hospitality outweigh the challenges. However, it is important to be aware of the challenges and to take steps to address them.

Hotels And Restaurants Are Addressing Service Robot Challenges Through Various Strategies And Innovations.

Cost: Some hotels and restaurants are leasing robots rather than purchasing them outright. This can help to reduce the upfront cost of deploying robots.

Acceptance: Some hotels and restaurants are using robots in a limited capacity, such as in back-of-house areas or in areas that are not open to the public. This can help guests to become more familiar with robots and to build trust.

Safety: Some hotels and restaurants are using robots that have been certified by safety organizations. This helps to ensure that robots are safe for guests and staff.

Regulation: Some hotels and restaurants are working with government agencies to develop clear regulations governing the use of service robots in hospitality. This helps to ensure that robots are deployed safely and responsibly.

As the use of service robots in hospitality continues to grow, we can expect to see even more innovative and effective solutions to the challenges that these robots present.

10.2 Implementation And Case Studies

Implementation Of Service Robots In Hospitality

The implementation of service robots in hospitality can vary depending on the specific needs of the hotel or restaurant. However, there are some general steps that can be followed: Identify the tasks that robots can be used to automate. This could include tasks such as cleaning, delivery, customer service, and concierge services.

- Choose the right robots for the job. There are a variety of service robots available, so it is important to choose the robots that are best suited for the specific tasks that need to be automated.

- Plan the deployment of the robots. This includes determining where the robots will be deployed, how they will be used, and how guests and staff will interact with them.

- Train staff on how to use and interact with the robots. This is important to ensure that staff are able to use the robots effectively and that guests have a positive experience.

- Monitor the performance of the robots and make adjustments as needed. This is important to ensure that the robots are meeting the needs of the hotel or restaurant and that guests are satisfied with the service they are receiving.

Implementation of service robots in hospitality

Case Studies Of Service Robots Revolutionizing Hospitality Industry:

Henn-na Hotel, Japan: Henn-na Hotel, located in Nagasaki, Japan, gained global attention for its innovative use of robots. The hotel implemented robots at various touchpoints, including robotic receptionists, luggage carriers, and in-room assistants. Their approach aimed to create a futuristic, efficient, and unique guest experience. While it received mixed reviews initially, Henn-na Hotel's initiative highlighted the potential of robots in hospitality, especially in roles like check-in, check-out, and concierge services.

Aloft Hotel, Cupertino, California: The Aloft Hotel in Cupertino, California, introduced a robot known as "Botlr" to handle room service deliveries. This robot can navigate through the hotel's corridors, use elevators, and autonomously deliver requested items to guest rooms. By implementing Botlr, the Aloft Hotel improved efficiency, reduced delivery times, and added a tech-savvy, memorable element to the guest experience.

Yotel, New York City: Yotel, a chain of micro-hotels, incorporated a luggage-storing robot named "Yobot" at its New York City location. Yobot can securely store guests' luggage and retrieve it upon request. The use of Yobot has not only reduced the need for human staff to manage luggage but has also become a talking point among guests for its novelty and convenience.

M Social Hotel, Singapore:M Social Hotel in Singapore implemented a robot named "Aura" to enhance in-room experiences. Aura can provide guests with information, answer questions, and even perform basic room service tasks like delivering amenities. This personal assistant robot adds an element of modernity to the guest's stay, providing useful information while reducing the workload on human staff.

These are just a few examples of how service robots are being used in hospitality. As the technology continues to develop, we can expect to see even more innovative and groundbreaking applications of service robots in the hospitality industry.

Efficiency and Excellence: The Varied Roles of Robots in Hotel Operations:

Concierge Robots: Some hotels employ concierge robots to provide guests with information about the hotel's facilities, local attractions, and dining options. These robots can offer recommendations and directions, enhancing the overall guest experience.

Security Robots: Security robots are used in hotels to enhance safety and surveillance. Equipped with cameras and sensors, these robots can patrol the premises, detect anomalies, and alert human security staff when necessary.

Housekeeping Robots: In addition to traditional cleaning robots, some hotels use housekeeping robots for tasks such as making beds, restocking amenities, and tidying up guest rooms. These robots can work in conjunction with human housekeeping staff to streamline the cleaning process.

Bellhop Robots: While porter robots are primarily responsible for transporting luggage, bellhop robots go a step further by escorting guests to their rooms and offering information about hotel services. These robots aim to create a unique and engaging arrival experience.

Room Service Robots: Beyond delivering meals, room service robots can provide guests with room amenities, toiletries, and other requests. They are equipped with compartments for different items and can navigate the hotel's corridors autonomously.

Maintenance and Repair Robots: Maintenance robots are used for tasks such as inspecting and repairing equipment, performing routine maintenance on HVAC systems, and addressing technical issues in guest rooms. Their ability to identify and address problems promptly contributes to overall guest satisfaction.

These examples showcase the diverse roles that service robots play in hotels, going beyond the traditional porter, cleaning, reception, delivery, and customer service functions. As technology continues to advance, hotels are finding creative ways to leverage robots to enhance the guest experience and improve operational efficiency.

Successful Integration Strategies for Service Robots in Hotels:

Enhancing Hotels with Service Robots: Key Integration Strategies

Clearly Defined Objectives: Before implementing service robots, hotels should identify their specific goals. These could include improving guest satisfaction, reducing labor costs, or enhancing operational efficiency. Having clear objectives helps in selecting the right technology and measuring success.

Guest Engagement and Personalization: Service robots can offer personalized interactions with guests. They can greet guests by name, offer information about local attractions, and provide assistance with requests. The key is to strike a balance between automation and maintaining a personal touch.

Training and Guest Awareness: Ensure that staff and guests are aware of how to interact with service robots. Training staff to work alongside robots and educating guests on their capabilities will lead to a smoother integration.

Complementary Human-Technology Interaction: Robots should complement human staff, not replace them. They can handle routine tasks like delivering amenities, room service, or providing directions, allowing human employees to focus on more complex and guest-centric tasks.

Customization and Localization: Customize the robot's appearance and functionality to suit the hotel's brand and guest expectations. Consider language options and regional customs to make guests feel at home.

Maintenance and Technical Support: Regular maintenance and technical support are essential. Establish a plan to address technical issues promptly and provide staff with troubleshooting procedures.

Data Security and Privacy: Ensure that any data collected by the robots, such as guest preferences, remains secure. Guests should have the option to opt-in or out of data collection.

Seamless Integration with Hotel Systems: Integrate robots with the hotel's existing systems, such as property management, room service, and customer relationship management. This ensures a smooth workflow and efficient communication.

Feedback Mechanisms: Create channels for guests and staff to provide feedback on their experiences with service robots. This feedback can help improve the robot's performance and guest satisfaction.

Cost-Benefit Analysis: Evaluate the cost-effectiveness of implementing service robots by considering factors like initial investment, maintenance costs, labor savings, and increased revenue from improved guest experiences.

Regulatory Compliance: Ensure that the use of service robots complies with local and national regulations. Keep abreast of legal and ethical considerations, especially in data privacy and safety.

Marketing and Promotion: Promote the use of service robots as a unique feature of the hotel. Highlight their benefits in marketing materials and online platforms to attract tech- savvy travelers.

Continuous Improvement: Continuously monitor and adapt the use of service robots based on changing guest expectations and technological advancements.

Sustainability and Eco-Friendly Features: Consider the environmental impact of service robots and opt for eco-friendly designs and features where possible.

Accessibility: Ensure that the use of service robots is accessible to all guests, including those with disabilities. Provide alternative options for those who may not be able to use or interact with robots.

Successful integration of service robots in hotels involves a holistic approach, focusing on guest satisfaction, staff training, and the efficient use of technology to enhance the overall hotel experience while maintaining a personal touch.

Successful Integration Strategies for Service Robots in Hotels

Cost-Benefit Analysis

The cost-benefit analysis of service robots in hotels is complex and depends on a number of factors, such as the specific tasks that the robots are being used for, the type of robots that are being used, and the cost of the robots.

Potential Costs Of Service Robots In Hotels:

Initial investment: Service robots can be expensive to purchase and maintain. This can be a barrier for some hotels.

Acceptance: Some guests may be hesitant to interact with service robots. This is because they may feel that robots are impersonal or untrustworthy.

Safety: It is important to ensure that service robots are safe for guests and staff. This means that robots need to be properly designed and tested to avoid accidents.

Regulation: There is currently a lack of clear regulation governing the use of service robots in hospitality. This can make it difficult for hotels to deploy robots safely and responsibly.

Potential Benefits Of Service Robots In Hotels:

Improved efficiency: Service robots can help hotels to improve efficiency by automating tasks such as cleaning, delivery, and customer service. This can free up staff to focus on other tasks, such as providing personalized service to guests.

Reduced costs: Service robots can help hotels to reduce costs by automating tasks that are currently performed by human workers. This can lead to savings on labor costs, as well as other costs such as training and benefits.

Improved guest experience: Service robots can help to improve the guest experience by providing a more personalized and efficient service. For example, robots can be used to greet guests, answer questions, and provide directions. Robots can also be used to deliver food and drinks to guests' rooms, which can save guests time and hassle.

Overall, the cost-benefit analysis of service robots in hotels is positive. Service robots can help hotels to improve efficiency, reduce costs, and improve the guest experience. However, it is important to be aware of the potential costs of service robots before deploying them.

If the benefits of service robots outweigh the costs, then it is likely that a hotel will see a positive return on investment. However, it is important to carefully consider all of the factors involved before making a decision about whether or not to deploy service robots.

It is also important to note that the cost-benefit analysis of service robots in hotels is constantly evolving. As the technology continues to develop and become more affordable, the cost-benefit analysis is likely to become more favorable for hotels.

10.3. Human-Robot Interaction And Customer Experience

The human-robot interaction (HRI) and customer experience (CX) of service robots in hotels is a complex topic. There are a number of factors that can affect the HRI and CX, including the specific tasks that the robots are being used for, the type of robots that are being used, and the way that the robots are integrated into the hotel environment.

Potential Benefits of HRI And CX in Hotels:

Personalized service: Robots can be programmed to provide personalized service to guests. For example, a robot could greet a guest by name and ask them about their preferences.

Efficiency: Robots can help to improve the efficiency of the hotel by automating tasks such as cleaning, delivering food and drinks, and providing customer service. This can free up staff to focus on other tasks, such as providing personalized service to guests.

Convenience: Robots can provide a more convenient experience for guests. For example, a robot could deliver food and drinks to a guest's room at their request.

Potential Challenges Of HRI And CX in Hotels:

Trust: Some guests may not trust robots to provide them with good service. This is because robots are not human and may not be able to understand or respond to human emotions.

Acceptance: Some guests may be hesitant to interact with robots. This is because they may feel that robots are impersonal or even threatening.

Safety: It is important to ensure that service robots are safe for guests and staff. This means that robots need to be properly designed and tested to avoid accidents.

Human-Robot Interaction and Customer Experience

Overall, the HRI and CX of service robots in hotels can be positive. However, it is important to be aware of the potential challenges and to take steps to address them.

Enhancing Service Robot HRI And CX in Hotel Environments:

- Design robots that are friendly and approachable. Robots should be designed with human- like features, such as facial expressions and body language.

- Program robots to be polite and respectful. Robots should be programmed to greet guests by name and to ask them about their preferences.

- Provide guests with information about how to interact with the robots. Hotels should provide guests with information about how to use the robots and how to get help if they need it.

- Train staff on how to use and interact with the robots. Staff should be trained on how to use the robots and how to help guests interact with them.

- Monitor the performance of the robots and make adjustments as needed. Hotels should monitor the performance of the robots and make adjustments as needed to improve the HRI and CX.

By following these tips, hotels can improve the HRI and CX of service robots and provide a better experience for their guests.

Designing User-Friendly Interfaces

Designing user-friendly interfaces for service robots in hotels is important for a number of reasons. First, it can help to improve the HRI and CX of the robots. Second, it can help to reduce the amount of training that guests need to interact with the robots. Third, it can help to make the robots more efficient and productive.

Tips For Designing User-Friendly Hotel Service Robot Interfaces:

- Use clear and concise language. The language used in the interface should be clear and concise, so that guests can easily understand what they need to do.

- Use simple and intuitive controls. The controls in the interface should be simple and intuitive, so that guests can easily use them without having to read any instructions.

- Provide feedback to users. The interface should provide feedback to users so that they know what is happening and whether their actions are successful.

- Use visual cues. Visual cues can be used to help users understand what they are doing and to provide feedback.

- Test the interface with users. It is important to test the interface with users to get feedback and make sure that it is easy to use.

By following these tips, hotels can design user-friendly interfaces for service robots that will improve the HRI and CX, reduce training requirements, and make the robots more efficient and productive.

User-Friendly Service Robot Interfaces For Hotels:

- A robot that delivers food and drinks to guests' rooms could have a simple interface that allows guests to select the food and drinks they want and to specify their room number.

- A robot that provides customer service could have an interface that allows guests to ask questions and get help.

- A robot that cleans rooms could have an interface that allows staff to control the robot and to set cleaning schedules.

These are just a few examples of user-friendly interfaces for service robots in hotels. As the technology continues to develop, we can expect to see even more innovative and user-friendly interfaces.

Ensuring Safety And Comfort

Ensuring the safety and comfort of guests and staff is paramount when using service robots in hotels.

Ensuring Safety and Comfort

Here are some tips:

Choose robots that are designed for safety. Robots should be equipped with safety features such as sensors to detect obstacles and to prevent collisions.

Test the robots thoroughly before deploying them. Robots should be tested in a variety of environments to make sure that they are safe to operate.

Monitor the robots closely. Robots should be monitored by staff to ensure that they are operating safely.

Provide guests with information about the robots. Guests should be informed about the robots and how to interact with them safely.

Train staff on how to use and interact with the robots. Staff should be trained on how to use the robots and how to help guests interact with them safely.

Enhancing Safety and Comfort with Service Robots in Hotels:

- A robot that delivers food and drinks to guests' rooms could be equipped with sensors to detect obstacles and to prevent collisions. The robot could also be equipped with a safety feature that stops the robot if it detects a person in its path.

- A robot that provides customer service could be placed in a safe area where it is unlikely to come into contact with guests or staff. The robot could also be equipped with a safety feature that allows staff to remotely stop the robot if necessary.

- A robot that cleans rooms could be programmed to avoid areas where guests are likely to be present. The robot could also be equipped with a safety feature that stops the robot if it detects a person in its path.

By following these tips, hotels can ensure the safety and comfort of guests and staff when using service robots.It is also important to note that the safety and comfort of guests and staff may vary depending on the specific type of robot being used and the specific tasks that the robot is being used for.

For example, a robot that delivers food and drinks to guests' rooms may pose a different set of safety and comfort concerns than a robot that cleans rooms.

It is important for hotels to carefully consider the safety and comfort of guests and staff when deploying service robots. By doing so, hotels can help to ensure a positive experience for everyone involved.

Measuring Customer Satisfaction

There are a number of ways to measure customer satisfaction with service robots in hotels. Some common methods include:

- **Customer surveys:** Customer surveys can be used to collect feedback on guests' experiences with service robots. Surveys can be administered online, in person, or over the phone.
- **Focus groups:** Focus groups can be used to gather in-depth feedback from a small group of guests. Focus groups can be conducted in person or online.
- **Observational studies:** Observational studies can be used to observe guests' interactions with service robots. This can be done by using hidden cameras or by having staff observe guests.
- **Social media:** Social media can be used to monitor guest feedback about service robots. Hotels can use social media listening tools to track mentions of service robots on social media.

By using a combination of these methods, hotels can get a comprehensive understanding of customer satisfaction with service robots.

Measuring Customer Satisfaction With Service Robots In Hotels:

- A hotel could ask guests to complete a survey after they check out. The survey could ask guests about their overall satisfaction with their stay, as well as their specific experiences with service robots.

- A hotel could conduct a focus group with guests to get their feedback on service robots. The focus group could discuss the benefits and drawbacks of service robots, as well as their suggestions for improvement.

- A hotel could observe guests interacting with service robots. This could be done by using hidden cameras or by having staff observe guests. The observations could be used to identify areas where service robots could be improved.

- A hotel could monitor guest feedback about service robots on social media. This could be done by using social media listening tools to track mentions of service robots on social media. The feedback could be used to identify areas where service robots are meeting or not meeting guest expectations.

- By measuring customer satisfaction with service robots, hotels can identify areas where they can improve the customer experience. This can help hotels to make better decisions about how to deploy and use service robots in the future.

It is also important to note that customer satisfaction with service robots may vary depending on the specific type of robot being used and the specific tasks that the robot is being used for. For example, a robot that delivers food and drinks to guests' rooms may be more popular than a robot that cleans rooms.

Measuring Customer Satisfaction with Service Robots in Hotels

It is important for hotels to track customer satisfaction with service robots over time. This will help hotels to identify trends and to make necessary adjustments.

Chapter 11: Service Robotics Products And Leading Companies

11.1. Introduction To Service Robotics Market

The service robotics market is the market for robots that are designed to perform tasks that are traditionally performed by humans. Service robots can be used in a wide variety of industries, including healthcare, hospitality, logistics, and manufacturing.

Introduction to Service Robotics Market

The Service Robotics Market Is Growing Rapidly, Driven By A Number Of Factors, Including:

The aging population: As the population ages, there is a growing need for robots to help with tasks such as eldercare and rehabilitation.

The increasing adoption of automation: Businesses are increasingly adopting automation to improve productivity and reduce costs. Service robots can play a major role in this trend.

The technological advancements in robotics: Advances in artificial intelligence, machine learning, and robotics hardware are making service robots more sophisticated and capable.

The global service robotics market is expected to reach $113.06 billion by 2030, according to MarketsandMarkets Research. The Asia Pacific region is expected to be the largest market for service robots, followed by North America and Europe.

Some of the key trends in the service robotics market:

AI and ML are enhancing service robots' intelligence and capabilities.

- For example, AI-powered service robots can be used to navigate dynamic environments, interact with humans in a natural way, and learn to perform new tasks on their own.

Advancements in Service Robotics: A Look at Emerging Robot Types

The landscape of service robotics is continually evolving, witnessing the emergence of novel robotic solutions. Among these innovations are robots designed to fulfill diverse roles, such as food delivery, household cleaning, and surgical assistance. These robots exemplify the ever-expanding realm of possibilities within the service robotics sector.

Moreover, the service robotics market has been extending its reach into previously unexplored industries. Industries like agriculture, construction, and mining are now harnessing the power of service robots to improve efficiency, safety, and productivity. These robots are proving to be indispensable tools in challenging environments and tasks.

This dynamic market is characterized by rapid growth and immense potential. A notable trend driving this growth is the pervasive integration of artificial intelligence and machine learning. These technologies are facilitating the development of smarter, more adaptive service robots, enabling them to handle increasingly complex tasks and interact seamlessly with humans.

Advancements in Service Robotics

In summary, the service robotics market is experiencing continuous expansion, both in terms of the capabilities of the robots themselves and their penetration into diverse industries. With the integration of cutting-edge technologies, service robots are poised to redefine the way we work and live, making them a fascinating and promising area of innovation.

Growth Trends And Market Drivers:

The service robotics market is expected to grow from USD 36.2 billion in 2021 to USD 103.3 billion by 2026, at a CAGR of 23.3% from 2021 to 2026.

The growth of the market is driven by a number of factors, including:

The aging population: As the population ages, there is a growing demand for service robots to assist with tasks such as eldercare and rehabilitation.

The increasing adoption of automation: Businesses are increasingly adopting automation to improve productivity and reduce costs. Service robots can play a major role in this trend.

The technological advancements in robotics: Advances in artificial intelligence, machine learning, and robotics hardware are making service robots more sophisticated and capable.

Other factors driving the growth of the service robotics market include:

The growing demand for service robots in developing countries: Developing countries are expected to be a major driver of growth in the service robotics market in the coming years. This is due to factors such as the growing middle class, the increasing urbanization, and the rising demand for healthcare and other services.

The increasing government support for the development and use of service robots: Governments around the world are providing support for the development and use of service robots. This support comes in the form of funding, tax breaks, and other incentives.

Catalysts for the Service Robotics Market Growth :

The increasing adoption of artificial intelligence and machine learning: Artificial intelligence and machine learning are being used to make service robots more intelligent and capable. For example, AI-powered service robots can be used to navigate dynamic environments, interact with humans in a natural way, and learn to perform new tasks on their own.

The development of new types of service robots: New types of service robots are being developed all the time. For example, there are now robots that can deliver food, clean homes, and assist with surgeries.

The expansion of the service robotics market into new industries: Service robots are being used in a wider range of industries than ever before. For example, service robots are now being used in agriculture, construction, and mining.

Overall, the service robotics market is a rapidly growing market with a lot of potential. Service robots are being used in a wide range of industries and are becoming more sophisticated and capable. The increasing adoption of artificial intelligence and machine learning is one of the key trends in the service robotics market.

Understanding popular service robotics products and companies is crucial for various reasons.

To stay informed about the latest developments in the field: The service robotics field is constantly evolving, with new products and companies emerging all the time. By staying informed about the latest developments, you can ensure that you are up-to-date on the latest technologies and trends.

To make informed decisions: If you are considering purchasing a service robot or working with a service robotics company, it is important to have a good understanding of the different options available to you. By knowing about popular service robotics products and companies, you can make more informed decisions based on your needs and budget.

To identify potential investment opportunities: The service robotics market is a rapidly growing market with a lot of potential. By knowing about popular service robotics products and companies, you can identify potential investment opportunities.

Latest Developments in the Field of Service Robotics

Unlocking Opportunities: The Benefits of Staying Informed about Service Robotics:

For individuals considering a career in robotics and entrepreneurs and stakeholders, staying informed about prominent service robotics products and companies holds significant advantages. This knowledge empowers them to make informed decisions about investments and career choices.Aspiring professionals can align their skills and aspirations with market trends, potentially securing rewarding positions in innovative companies.

Business owners, on the other hand, can leverage this information to identify service robots that enhance productivity, reduce operational costs, and gain a competitive edge. In essence, staying abreast of developments in the service robotics sector is pivotal for those pursuing a robotics- focused career or striving to optimize their business operations. This knowledge equips them with the insights needed to make strategic choices that lead to improved efficiency and profitability.

For example, you could use a service robot to automate tasks such as cleaning, delivering goods, or providing customer service.

If you are a healthcare provider, you can use this knowledge to identify service robots that can help you improve patient care. For example, you could use a service robot to assist with surgery, rehabilitation, or eldercare.

If you are a researcher, you can use this knowledge to identify service robots that you can use in your research. For example, you could use a service robot to collect data, conduct experiments, or perform tasks that are dangerous or difficult for humans to perform.

Overall, knowing about popular service robotics products and companies is important for anyone who wants to stay informed about the latest developments in the field, make informed decisions, or identify potential investment opportunities.

Some of the key players in the service robotics market include:

Boston Dynamics: Known for its advanced robotics, including the famous Spot and Atlas robots.

Fetch Robotics: A company specializing in autonomous mobile robots for various applications, particularly in logistics and warehouses.

Universal Robots: A leader in collaborative robot (cobots) technology used in various industries.

Clearpath Robotics: Focuses on providing industrial robots for research and development purposes.

Aethon: Specializes in autonomous mobile robots for hospital and warehouse applications.

Rethink Robotics: Known for its collaborative robot, Baxter, designed for manufacturing and research purposes.

PAL Robotics: A Barcelona-based company that produces humanoid and mobile robots for various applications.

Savioke: Known for its service robots used in hotels and healthcare facilities.

Soft Robotics: Develops soft robotic grippers and automation solutions for handling delicate and irregularly shaped objects.

Blue Ocean Robotics: An innovative company that develops various robotic solutions and focuses on robot-as-a-service models.

Techmetics: Specializes in service robots for hospitality, including room service and concierge robots.

Ava Robotics: Known for its telepresence robots designed for remote collaboration and communication.

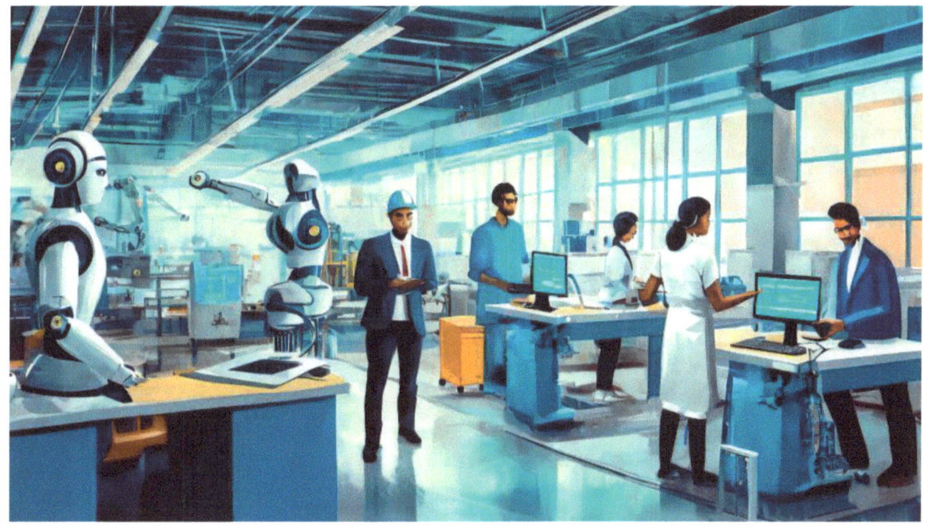

The Increasing Adoption of Automation

Top 10 Indian Robotics Companies Making Waves in 2023:

Simelabs: Simelabs is a cutting-edge robotics company known for its innovative solutions in automation and artificial intelligence.

Queppelin: Queppelin specializes in robotics and AR/VR technologies, offering unique solutions for a wide range of industries.

DiFACTO Robotics: DiFACTO Robotics is a prominent player in the field of industrial robotics, providing automation solutions for manufacturing and logistics.

Kuka Robotics Private Limited: Kuka Robotics is a global leader with a strong presence in India, offering high-quality industrial robots and automation solutions.

ABB: ABB is a multinational corporation, well-known for its comprehensive range of robotics and automation products and services.

Techasoft: Techasoft is an emerging robotics company, focusing on delivering robotic process automation (RPA) and AI-driven solutions.

Gridbots: Gridbots is recognized for its advanced robotic vision systems, catering to applications in agriculture, healthcare, and more.

Wipro PARI Robotics: Wipro PARI Robotics is a collaborative venture providing robotic solutions across various industries, including healthcare and manufacturing.

Hi-Tech Robotic Systemz: Hi-Tech Robotic Systemz specializes in autonomous navigation technology for robotics, contributing to the development of self-driving vehicles and industrial robots.

Solution Analysts: Solution Analysts is an Indian robotics company that offers custom robotics and automation solutions for businesses in various sectors, including healthcare and logistics.

Some examples of popular service robotics products and companies:

iRobot Roomba: A vacuum-cleaning robot (https://www.irobot.com/)

Amazon Astro: A home robot that can provide companionship, security, and entertainment (https://www.amazon.in/)

Diligent Robotics Relay: A healthcare robot that can transport patients, supplies, and equipment (https://www.diligentrobots.com/)

SoftBank Robotics Pepper: A humanoid robot that can provide customer service and information (https://us.softbankrobotics.com/pepper)

Intuitive Surgical: A computer-assisted surgical system that allows surgeons to perform complex procedures through small incisions. (https://www.intuitive.com/en-us)

Intel: It helps robots to perceive their surroundings, understand objects, and act intelligently. (https://www.intel.com/content/www/us/en/homepage.html)

LG Electronics: It uses sensors and AI to navigate busy environments and perform tasks such as serving food, delivering goods, and guiding customers. (https://www.lgcorp.com/)

Samsung Electronics: It uses AI and robotics to perform tasks such as delivery, customer service, and household chores in a variety of settings.

(https://www.samsung.com/in/?cid=in_pd_affiliate_Fanuc)

Kawasaki Heavy Industries: It uses advanced technology to perform tasks such as serving meals, performing medical check-ups, and assisting with manufacturing.
(https://global.kawasaki.com/)

Yaskawa Electric: It use advanced sensors and software to perform tasks such as delivery, inspection, and customer service, in a safe and efficient manner.
(https://www.yaskawa-global.com/)

The service robotics market is still in its early stages of development, but it has the potential to revolutionize many aspects of our lives. Service robots can help to improve productivity and efficiency in businesses, and they can also help to improve the quality of life for people in many ways.

This is just a small sample of the many popular service robotics products and companies that are available. To learn more, I recommend that you research the specific needs and applications that you are interested in.

11.2. Home Assistant Robots And Their Leading Brands

1. Home Assistant Robots Like Smart Speakers.

Home assistant robots like smart speakers are a type of service robot that can be used to control smart home devices, provide information, and entertain users. They are typically voice-activated and can be controlled using natural language commands.

Benefits Of Using Home Assistant Robots Like Smart Speakers:

Convenience: Home assistant robots like smart speakers can make our lives more convenient by making it easy to control smart home devices and access information. For example, we can use a home assistant robot to turn on the lights, adjust the thermostat, and play music without having to get up from the couch.

Productivity: Home assistant robots like smart speakers can also help us to be more productive by automating tasks such as setting alarms, reminders, and appointments. For example, we can tell our home assistant robot to wake us up at a certain time, remind us to take our medication or schedule a meeting.

Accessibility: Home assistant robots like smart speakers can also make our homes more accessible to people with disabilities. For example, people with mobility issues can use a home assistant robot to control smart home devices and access information without having to get up or down.

Companionship: Home assistant robots like smart speakers can also provide companionship to people who live alone or who are isolated. For example, we can talk to our home assistant robot about our day, ask it to play games, or simply have a conversation.

Overall, home assistant robots like smart speakers have the potential to improve our lives in many ways. They can make our lives more convenient, productive, accessible, and enjoyable.

Home Assistant Robots

How Home Assistant Robots Like Smart Speakers Work

Home assistant robots like smart speakers typically use a combination of artificial intelligence (AI), natural language processing (NLP), and machine learning to understand and respond to user requests.

AI: AI allows home assistant robots to learn and adapt over time. For example, an AI- powered home assistant robot can learn your preferences and habits, and then use that information to provide you with more personalized recommendations.

NLP: NLP allows home assistant robots to understand the meaning of human language. For example, an NLP-powered home assistant robot can understand the difference between "turn on the lights" and "turn off the lights."

Smart Speakers

Machine learning: Machine learning allows home assistant robots to improve their performance over time. For example, a machine learning-powered home assistant robot can learn to better understand your voice and to better anticipate your needs.

Examples of Home assistant robots like smart speakers:

Amazon Echo: A smart speaker powered by Amazon's Alexa voice assistant, offering hands-free control of smart devices, music streaming, and information retrieval.

Google Nest Audio: A high-quality smart speaker with Google Assistant, known for its exceptional sound quality and seamless integration with Google ecosystem.

Apple HomePod mini: A compact and powerful smart speaker by Apple, featuring Siri integration and impressive audio performance in a small package.

Sonos One: A versatile smart speaker compatible with both Amazon Alexa and Google Assistant, offering excellent audio quality and multi-room audio capabilities.

JBL Link 500: A robust and voice-activated smart speaker with Google Assistant, delivering powerful sound and a wide range of streaming options.

Amazon Astro: An innovative home robot by Amazon, designed to assist with various tasks, equipped with AI capabilities for navigation and interaction.

LG CLOi: A series of service robots by LG designed for diverse roles, including customer service, cleaning, and delivery, enhancing efficiency and convenience in various industries.

Samsung Bot Handy: A versatile robotic assistant developed by Samsung, capable of recognizing and handling various household tasks, from setting the table to pouring drinks.

SoftBank Robotics Pepper: A humanoid robot designed to engage with humans in social and customer service settings, equipped with advanced sensors and software for natural interactions.

Jibo: An interactive social robot designed for companionship, offering personalized interactions and emotional engagement through conversation and expressive movements.

Future of home assistant robots

Home assistant robots like smart speakers are still under development, but they have the potential to revolutionize the way we interact with our homes. In the future, we can expect to see home assistant robots that are more intelligent, more capable, and more affordable than ever before.

Some of the potential future applications of home assistant robots like smart speakers include:

Home automation: Home assistant robots will be able to automate more and more tasks in our homes, such as cooking, cleaning, and childcare. This will free up our time and energy for other things.

Eldercare: Home assistant robots can provide companionship and assistance to elderly people, helping them to live independently for longer.

Education: Home assistant robots can be used to educate children and adults on a variety of topics.

Entertainment: Home assistant robots can be used to provide entertainment, such as playing games and telling stories.

Overall, the future of home assistant robots like smart speakers is very bright. These robots have the potential to make our lives more convenient, productive, accessible, and enjoyable.Home assistant robots like smart speakers are a type of service robot that can be used to control smart home devices, provide information, and entertain users. They are typically voice-activated and can be controlled using natural language commands.Smart speakers are one of the most popular types of home assistant robots. They are relatively inexpensive and easy to use, and they can be used to control a wide range of smart home devices, such as lights, thermostats, and locks. Smart speakers can also be used to play music, get the news and weather, and set alarms and timers.

2. Kitchen And Cooking Robots

Kitchen and cooking robots are a type of service robot that can be used to automate tasks in the kitchen, such as cooking, cleaning, and food preparation. These robots are typically equipped with a variety of sensors and actuators, which allow them to interact with their environment and perform tasks in a safe and efficient manner.

Examples of kitchen and cooking robots:

Cooking robots: These robots can be used to prepare meals from scratch, or to cook pre- made meals. Some cooking robots can even adjust recipes based on the ingredients that are available.

Food preparation robots: These robots can be used to chop vegetables, dice fruits, and prepare other food items. Some food preparation robots can even mix and blend ingredients to create sauces and dressings.

Cleaning robots: These robots can be used to clean kitchen counters, floors, and appliances. Some cleaning robots can even wash dishes and silverware.

Benefits of using kitchen and cooking robots:

Convenience: Kitchen and cooking robots can make our lives more convenient by automating tasks that we would otherwise have to do ourselves. This can free up our time so that we can spend it on other things, such as spending time with family and friends or pursuing our hobbies.

Productivity: Kitchen and cooking robots can also help us to be more productive by allowing us to cook meals more quickly and efficiently. This can be especially helpful for people who are busy and don't have a lot of time to spend in the kitchen.

Health: Kitchen and cooking robots can also help us to eat healthier by allowing us to prepare meals that are more nutritious and balanced. For example, a cooking robot can be programmed to follow a specific diet plan or to avoid certain ingredients.

Accessibility: Kitchen and cooking robots can also make the kitchen more accessible to people with disabilities. For example, a robot can be used to prepare meals for people who have difficulty using their hands or who are unable to stand for long periods of time.

Overall, kitchen and cooking robots have the potential to make our lives more convenient, productive, healthy, and accessible.

Benefits of Using Kitchen and Cooking Robots

How Kitchen And Cooking Robots Work:

Kitchen and cooking robots typically use a combination of sensors, actuators, and artificial intelligence (AI) to perform their tasks.

Sensors: Sensors allow the robot to perceive its environment and to understand what is happening around it. For example, a cooking robot might use sensors to detect the temperature of food or to identify the ingredients that are available.

Actuators: Actuators allow the robot to interact with its environment and to perform tasks. For example, a cooking robot might use actuators to move its arms and legs, to open and close doors, and to operate kitchen appliances.

AI: AI allows the robot to learn and adapt over time. For example, an AI-powered cooking robot can learn your preferences and habits, and then use that information to prepare meals that you are more likely to enjoy.

Top 10 popular products of Kitchen and cooking robots

Moley Robotics: Robotic kitchen system that can cook and prepare a wide variety of meals. **Nymble:** Home kitchen robot that can cook a variety of meals, from simple snacks to complex dishes.

Chef iQ: Smart pressure cooker that can cook a variety of meals with the touch of a button. **Instant Pot:** Popular multi-cooker that can be used to cook a variety of meals, including rice, soup, stew, and meat.

Philips Air Fryer: Kitchen appliance that cooks food using hot air.

Anova Sous Vide Machine: Kitchen appliance that cooks food by sealing it in a vacuum- sealed bag and then cooking it in a water bath.

Panasonic Bread Maker: Kitchen appliance that automatically mixes, kneads, bakes, and keeps warm bread.

Cuisinart Ice Cream Maker: Kitchen appliance that makes homemade ice cream. **Cuisinart Food Processor:** Kitchen appliance that chops, slices, dices, and blends food. **KitchenAid Stand Mixer:** Kitchen appliance that mixes, kneads, and whips ingredients.

Future Of Kitchen And Cooking Robots:

Kitchen and cooking robots are still under development, but they have the potential to revolutionize the way we cook and eat. In the future, we can expect to see kitchen and cooking robots that are more intelligent, more capable, and more affordable than ever before.

Personalized cooking: Kitchen and cooking robots can be used to cook meals that are tailored to our individual preferences and dietary needs.

Sustainable cooking: Kitchen and cooking robots can be used to cook meals that are more environmentally friendly. For example, a robot can be programmed to use less energy and waste less food.

Future of Kitchen and Cooking Robots

Automated restaurants: Kitchen and cooking robots can be used to automate the entire cooking process in restaurants. This could lead to lower prices and more efficient service.

Overall, the future of kitchen and cooking robots is very bright. These robots have the potential to make our lives easier, healthier, and more sustainable.

3. Security And Surveillance Robots For Homes

Security and surveillance robots for homes are a type of service robot that can be used to monitor and protect homes from intruders. These robots are typically equipped with a variety of sensors and cameras, which allow them to detect and track movement in and around the home.

Examples of security and surveillance robots for homes:

Patrolling robots: These robots can be used to patrol the perimeter of a home and to detect any unauthorized activity.

Monitoring robots: These robots can be used to monitor specific areas of a home, such as the front door or the backyard.

Surveillance robots: These robots can be used to record video footage of the home and its surroundings

Benefits of using security and surveillance robots for homes:

Increased security: Security and surveillance robots can help to increase the security of a home by deterring intruders and by providing early warning of any unauthorized activity.

Peace of mind: Security and surveillance robots can provide homeowners with peace of mind knowing that their home is being monitored and protected.

Convenience: Security and surveillance robots can be monitored remotely, which gives homeowners the ability to check on their homes even when they are not there.

Overall, security and surveillance robots for homes have the potential to make homes safer and more secure.

How Security And Surveillance Robots For Homes Work

Security and surveillance robots for homes typically use a combination of sensors, cameras, and artificial intelligence (AI) to perform their tasks.

Sensors: Sensors allow the robot to perceive its environment and to understand what is happening around it. For example, a security robot might use sensors to detect movement, sound, or heat.

Cameras: Cameras allow the robot to record video footage of its surroundings. This footage can then be used to identify intruders or to monitor suspicious activity.

AI: AI allows the robot to learn and adapt over time. For example, an AI-powered security robot can learn the patterns of normal activity in a home and then use that information to identify unusual or suspicious activity.

Popular Service Robotics Product in Security and Surveillance Domain :

Knightscope K5 Autonomous Security Robot: This robot is equipped with a variety of sensors, including cameras, thermal imaging, and lidar, which allow it to autonomously patrol and monitor large areas. It can also detect and deter intruders, and report incidents to security personnel.

Spot 4-Legged Robot: This robot is highly mobile and can navigate difficult terrain, making it ideal for use in security and surveillance applications. It can be equipped with a variety of sensors, including cameras, lidar, and gas sensors, and can be programmed to follow specific patrol routes or respond to alarms.

Cobalt Robotics Security Robot: This robot is designed to be deployed in indoor environments, such as warehouses and retail stores. It can autonomously patrol and monitor areas, and detect and deter intruders. It can also be used to collect inventory data and perform other tasks.

Flyability Elios 2 Drone: This drone is equipped with a high-resolution camera and can be used to inspect and monitor areas that are difficult or dangerous to reach. It is also equipped with a collision-tolerant cage, making it safe to use in indoor environments.

Airobotic Quadrotors: Airobotic quadrotors are a versatile platform that can be used for a variety of security and surveillance applications. They can be equipped with a variety of sensors, including cameras, thermal imaging, and lidar, and can be programmed to follow specific patrol routes or respond to alarms.

GuardBot Security Robot: This robot is designed to be deployed in outdoor environments, such as construction sites and perimeter security. It can autonomously patrol and monitor areas, and detect and deter intruders. It is also equipped with a variety of sensors, including cameras, thermal imaging, and lidar.

Robotex Guardbot X4 Robot: This robot is a versatile platform that can be used for a variety of security and surveillance applications. It can be equipped with a variety of sensors, including cameras, thermal imaging, and lidar, and can be programmed to follow specific patrol routes or respond to alarms.

Orasis Guardbot X2 Robot: This robot is designed to be deployed in indoor environments, such as warehouses and retail stores. It can autonomously patrol and monitor areas, and detect and deter intruders. It is also equipped with a variety of sensors, including cameras, thermal imaging, and lidar.

Dahua Technology WizMind AI-Powered Surveillance Camera: This camera is equipped with AI capabilities that allow it to detect and track objects and people of interest. It can also be used to generate alerts and notifications when suspicious activity is detected.

Hanwha Techwin Wisenet AI-Powered Surveillance Camera: This camera is equipped with AI capabilities that allow it to detect and track objects and people of interest. It can also be used to generate alerts and notifications when suspicious activity is detected.

Future Of Security And Surveillance Robots For Homes:

Security and surveillance robots for homes are still in their early stages of development, but they have the potential to revolutionize the way we protect our homes. In the future, we can expect to see security and surveillance robots that are more intelligent, more capable, and more affordable than ever before.

Some of the potential future applications of security and surveillance robots for homes include:

More personalized security: Security and surveillance robots can be programmed to learn the specific needs and preferences of homeowners. This could lead to more effective and efficient security measures.

More sustainable security: Security and surveillance robots can be designed to be more environmentally friendly. For example, a robot could be powered by solar energy or other renewable energy sources.

More integrated security: Security and surveillance robots can be integrated with other smart home devices, such as lights, locks, and thermostats. This could create a more holistic and secure home environment.

Overall, the future of security and surveillance robots for homes is very bright. These robots have the potential to make our homes safer and more secure in ways that were not possible before.

11.3. Robot Concierges In Hotels

Robot concierges in hotels are a type of service robot that can be used to provide information and assistance to hotel guests. These robots are typically equipped with a variety of sensors and actuators, which allow them to navigate the hotel and to interact with guests in a natural way.

Examples of robot concierges in hotels:

Information robots: These robots can provide guests with information about the hotel, such as its amenities, services, and room rates.

Assistance robots: These robots can provide guests with assistance with tasks such as checking in and out, finding their rooms, and getting directions.

Entertainment robots: These robots can provide guests with entertainment, such as playing games and telling stories.

Benefits of using robot concierges in hotels:

Convenience: Robot concierges can provide guests with convenient and efficient service. For example, a guest can simply ask a robot concierge for directions to their room, and the robot can lead them there directly.

Personalization: Robot concierges can be programmed to provide personalized service to guests. For example, a robot can remember a guest's preferences and then suggest activities or amenities that the guest might enjoy.

Sustainability: Robot concierges can help hotels to be more sustainable by reducing the need for human staff.

Robot Concierges in Hotels

Overall, robot concierges in hotels have the potential to make hotel stays more convenient, personalized, and sustainable.

How robot concierges in hotels work:

Robot concierges in hotels typically use a combination of sensors, actuators, and artificial intelligence (AI) to perform their tasks.

Sensors: Sensors allow the robot to perceive its environment and to understand what is happening around it. For example, a robot concierge might use sensors to detect movement, sound, or light.

Actuators: Actuators allow the robot to interact with its environment and to perform tasks. For example, a robot concierge might use actuators to move its arms and legs, to open and close doors, and to operate electronic devices.

AI: AI allows the robot to learn and adapt over time. For example, an AI-powered robot concierge can learn the names of guests and their preferences, and then use that information to provide personalized service.

Popular products of Robot Concierges in Hotels:

Bellhop - A robotic luggage delivery service that can transport guests' luggage to their rooms.

Connie - A robot concierge that can check guests in and out, answer questions, and provide recommendations.

Jibo - A social robot that can interact with guests, provide information, and entertain them.

Relay - A robotic waiter that can deliver food and drinks to guests' rooms.

Savioke Relay - A robotic butler that can deliver room service, amenities, and laundry to guests' rooms.

Temi - A telepresence robot that can allow guests to connect with friends and family members who are not at the hotel.

Ubtech Walker - A robotic security guard that can patrol hotel grounds and deter crime.

Xenex Germ-Zapping Robots - Robots that can disinfect hotel rooms and other public spaces to kill germs and reduce the risk of infection.

Yobot - A robotic bellhop that can deliver luggage to guests' rooms and provide information about the hotel and the surrounding area.

Zoe - A robotic concierge that can check guests in and out, answer questions, and provide recommendations.

Future of robot concierges in hotels

Robot concierges in hotels are still in their early stages of development, but they have the potential to revolutionize the way that hotels operate. In the future, we can expect to see robot concierges that are more intelligent, more capable, and more affordable than ever before.

Some of the potential future applications of robot concierges in hotels include:

More personalized service: Robot concierges can be programmed to learn the specific needs and preferences of guests. This could lead to more effective and efficient service.

More sustainable service: Robot concierges can be designed to be more environmentally friendly. For example, a robot could be powered by solar energy or other renewable energy sources.

More integrated service: Robot concierges can be integrated with other hotel systems, such as the reservation system and the room service system. This could create a more seamless and convenient experience for guests.

Overall, the future of robot concierges in hotels is very bright. These robots have the potential to make hotel stays more convenient, personalized, sustainable, and integrated.

Automated check-in/check-out services

Automated check-in/check-out services are a type of service robot that can be used to automate the check-in and check-out process for hotel guests. These robots are typically equipped with a variety of sensors and actuators, which allow them to interact with guests and to perform the check-in and check-out process without the need for human staff.

Automated check-in/check-out services

Examples of automated check-in/check-out services

Robotic concierges: These robots can be used to greet guests and to guide them through the check-in and check-out process.

Future of automated check-in/check-out services:

Automated check-in/check-out services are still in their early stages of development, but they have the potential to revolutionize the way that hotels operate. In the future, we can expect to see automated check-in/check-out services that are more intelligent, more capable, and more affordable than ever before.

Overall, the future of automated check-in/check-out services is very bright. These services have the potential to make hotel stays more convenient, personalized, sustainable, and integrated.

Future of automated check-in/check-out services

11.4 Restaurant And Food Service Robots

Restaurant and food service robots are a type of service robot that can be used to automate tasks in restaurants and other food service establishments. These robots are typically equipped with a variety of sensors and actuators, which allow them to interact with their environment and to perform tasks in a safe and efficient manner.

Restaurant and food service robots

Examples of restaurant and food service robots:

Cooking robots: These robots can be used to prepare meals from scratch, or to cook pre- made meals. Some cooking robots can even adjust recipes based on the ingredients that are available.

Food preparation robots: These robots can be used to chop vegetables, dice fruits, and prepare other food items. Some food preparation robots can even mix and blend ingredients to create sauces and dressings.

Serving robots: These robots can be used to deliver food and drinks to tables, and to collect dirty dishes.

Cleaning robots: These robots can be used to clean tables, floors, and other surfaces in restaurants.

Benefits of using restaurant and food service robots:

Convenience: Restaurant and food service robots can make it more convenient for customers to order and receive food. For example, customers can order food from a self- service kiosk and then have it delivered to their table by a robot.

Efficiency: Restaurant and food service robots can help restaurants to be more efficient. For example, restaurants can reduce the number of staff members needed to prepare and serve food.

Cost savings: Restaurant and food service robots can help restaurants to save money. For example, restaurants can reduce the cost of labor and the cost of providing space for human staff.

Improved quality: Restaurant and food service robots can help to improve the quality of food and service. For example, robots can be programmed to follow precise recipes and to ensure that food is cooked to perfection.

Benefits of using restaurant and food service robots

Overall, restaurant and food service robots have the potential to make dining out more convenient, efficient, cost-effective, and enjoyable.

How restaurant and food service robots work:

Restaurant and food service robots typically use a combination of sensors, actuators, and artificial intelligence (AI) to perform their tasks.

Sensors: Sensors allow the robot to perceive its environment and to understand what is happening around it. For example, a robot might use sensors to detect movement, sound, or light.

Actuators: Actuators allow the robot to interact with its environment and to perform tasks. For example, a robot might use actuators to move its arms and legs, to open and close doors, and to operate electronic devices.

AI: AI allows the robot to learn and adapt over time. For example, an AI-powered robot can learn the recipes that are used in a restaurant and then use that information to prepare food.

Popular Products For Restaurant And Food Service Robots:

Flippy (Miso Robotics): A robotic arm that can cook, fry, and flip food, including burgers, fries, chicken wings, and onion rings.(https://misorobotics.com/)

Sally (Basil Robotics): A robotic pizza maker that can produce up to 100 pizzas per hour. (https://www.saladsbycasey.com/sally-the-robot)
Kiwibot (Kiwibot): A self-driving delivery robot that can bring food from restaurants to customers' doorsteps.(https://www.kiwibot.com/)
Starship Technologies: A company that develops and operates a fleet of self-driving delivery robots.(https://www.starship.xyz/)
Bellabot (Pudu Robotics): A delivery robot that can navigate through crowded restaurants and deliver food to customers' tables. (https://www.pudurobotics.com/product/detail/bellabot)
Pepper (SoftBank Robotics): A humanoid robot that can greet customers, provide information, and take orders. (https://robotsguide.com/robots/pepper)
Cecilia (Cecilia.ai): A robotic barista that can make coffee drinks, smoothies, and other beverages. (https://cecilia.ai/)
Piestro (Piestro): A robotic pizza maker that can produce up to 400 pizzas per hour. (https://www.piestro.com/)
Makr Shakr (Makr Shakr): A robotic bartender that can make cocktails, mocktails, and other drinks. (https://www.makrshakr.com/)

These are just a few of the many companies that are developing and deploying robots for restaurant and food service applications. As the use of robots in the food industry continues to grow, we can expect to see even more innovative and sophisticated products in the years to come.

Future of restaurant and food service robots:

Restaurant and food service robots are still in their early stages of development, but they have the potential to revolutionize the way that we eat out. In the future, we can expect to see restaurant and food service robots that are more intelligent, more capable, and more affordable than ever before.

Some of the potential future applications of restaurant and food service robots include:

More personalized service: Restaurant and food service robots can be programmed to learn the specific needs and preferences of customers. This could lead to more effective and efficient service.

More sustainable service: Restaurant and food service robots can be designed to be more environmentally friendly. For example, a robot could be powered by solar energy or by other renewable energy sources.

More integrated service: Restaurant and food service robots can be integrated with other restaurant systems, such as the point-of-sale system and the inventory management system. This could create a more seamless and convenient experience for customers.

Overall, the future of restaurant and food service robots is very bright. These robots have the potential to make dining out more convenient, personalized, sustainable, and integrated.

11.5. Retail Robots For Customer Assistance

Retail robots for customer assistance are a type of service robot that can be used to provide assistance to customers in retail stores. These robots are typically equipped with a variety of sensors and actuators, which allow them to navigate the store and to interact with customers in a natural way.

Examples of retail robots for customer assistance:

Greeting robots: These robots can be used to greet customers and to provide them with information about the store.

Navigation robots: These robots can be used to help customers find the products they are looking for.

Product recommendation robots: These robots can recommend products to customers based on their past purchases and preferences.

Payment robots: These robots can be used to process customer payments.

Cleaning robots: These robots can be used to clean store floors and other surfaces.

Retail robots for customer assistance

Benefits of using retail robots for customer assistance:

Convenience: Retail robots for customer assistance can make it more convenient for customers to shop. For example, customers can ask a robot for help finding a product or for recommendations.

Efficiency: Retail robots for customer assistance can help stores to be more efficient. For example, stores can reduce the number of staff members needed to provide customer assistance.

Cost savings: Retail robots for customer assistance can help stores to save money. For example, stores can reduce the cost of labor and the cost of providing space for human staff.

Improved customer satisfaction: Retail robots for customer assistance can help to improve customer satisfaction. For example, customers may appreciate the convenience and efficiency of being able to get help from a robot.

Overall, retail robots for customer assistance have the potential to make shopping more convenient, efficient, cost-effective, and enjoyable.

How retail robots for customer assistance work:

Retail robots for customer assistance typically use a combination of sensors, actuators, and artificial intelligence (AI) to perform their tasks.

Sensors: Sensors allow the robot to perceive its environment and to understand what is happening around it. For example, a robot might use sensors to detect movement, sound, or light.

Actuators: Actuators allow the robot to interact with its environment and to perform tasks. For example, a robot might use actuators to move its arms and legs, to open and close doors, and to operate electronic devices.

AI: AI allows the robot to learn and adapt over time. For example, an AI-powered robot can learn the layout of the store and the names of products.

Popular products for Retail robots for customer assistance

Buddy by Blue Frog Robotics: Buddy is a friendly, interactive robot that can help customers with a variety of tasks, such as finding products, providing information, and collecting feedback.

Pepper by SoftBank Robotics: Pepper is a humanoid robot that can greet and engage customers, answer their questions, and provide directions. It is also used for product demonstrations and promotions.

Giraff by Giraff Technologies: Giraff is a tall, mobile robot that can provide customers with a bird's-eye view of the store, helping them to find products and navigate more easily. It can also be used to collect inventory data and provide real-time updates to staff.

Marty by Simbe Robotics: Marty is a shelf-scanning robot that can help retailers to keep track of their inventory and identify out-of-stock items. It can also be used to collect data on customer behavior and product trends.

Xenex LightStrike Germ-Zapping Robots: Xenex robots use ultraviolet (UV) light to disinfect surfaces and kill harmful bacteria and viruses. They are used in a variety of healthcare settings, but they can also be used in retail stores to help keep customers and staff safe.

Promobot: Promobot is a Russian company that produces a variety of humanoid robots, including V4 and PromoBot GO. These robots can be used in retail stores to greet customers, answer their questions, and provide directions. They can also be used for product demonstrations and promotions.

Loomo by Segway-Ninebot: Loomo is a self-balancing robot that can be used to deliver products to customers, collect inventory data, and provide customer assistance. It is still under development, but it has the potential to revolutionize the retail industry.

Aeolus by Aeolus Robotics: Aeolus is a mobile robot that can be used to disinfect and clean retail stores. It uses UV light to kill harmful bacteria and viruses, and it can also be used to dispense disinfectants and other cleaning products.

Kiwibot by Kiwi Campus: Kiwibot is a self-driving delivery robot that can be used to deliver products to customers in retail stores. It is still under development, but it has the potential to make shopping more convenient and efficient for customers.

Relay by Starship Technologies: Relay is a self-driving delivery robot that can be used to deliver products to customers in retail stores and other locations. It is still under development, but it has the potential to revolutionize the way that products are delivered.

These are just a few of the many retail robots that are available on the market today. As technology continues to advance, we can expect to see even more innovative and sophisticated robots being used in retail stores in the future.

Future Of Retail Robots For Customer Assistance:

Retail robots for customer assistance are still in their early stages of development, but they have the potential to revolutionize the way that we shop. In the future, we can expect to see retail robots that are more intelligent, more capable, and more affordable than ever before.

Future of retail robots for customer assistance

Some of the potential future applications of retail robots for customer assistance include:

More personalized service: Retail robots for customer assistance can be programmed to learn the specific needs and preferences of customers. This could lead to more effective and efficient service.

More sustainable service: Retail robots for customer assistance can be designed to be more environmentally friendly. For example, a robot could be powered by solar energy or by other renewable energy sources.

More integrated service: Retail robots for customer assistance can be integrated with other store systems, such as the point-of-sale system and the inventory management system. This could create a more seamless and convenient experience for customers.

Overall, the future of retail robots for customer assistance is very bright. These robots have the potential to make shopping more convenient, personalized, sustainable, and integrated.

11.6. Agricultural And Industrial Service Robots

Agricultural Service Robots : Agricultural service robots are a type of service robot that can be used to automate tasks in agriculture. These robots are typically equipped with a variety of sensors and actuators, which allow them to interact with their environment and to perform tasks in a safe and efficient manner.

Popular products of Agricultural Service Robots:

Milking robots: These robots automatically milk cows, freeing up farmers to focus on other tasks.

Harvesting robots: These robots can harvest a variety of crops, including fruits, vegetables, and nuts.

Weeding robots: These robots can identify and remove weeds from fields, reducing the need for herbicides.

Planting robots: These robots can plant seeds and seedlings with precision, improving yields and reducing costs.

Spraying robots: These robots can apply pesticides and fertilizers to crops in a targeted and efficient manner.

Pruning robots: These robots can prune trees and vines, improving crop quality and reducing labor costs.

Livestock monitoring robots: These robots can monitor the health and welfare of livestock, helping farmers to identify and address problems early on.

Dairy farm robots: These robots can perform a variety of tasks on dairy farms, such as milking cows, cleaning stalls, and feeding animals.

Greenhouse automation robots: These robots can help to automate tasks in greenhouses, such as watering, fertilizing, and harvesting plants.

Field mapping robots: These robots can create detailed maps of fields, which can be used to improve crop management and yields.

Pest control robots: These robots can be used to identify and control pests in crops.

In addition to these specific products, there is a growing trend towards the development of agricultural service robots that can perform multiple tasks. These robots are often equipped with artificial intelligence and machine learning capabilities, which allow them to learn and adapt to their environment.

Agricultural service robots have the potential to revolutionize the agriculture industry by improving efficiency, reducing costs, and making it easier to produce food in a sustainable manner.

Benefits Of Using Agricultural Service Robots:

Increased productivity: Agricultural service robots can help farmers to increase their productivity by automating tasks that are currently performed by human labor.

Reduced costs: Agricultural service robots can help farmers to reduce their costs by reducing the need for human labor.

Improved quality: Agricultural service robots can help farmers to improve the quality of their crops by performing tasks more accurately and consistently than human workers.

Sustainability: Agricultural service robots can help farmers to be more sustainable by reducing their environmental impact.

Overall, agricultural service robots have the potential to revolutionize the way that agriculture is practiced.

How Agricultural Service Robots Work:

Agricultural Service Robots Typically Use A Combination Of Sensors, Actuators, And Artificial Intelligence (Ai) To Perform Their Tasks.

Sensors: Sensors allow the robot to perceive its environment and to understand what is happening around it. For example, a robot might use sensors to detect the presence of weeds, pests, or ripe fruit.

Actuators: Actuators allow the robot to interact with its environment and to perform tasks. For example, a robot might use actuators to move its arms and legs, to apply fertilizer, or to pick fruit.

AI: AI allows the robot to learn and adapt over time. For example, an AI-powered robot can learn to identify weeds more accurately or to pick fruit more efficiently.

Popular Products For Agricultural Robots For Farming And Harvesting

Tortuga AgTech Field Robot: A versatile robot that can be used for a variety of tasks, including planting, weeding, and harvesting. (Website: https://www.tortugaagtech.com/)

Harvest CROO Strawberry Harvester: A robotic strawberry picker that can harvest up to 24,000 strawberries per hour. (Website: https://www.harvestcroorobotics.com/)

FFRobotics Fruit Harvester: A robotic fruit harvester that can be used to harvest a variety of fruits, including apples, oranges, and pears. (Website: https://www.ffrobotics.com/)

Lely Discovery Collector: A robotic manure collector that helps to keep dairy farms clean and hygienic. (Website: https://www.lely.com/solutions/housing-and-caring/discovery-collector/)

Agricultural robots for farming and harvesting

GEA DairyRobot R9500: A robotic milking system that allows cows to be milked on their own schedule.
(Website: https://www.gea.com/en/products/milking-farming-barn/dairyrobot-automated-milking/dairyrobot-r9500-robotic-milking- system.jsp)
BouMatic Gemini Robotic Milker: Another robotic milking system that is popular among dairy farmers.
 (Website: https://boumatic.com/eu_en/product-categories/europe/products/robotics/)
DeLaval VMS V300 Robotic Milker: A third robotic milking system on the list, this one from DeLaval. (Website:https://www.delaval.com/en-us/explore-our-farm-solutions/milking/delaval-vms-series/)
Halter Hexoskin Livestock Monitoring System: A system that uses sensors to monitor the health and well-being of livestock.
 (Website: https://www.halterhq.com/health-and-heat-detection)
FarmWise Carrot Harvester: A robotic carrot harvester that can harvest up to 1 acre of carrots per hour. (Website: https://farmwise.io/)
Lely Vector Automatic Feeding System: A robotic feeding system that automatically delivers feed to livestock. (Website: https://www.lely.com/us/solutions/feeding/vector/)

Future Of Agricultural Service Robots:

Agricultural service robots are still in their early stages of development, but they have the potential to revolutionize the way that agriculture is practiced. In the future, we can expect to see agricultural service robots that are more intelligent, more capable, and more affordable than ever before.

Some of the potential future applications of agricultural service robots include:

More personalized farming: Agricultural service robots can be programmed to learn the specific needs of each farm. This could lead to more efficient and sustainable farming practices.

More sustainable farming: Agricultural service robots can be designed to be more environmentally friendly. For example, a robot could be powered by solar energy or by other renewable energy sources.

More integrated farming: Agricultural service robots can be integrated with other agricultural systems, such as irrigation systems and crop monitoring systems. This could create a more seamless and efficient farming operation.

Overall, the future of agricultural service robots is very bright. These robots have the potential to make agriculture more productive, sustainable, and efficient.

Future of agricultural service robots

Chapter 12: Robotics Case Studies: Transforming Industries And Enriching Lives

12.1. Case Studies And Success Stories

1. Healthcare Revolution: Robots in Medical Settings

Description: This section focuses on the innovative integration of robots in healthcare, transforming the industry with advanced technology.

The Healthcare Frontier: Surgical and Assistive Robots:

- Service robots are revolutionizing healthcare by assisting in surgeries and providing aid to healthcare professionals.
- Surgical robots like the da Vinci Surgical System offer enhanced precision and dexterity, aiding in complex surgical procedures.
- Assistive robots help in patient care, rehabilitation, and tasks that improve the quality of life for individuals with disabilities or the elderly.

Use Cases:

Robots in Surgery: Discuss how robots are assisting surgeons in performing precise and minimally invasive procedures, leading to improved outcomes and faster recovery times.

Patient Care and Assistance: Explain how robots are aiding in patient monitoring, medication delivery, and rehabilitation exercises, enhancing overall healthcare delivery.

Telemedicine and Remote Consultations: Detail how robots facilitate remote consultations and diagnostics, providing healthcare access to remote or underserved areas.

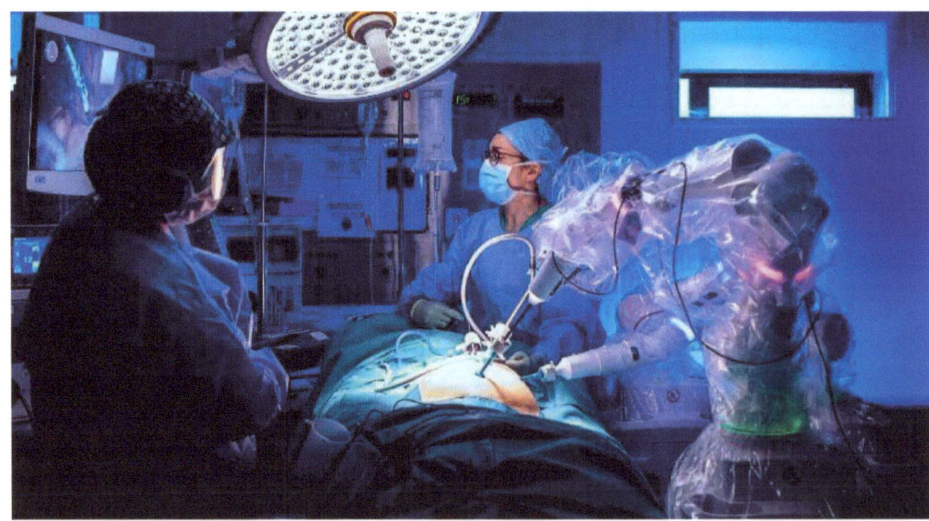

Healthcare Revolution: Robots in Medical Settings

2. Streamlining Logistics: Robots in Warehouses

Description: This section highlights how robots are optimizing warehouse operations and revolutionizing the logistics industry.

Use Cases:
Inventory Management: Discuss how robots are efficiently managing inventory, tracking stock levels, and ensuring timely restocking, reducing delays and improving productivity. **Order Fulfillment and Sorting:** Explain how robots are automating order processing, sorting packages, and expediting the delivery process in warehouses.
Autonomous Guided Vehicles (AGVs): Describe the role of AGVs in moving goods within warehouses, streamlining material flow and minimizing human intervention.

Streamlining Logistics: Robots in Warehouses

3. Transforming Hospitality: Robots in Customer Service

Description: This section delves into the impact of robots on the hospitality sector, enhancing customer experiences.

Hospitality at Its Best: Service Robots in Hotels and Restaurants

- Robots are being employed in the hospitality industry to enhance guest experiences and optimize operations.
- In hotels, robots assist with check-ins, room service, and concierge services, providing a seamless and efficient experience for guests.
- In restaurants, robots can take orders, deliver food, and even assist in the kitchen, augmenting the overall dining experience.

Reception and Concierge Services: Discuss how robots are being used to greet guests, provide information, and enhance hospitality experiences in hotels and resorts.

Room Service and Delivery: Explain how robots are delivering food, amenities, and other services to guest rooms, optimizing efficiency and service quality.

Cleaning and Maintenance: Detail the role of robots in housekeeping, cleaning, and maintenance tasks, contributing to a cleaner and more pleasant environment.

Service Robots in Hotels and Restaurants

Retail Revolution: Robots in Stores and Shopping Centers

- Service robots are transforming the retail landscape by improving customer service and optimizing inventory management.
- Robots in retail stores can guide customers to specific products, provide information, and even facilitate transactions.
- Automated robots in warehouses streamline inventory handling, reducing errors and enhancing efficiency in the supply chain.

4. Cultivating Tomorrow: Robots in Agriculture

Description: This section explores the applications of robots in agriculture, revolutionizing farming practices.

Farming the Future: Agricultural Robots in Crop Management:
- Agricultural robots are reshaping farming practices by automating various tasks involved in crop management.
- Robots equipped with sensors and AI can monitor crop health, automate planting and harvesting, and optimize resource usage.
- Precision farming through robotic technology leads to increased productivity, cost- efficiency, and sustainable agriculture practices.

Use Cases:

Precision Farming: Discuss how robots are aiding in precise planting, fertilizing, and pest control, optimizing crop yield and reducing resource wastage.

Crop Monitoring and Harvesting: Explain how robots equipped with sensors are monitoring crop health and automating the harvesting process, improving efficiency and reducing labor costs.

Greenhouse Automation: Describe how robots are automating greenhouse operations, maintaining optimal conditions for plant growth and enhancing productivity.

5. Building the Future: Robots in Construction

Description: This section focuses on the integration of robots in the construction industry, transforming how we build structures.

Use Cases:

Automated Construction Tasks: Discuss how robots are involved in tasks like bricklaying, welding, and painting, speeding up construction processes while maintaining precision.

Site Safety and Inspection: Explain how robots are enhancing safety by inspecting construction sites, identifying potential hazards, and ensuring compliance with safety standards.

3D Printing and Modular Construction: Detail how robots are enabling 3D printing of building components and facilitating modular construction techniques, revolutionizing construction methodologies.

Building the Future: Robots in Construction

6. Driving Change: Autonomous Vehicles and Transportation:

Description: service robots are making remarkable contributions to the evolution of autonomous vehicles and transportation. These robots are instrumental in enhancing the safety and efficiency of self-driving cars and automated transit systems. They play key roles in tasks like monitoring and maintenance, ensuring that autonomous vehicles remain in optimal operating condition. Service robots also contribute to the broader infrastructure of smart cities, where they assist in traffic management, waste collection, and public transportation. Their presence in autonomous transportation not only improves overall system performance but also shapes a sustainable and convenient future for urban mobility.

Green and Efficient: The Impact of Autonomous Vehicles on Transportation

- Autonomous vehicles, including self-driving cars and drones, are a significant advancement in transportation.
- These vehicles use AI and sensor technology to navigate and operate without human intervention, potentially improving road safety and traffic flow.
- Autonomous transportation has the potential to revolutionize urban mobility and logistics, making transportation more efficient and environmentally friendly.

Autonomous Vehicles and Transportation

In summary, service robots are making a substantial impact across diverse sectors, enhancing efficiency, productivity, and the overall quality of services in the real world. From healthcare to hospitality, retail, agriculture, and transportation, these robots are driving transformative changes and shaping the future of various industries.

12.2. Impact Assessment And User Feedback For Service Robotics

Service robotics refers to the integration of autonomous robotic systems into various service sectors, aimed at improving efficiency, productivity, and the overall quality of service delivery. These robots are designed to perform tasks that traditionally require human intervention, ranging from healthcare and customer service to logistics and maintenance. Assessing the impact of introducing service robotics and gathering user feedback is crucial for understanding the implications on society, businesses, and individuals.

a. Societal Impact Assessment

Employment and Workforce Dynamics:
- Evaluate the effect of service robotics on employment, job displacement, and skills required for the workforce.
- Analyze the potential for new job creation in robot maintenance, programming, and oversight.

Economic Implications:
- Assess the economic impact, including cost savings, revenue generation, and market growth in industries utilizing service robotics.
- Consider the impact on small and medium-sized businesses and startups in terms of competition and market accessibility.

Ethical and Social Considerations:
- Investigate ethical concerns related to service robots, such as privacy, security, and potential biases in decision-making algorithms.
- Examine societal acceptance, trust, and attitudes toward service robots, considering cultural, demographic, and generational differences.

b. Business and Industry Impact Assessment:

Operational Efficiency and Productivity:
- Measure improvements in operational efficiency, accuracy, and speed achieved by integrating service robots into various industry processes.
- Explore cost reduction and resource optimization resulting from automated tasks and processes.

Customer Experience and Satisfaction:
- Evaluate the impact of service robotics on customer satisfaction, loyalty, and overall experience in different sectors, considering feedback and preferences.
- Analyze customer acceptance and adaptation to interacting with robots in service settings.

Market Penetration and Competitive Landscape:
- Study the penetration of service robotics in different industries and its influence on market dynamics, competition, and market share.

- Assess the strategic positioning of companies adopting service robotics in comparison to those relying solely on human-based services.

c. User Feedback and Acceptance

User Experience and Interaction:

- Gather user feedback on their experiences interacting with service robots, including ease of use, effectiveness, and reliability.
- Understand the factors that influence user acceptance or resistance to utilizing service robots.

Usability and Adaptability:

- Assess the adaptability of service robots to varying user needs, preferences, and environmental contexts.
- Identify usability challenges and areas for improvement to enhance user satisfaction and engagement.

Feedback Incorporation and Iterative Development:

- Incorporate user feedback into the iterative design and development of service robots to enhance their functionality, usability, and acceptance.
- Highlight the importance of an ongoing feedback loop to ensure continuous improvements aligned with user expectations and requirements.

By conducting a comprehensive assessment of the impact of introducing service robotics and actively seeking user feedback, we can better navigate the evolving landscape of robotics, ensuring a symbiotic integration of technology that benefits both society and businesses.

d. Precision and Efficiency: Assessing Operational Impact for Service Robotics

The integration of service robotics in various industries has significantly impacted operational processes, aiming to enhance precision and efficiency. This paper delves into the assessment of this impact, exploring the multifaceted aspects of service robotics and their potential to optimize operations.

Operational Challenges Faced by Industries:

- Identifying the operational inefficiencies and challenges faced by industries, which service robotics aims to address.

Common Issues:

- Discussing common problems such as labor shortages, operational errors, and cost management.

Benefits of Service Robotics in Precision and Efficiency:

Enhanced Precision: Detailing how service robots improve precision in tasks, minimizing errors and maximizing accuracy.

Efficiency Gains: Exploring the impact of service robotics on operational efficiency, including speed, resource allocation, and time management.

Cost Reduction: Analyzing how the use of service robots leads to cost savings through optimized operations.

Future Prospects and Recommendations:

Predicting future trends and advancements in service robotics that will further enhance precision and efficiency.

Recommendations: Providing insights and recommendations for organizations looking to implement service robotics to optimize their operations.

e. Financial Implications: Cost-effectiveness of Service Robots

In recent years, service robots have gained significant traction in various industries, revolutionizing the way businesses operate and serve their customers. The integration of service robots has raised pertinent questions regarding their financial implications and cost-effectiveness. This article aims to delve into various topics related to the financial aspects of employing service robots, offering a comprehensive understanding for readers.

Initial Investment and ROI Analysis: Service robots require an initial investment for acquisition, implementation, and customization. This section discusses how businesses evaluate the return on investment (ROI) by considering factors such as cost, operational efficiency, and potential revenue increase associated with the use of service robots.

Operational Costs and Maintenance: Delving into the ongoing operational costs and maintenance expenses is crucial for assessing the overall cost-effectiveness of service robots. It explores expenditures related to power consumption, repairs, software updates, and regular maintenance to ensure optimal functioning and durability of the robots.

Labor Savings and Workforce Optimization: Service robots can often replace or augment human labor, leading to potential cost savings by reducing the need for a large workforce. This section analyzes how service robots optimize labor costs and enhance workforce productivity, thus impacting the financial bottom line positively.

Efficiency and Productivity Gains: Service robots are designed to enhance efficiency and productivity in tasks they are assigned to. This part delves into how increased efficiency due to automation can lead to cost-effectiveness, reduced operational time, and improved output, ultimately resulting in financial benefits.

Quality of Service and Customer Satisfaction: Better quality of service and improved customer satisfaction are outcomes of employing service robots. This section explores how these factors contribute to customer retention, increased sales, and long-term profitability, establishing a connection between financial gains and customer-centric advantages.

Scalability and Flexibility for Growth: Service robots offer scalability and flexibility, allowing businesses to adapt to changing demands and scale their operations efficiently. Discussing how this scalability impacts cost-effectiveness and financial viability helps readers grasp the strategic implications for business growth.

Risk Assessment and Mitigation: Every financial decision involves risks. This section focuses on evaluating the potential risks associated with investing in service robots and discusses strategies to mitigate these risks, ensuring a sound financial approach.

Comparison with Alternative Solutions: Comparing service robots with alternative solutions or conventional approaches is essential for assessing their cost-effectiveness. This part provides a comparative analysis, highlighting the advantages and drawbacks of using service robots over other available options.

By exploring these topics in detail, readers will gain a comprehensive understanding of the financial implications and cost-effectiveness of integrating service robots into their respective industries. This knowledge will enable informed decision-making and encourage the adoption of service robots where financially viable.

f. User Experience and Satisfaction: The Human Perspective in Service Robotics

Service robotics has witnessed remarkable advancements, transforming various industries with the promise of improving efficiency and convenience. However, the success of service robots ultimately hinges on the user experience and satisfaction they provide. In this comprehensive guide, we delve into the various facets of user experience (UX) and user satisfaction in the realm of service robotics, offering insights and recommendations from a human perspective.

Understanding User Expectations:

User-Centered Design: The foundation of a great robotic service experience begins with understanding the needs and expectations of users. User-centered design methodologies, such as personas and user journeys, can be applied to create robots that truly meet user needs.

Human-Robot Interaction:

Natural Communication: Service robots must be capable of natural and intuitive interactions with humans. This includes speech recognition, gesture understanding, and even facial expressions to convey emotions.

Personalization: Customizing robot behavior and responses to individual users can enhance the feeling of personal connection and satisfaction.

Safety and Trust:

Safety Features: Users need to feel secure when interacting with robots. We explore the importance of safety features like obstacle detection, emergency stop mechanisms, and fail- safe protocols.

Trust Building: Strategies for building trust between users and robots, such as transparency in robot capabilities and intentions, are crucial for a positive UX.

User-Friendly Interfaces:

Intuitive Interfaces: The design of user interfaces plays a pivotal role in ensuring ease of interaction. We discuss the importance of intuitive controls and feedback mechanisms.

Accessibility: Addressing the needs of diverse user groups, including individuals with disabilities, is essential for creating inclusive service robotics.

Adaptability and Learning:

Adaptive Behavior: Robots that can adapt to changing situations and user preferences enhance overall satisfaction. This section explores machine learning and AI-driven adaptability.

User Training: Providing users with resources and training to effectively utilize the robot's capabilities can improve their experience.

Ethical Considerations:

Privacy: Maintaining user privacy and data security is a paramount concern. We discuss strategies for responsible data handling.

Ethical Dilemmas: Addressing ethical challenges, such as robot rights and the potential for job displacement, is essential for a holistic perspective.

Feedback and Improvement:

User Feedback Loops: Establishing channels for users to provide feedback helps in continuous improvement. We explore methods for collecting and utilizing user insights.

Iterative Development: Embracing agile development methodologies enables rapid refinement based on user experiences.

Real-World Case Studies:

Examining real-world examples of successful service robots and the factors that contribute to their positive user experiences.

Future Trends:

Emerging Technologies: A glimpse into the future, including trends like emotional intelligence in robots, enhanced sensory capabilities, and multi-robot collaboration.

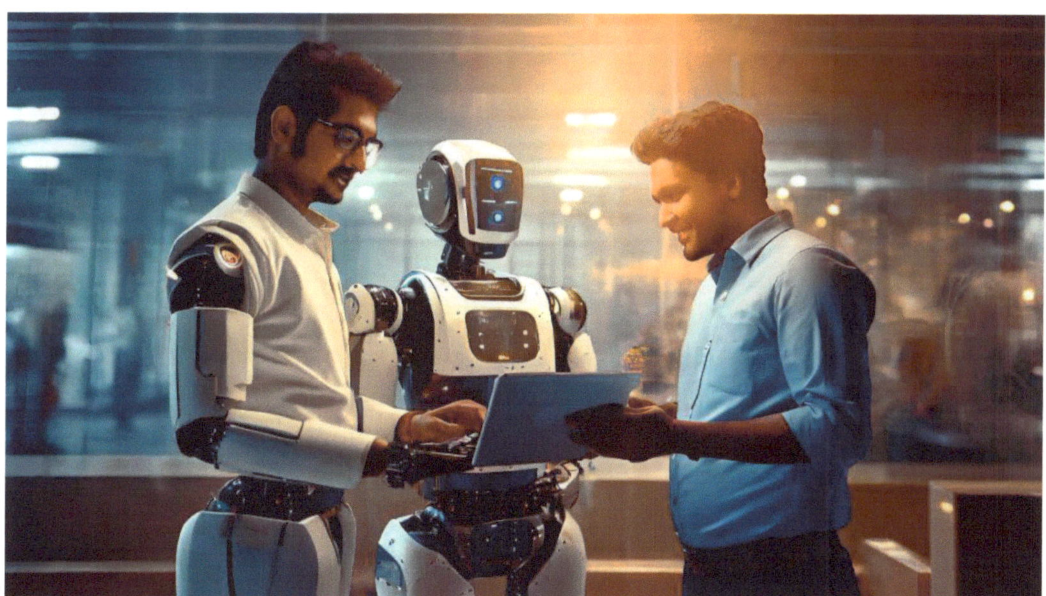

User Experience and Satisfaction: The Human Perspective in Service Robotics

Conclusion:

Service robotics holds immense potential to enhance various aspects of our lives, but this potential can only be fully realized when the human perspective takes center stage. Prioritizing user experience and satisfaction through user-centered design, natural interaction, safety, and ethical considerations will pave the way for a future where service robots truly benefit humanity.

By understanding the nuances of user expectations and continually striving to improve human-robot interaction, we can create a world where service robotics seamlessly integrate into our daily lives, making them more convenient, efficient, and enjoyable.

g. Safety First: Evaluating the Impact on Public and Workers

Service robotics have seen a significant advancement in recent years, offering a wide range of applications that enhance efficiency and convenience across various industries. However, as these technologies become more integrated into our daily lives, ensuring the safety of both the public and workers is of paramount importance. This section delves into the various aspects related to safety in service robotics and aims to evaluate its impact on society.

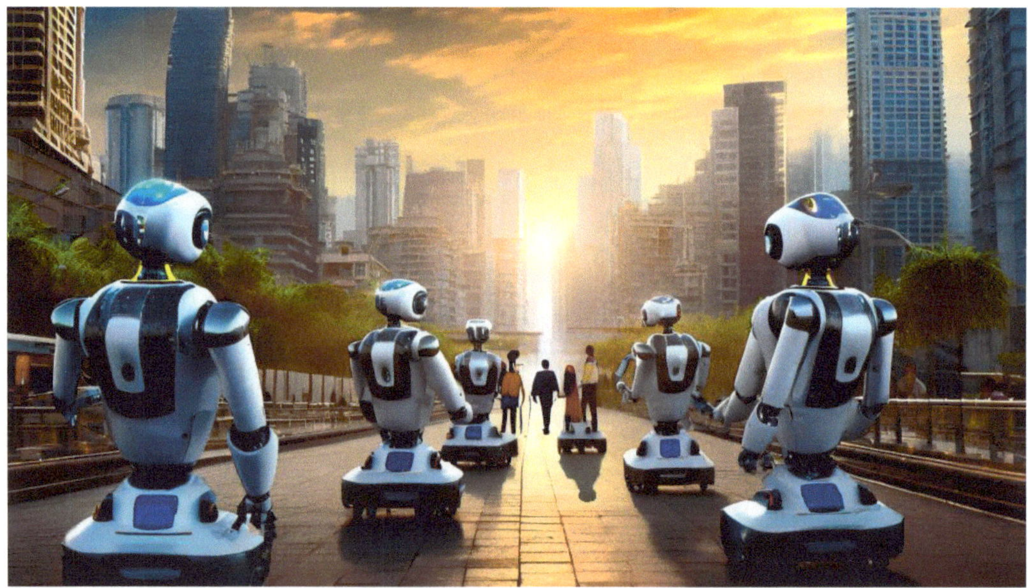
Evaluating the Impact on Public and Workers

Safety Challenges in Service Robotics:
Identify potential safety risks associated with service robots, including collision hazards, software malfunctions, and human-robot interaction issues.
Discuss how these challenges may impact both the public and workers.

Regulatory Framework and Standards:
Overview of existing regulations and standards related to safety in service robotics (e.g., ISO 13482:2014).

Discuss the importance of compliance with these standards to ensure safe deployment and operation of service robots.

Safety Measures and Protocols:

Detailed safety measures that can mitigate risks associated with service robots, such as sensors for object detection, emergency stop mechanisms, and fail-safes. Highlight the significance of proper maintenance, regular inspections, and ongoing training for operators to uphold safety protocols.

Human-Robot Interaction (HRI) Safety:

Explore the importance of designing service robots with safe interaction features to minimize risks to humans.

Discuss techniques like safety-rated monitored stops and speed limitations to ensure safe human-robot collaboration.

Case Studies and Success Stories:

Showcase real-world examples of how implementing robust safety measures has been successful in integrating service robots into various industries while ensuring the safety of the public and workers.

Public Perception and Acceptance:

Discuss how public perception of safety influences the adoption and acceptance of service robots.

Highlight strategies to build public trust in service robotics through transparent communication about safety measures and benefits.

Future Prospects and Challenges:

Discuss anticipated advancements in service robotics and their potential impact on safety. Address potential challenges and how researchers, manufacturers, and policymakers can proactively address them.

Summarize the importance of prioritizing safety in service robotics for the well-being of both the public and workers. Emphasize the need for a collaborative effort involving stakeholders to ensure responsible development and deployment of service robots.

h. Shaping Policies: Legal and Ethical Implications

Service robotics is a rapidly evolving field with significant potential to revolutionize various industries, from healthcare and manufacturing to hospitality and transportation. As these robots become increasingly integrated into our daily lives, it is imperative to address the legal and ethical implications surrounding their use. This article explores the key topics and considerations in shaping policies for service robotics from a reader's perspective.

Shaping Policies: Legal and Ethical Implications

Safety and Liability:

Product Liability: Manufacturers and developers must ensure their robots are safe for human interaction and use. This includes addressing issues like software glitches, hardware malfunctions, and potential harm to users.

Legal Framework: Policymakers need to establish clear liability frameworks that determine who is responsible in case of accidents or damage caused by service robots. This involves defining the roles of manufacturers, operators, and users in ensuring safety.

Privacy and Data Protection:

Data Collection: Service robots often collect vast amounts of data about their environment and users. Policies should address how this data is collected, stored, and used.

Consent: Users' consent for data collection should be obtained transparently, and they should have control over what data is shared and for what purposes.

Data Security: Regulations must require stringent data security measures to prevent unauthorized access or breaches that could compromise sensitive information.

Autonomy and Decision-Making:

Ethical Decision Algorithms: Service robots may make autonomous decisions, such as in healthcare or autonomous vehicles. Policies must ensure that these algorithms are programmed with ethical guidelines, prioritizing human well-being and safety.

Human Oversight: There should be a balance between robot autonomy and human control, particularly in critical situations where human judgment is essential.

Employment and Labor Market Impact:

Job Displacement: Policymakers should anticipate the potential displacement of workers due to automation and develop strategies for retraining and reskilling the workforce.

Labor Rights: Regulations should address the rights of workers who collaborate with or are supervised by robots, ensuring fair working conditions and compensation.

Ethical Considerations in Specific Domains:

Healthcare: Regulations should define the scope of robot involvement in healthcare, considering issues of patient consent, medical ethics, and the role of healthcare professionals.

Education: Policies should guide the integration of robots in education, ensuring that they enhance learning experiences while upholding ethical standards.

Military and Defense: The use of service robots in warfare raises questions about accountability, proportionality, and adherence to international humanitarian law.

Accessibility and Inclusivity:

Universal Design: Developers and policymakers should prioritize inclusive design principles to ensure that service robots are accessible to people with disabilities.

Affordability: Policies should encourage affordability to avoid exacerbating existing inequalities in access to technology.

International Collaboration and Standards:

Global Harmonization: Collaborative efforts on international standards are vital to avoid fragmented regulations and facilitate the global deployment of service robots. **Ethical Norms:** Policymakers should engage in international dialogues to establish ethical norms that guide the use of service robots on a global scale.

Public Awareness and Education:

Public Engagement: Policymakers should promote public awareness and engagement to ensure that the regulations reflect societal values and concerns.

Education: Initiatives to educate the public about the capabilities and limitations of service robots can help build trust and reduce fear.

The development and deployment of service robots hold great promise, but they also present complex legal and ethical challenges. Shaping policies in this field requires a multidisciplinary approach that considers safety, privacy, ethics, and the broader societal impact. Striking a balance between innovation and regulation is crucial to realizing the full potential of service robotics while safeguarding human well-being and rights.

i. Adaptive Behavior and Decision-making:

Context Awareness: Service robots should be designed to understand and adapt to different contexts and environments. They need to interpret real-time data, assess the situation, and make informed decisions to provide optimal service.

Dynamic Task Optimization: The ability of a robot to dynamically optimize its tasks based on feedback and changing circumstances is vital. This ensures efficient resource utilization and improved task completion rates.

Human-like Interaction: Striving for more human-like interaction involves understanding social cues, emotions, and gestures. Robots can adapt their behavior to match social norms, enhancing the user experience.

j. Incremental Learning and Model Updates:

Continuous Model Training: Service robots must undergo continuous training using new data and experiences. Incremental learning techniques ensure that the robot can adapt to evolving tasks and improve its performance based on changing requirements.

Update Integration: Regular software and hardware updates are essential for service robots. These updates can include bug fixes, security enhancements, and feature additions, ensuring that the robot remains up-to-date and capable of delivering high-quality service.

k. Robustness and Error Handling:

Feedback-based Error Correction: Utilizing feedback to identify errors and malfunctions allows for quick and targeted error correction. This approach enhances the robot's robustness and reliability in carrying out its tasks.

Failure Analysis and Recovery: Service robots need to analyze failures and implement recovery strategies autonomously. Learning from past failures helps in preventing similar issues in the future and improving the robot's overall performance.

l. Ethical and Privacy Considerations:

Data Privacy Safeguards: Implementing robust data privacy measures to ensure that the collection and utilization of user data are compliant with privacy regulations and respect individual privacy rights.

Bias Mitigation: Addressing biases in data and algorithms to ensure fair and unbiased interactions with users from diverse backgrounds and demographics, promoting inclusivity and equity.

m. Privacy in Service Robotics:

Data Collection and Storage: Service robots often collect and store personal data to perform tasks efficiently. However, ensuring privacy entails clear policies regarding what data is collected, how it is used, and who has access to it. It's crucial to anonymize and securely store sensitive information to mitigate potential breaches.

Informed Consent: Users should be well-informed about the data being collected by service robots and must give explicit consent. Transparent communication about the types of data collected, purpose, and potential risks involved is essential. Users should have the right to opt-out or have their data deleted.

Minimization of Data: Service robots should follow the principle of data minimization, collecting only necessary information for their intended functions. Unnecessary data collection should be avoided to reduce privacy risks and potential misuse.

Data Security: Implementing robust security measures to protect the data collected is fundamental. Encryption, access control, and regular security audits are necessary to safeguard sensitive information from unauthorized access or cyber-attacks.

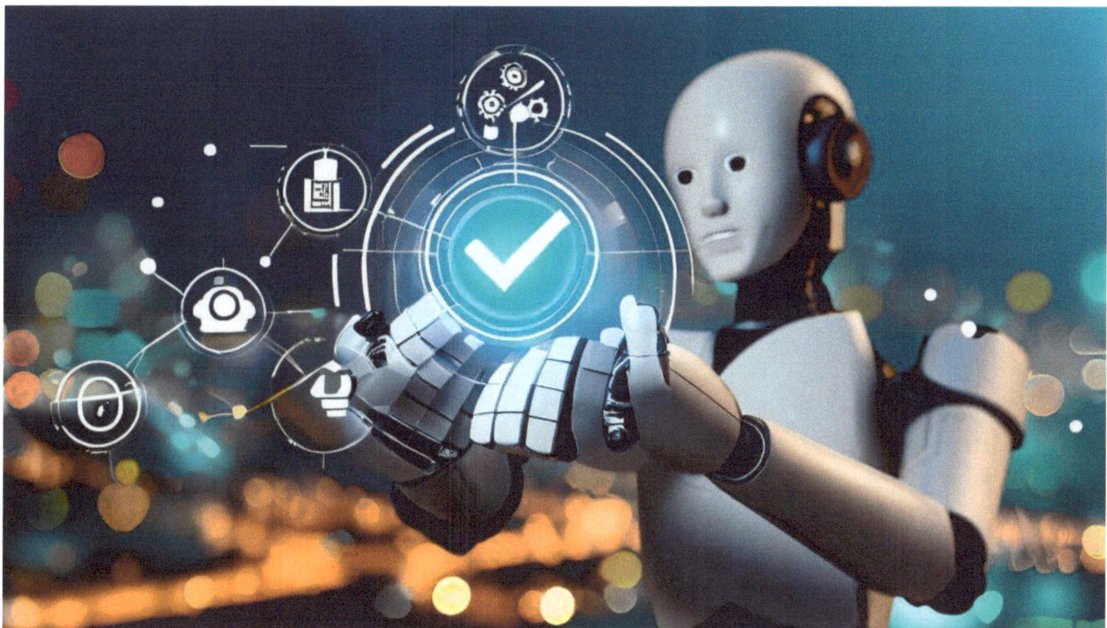

Ethical Considerations: Privacy, Consent, and Responsibility

n. Consent in Service Robotics:

Voluntary Participation:Users should willingly choose to engage with service robots, without any coercion or pressure. Consent should be given without manipulation and with a clear understanding of the implications.

Revocable Consent:Users should have the right to withdraw their consent at any time. Service robots should provide easy mechanisms for users to opt-out of services and have their data deleted. **Age and Capacity Consideration**:Special attention should be given to obtaining consent from minors or individuals with diminished capacity to understand the implications of their actions. Guardians or legal representatives should provide consent on their behalf.

o. Responsibility in Service Robotics:

Duty of Care: Developers and manufacturers of service robots have a duty to ensure that their creations are safe, reliable, and designed to minimize harm. Continuous monitoring and updates should be conducted to improve safety.

Ethical Algorithm Design: Algorithms governing the behavior of service robots should be designed with ethical considerations, ensuring they prioritize human well-being, inclusivity, and fairness. Bias mitigation should be a key focus.

Transparent Accountability: Stakeholders involved in the development, deployment, and use of service robots should be accountable for their actions and decisions. Transparency in operations, decision-making processes, and outcomes is essential to build trust.

The ethical integration of service robotics necessitates a careful balance between technological advancements and the protection of privacy, the assurance of consent, and the fulfillment of responsibilities. Adhering to clear guidelines and ethical principles is imperative for a society that benefits from the advantages of service robotics while upholding fundamental human rights and values.

12.3. Collaborative Innovation: The Power Of Interdisciplinary Teams

Service robotics is a rapidly evolving field that encompasses a wide range of applications, from healthcare and eldercare to logistics and customer service. To drive innovation in this dynamic sector, collaborative efforts among interdisciplinary teams are crucial. This article delves into the significance of interdisciplinary collaboration and how it fosters innovation in the realm of service robotics.

a. Understanding Interdisciplinary Collaboration

Interdisciplinary collaboration refers to the integration of diverse expertise, perspectives, and methodologies from multiple fields to tackle complex problems. In the context of service robotics, this involves pooling together knowledge from various disciplines like engineering, computer science, psychology, design, and more.

Benefits of Interdisciplinary Collaboration

Diverse Perspectives: Integrating insights from different disciplines provides a comprehensive understanding of the problem at hand.

Enhanced Creativity and Innovation: Various viewpoints stimulate creativity, leading to innovative solutions that might not be possible within a single discipline.

Holistic Problem Solving: Interdisciplinary collaboration enables a holistic approach to problem-solving, considering technical, social, and ethical aspects.

Fusion of Expertise: Key Players in Interdisciplinary Teams

Engineers and Technologists: Engineers contribute technical skills vital for developing the hardware, software, and mechatronic systems that underpin service robotics.

Computer Scientists: Computer scientists bring expertise in algorithms, artificial intelligence, and machine learning essential for the intelligent functioning of service robots.

Behavioral Scientists and Psychologists: Understanding human behavior, emotions, and responses is critical for designing robots that can interact with people in a natural and effective manner.

Designers and UX Experts: Designers play a crucial role in making robots user-friendly, aesthetically pleasing, and functional, ensuring a positive user experience.

Designers and UX Experts

Challenges and Solutions in Interdisciplinary Collaboration

Communication Barriers: Effective communication is essential; establishing a shared language and fostering a culture of open communication can mitigate this challenge.

Integration of Diverse Ideas: Encouraging active participation and providing platforms for idea sharing and feedback helps in integrating diverse ideas seamlessly.

Project Management and Coordination: Implementing efficient project management strategies, defining roles clearly, and ensuring a coordinated effort are vital for success.

Case Studies: Successful Collaborative Innovations

Assistive Robotics for Elderly Care: An interdisciplinary team developed a robotic companion that enhances the quality of life for the elderly by providing assistance in daily activities and companionship.

Autonomous Delivery Robots for E-commerce: Engineers, computer scientists, and designers collaborated to create autonomous delivery robots, streamlining last-mile delivery for e-commerce companies.

b. Future Outlook and Conclusion

Anticipated Trends:

- The future of service robotics will witness deeper integration of AI, improved natural language processing, enhanced mobility, and a focus on ethical and societal considerations.
- Collaborative innovation through interdisciplinary teams is the bedrock of progress in service robotics. Embracing diverse expertise and fostering effective collaboration will drive the industry towards solutions that revolutionize the way robots interact with and serve humanity. The key lies in a collective commitment to advancement and the power of collaboration.

Future Trends and Innovations AI-Powered Personalization:

AI-powered personalization in education harnesses artificial intelligence algorithms to tailor learning experiences for individual students. By analyzing students' preferences, learning styles, and progress, it optimizes curriculum and provides targeted recommendations. This approach ensures a customized educational journey, enhancing engagement and learning outcomes. It marks a shift towards adaptive and efficient learning, catering to the unique needs of each learner. Ultimately, AI-powered personalization strives to maximize student potential and foster a more effective educational environment.

AI-Powered Personalization

Intelligent Curriculum Adaptation: AI-powered personalization involves creating adaptive learning experiences for each student. Intelligent curriculum adaptation uses algorithms to tailor the educational material, pace, and difficulty level to match the individual student's learning style, preferences, and capabilities. This ensures a more effective and efficient learning process.

Learning Analytics for Individual Students: Utilizing AI and data analytics, educators can gain insights into each student's learning patterns, strengths, weaknesses, and engagement levels. Learning analytics help in understanding how students interact with educational content, enabling educators to make data-driven decisions to enhance the learning experience and provide personalized support.

Predictive Learning Pathways: AI algorithms can predict a student's future learning trajectory based on their past performance, preferences, and behaviors. By analyzing historical data and patterns, predictive learning pathways can suggest optimal educational routes, resources, and activities to maximize learning outcomes and help students reach their full potential.

12.4. Enhancing Education With Humanoid And Social Robots

Humanoid and social robots represent a significant advancement in robotics, emulating human-like features and behaviors. These robots are designed to interact, engage, and communicate with humans in a socially intuitive manner. They possess capabilities such as facial expressions, gestures, and verbal communication that enable natural interactions. Humanoid robots are being increasingly integrated into various domains, including education to enhance human-machine interactions and provide a more relatable and empathetic experience. Their potential to bridge the gap between technology and human emotion holds promise for a wide range of applications and societal impact.

Robot Companions for Students: Humanoid and social robots designed as companions for students can provide emotional support, motivation, and engagement. These robots can assist in learning activities, answer questions, and even engage in casual conversations. Their presence can create a friendly and comfortable learning environment, encouraging active participation and collaboration.

Emotional Intelligence in Robots: Robots with emotional intelligence are capable of recognizing, understanding, and responding to human emotions. This trait allows them to tailor their interactions based on the emotional state of the student, providing appropriate encouragement, empathy, or assistance. Emotional intelligence in robots enhances their ability to establish rapport and form meaningful connections with students.

Social Skills Development: Humanoid and social robots can be utilized to teach and improve social skills such as communication, teamwork, and conflict resolution. Through interactive scenarios and role-playing exercises, these robots can guide students in understanding social norms, etiquettes, and effective interpersonal interactions, preparing them for real-world social environments.

a. Remote Learning and Telepresence:

Remote Learning and Telepresence in service robotics revolutionize education by enabling virtual classroom experiences. Through robotic telepresence, students can participate in real-time classes, discussions, and activities from a distance. This innovative approach bridges geographical gaps, enhances engagement, and accommodates various learning styles, ensuring inclusivity and accessibility in education. Teachers can interact seamlessly with remote students, creating an interactive and collaborative learning environment. Ultimately, remote learning and telepresence hold immense potential to reshape the future of education.

Tele-Education Using Robots: Tele-education using robots involves employing robots as a medium for remote learning. These robots can facilitate live-streamed lessons, virtual classroom participation, and interactive discussions, bringing the classroom experience to remote or home-bound students. They enable real-time engagement and interaction, bridging the physical distance between educators and learners.

Remote Learning and Telepresence

Remote Presence for Absent Students: In cases where students cannot attend school physically due to illness or other reasons, robots can act as their remote presence in the classroom. Equipped with cameras, microphones, and screens, these robots allow absent students to virtually attend classes, interact with peers, and actively participate in the learning process, minimizing the impact of their absence.

Global Collaboration in Education: Robotic telepresence can facilitate global collaboration and cultural exchange in education. Students from different geographical locations can connect and collaborate on projects, share experiences, and gain diverse perspectives through virtual interactions using robots. This fosters a broader understanding of the world and promotes cultural awareness and tolerance.

These future trends and innovations in educational service robotics have the potential to significantly transform the learning landscape, making education more personalized, engaging, and accessible to students, regardless of their location or learning abilities. AI-powered personalization, humanoid and social robots, and remote learning technologies are paving the way for an inclusive and technologically enriched educational experience.

b. Case Studies and Success Stories - Introduction:

Case studies and success stories in the field of service robotics in education serve as valuable real- world illustrations of the transformative potential of robotic technology.

These narratives showcase how service robots are making a significant impact on the educational landscape, benefiting both students and educators alike. In this section, we will explore various case studies and success stories that highlight the practical applications, outcomes, and positive experiences associated with integrating service robots into educational settings.

Robotic Learning Labs: Robotic learning labs represent a cutting-edge approach to education, where universities harness the power of service robots to enhance the learning experience. These labs serve as innovation hubs, where research and development efforts converge to create state-of-the-art robotic systems designed to assist and educate students. In this section, we will delve into the world of robotic learning labs, examining how universities are leveraging service robots, the research they are conducting, and the profound impact these initiatives are having on education. We will also hear from students who have benefited from these pioneering educational technologies.

Universities using service robots:Universities across the globe have embraced service robots as an integral part of their educational infrastructure. These robots serve various roles, from providing on- demand tutoring to assisting in research projects. We will explore specific universities and institutions that have adopted service robots and the innovative ways in which they are being utilized to support learning and research.

Research outcomes and impact:The integration of service robots in universities has led to groundbreaking research outcomes. These robots facilitate data collection, experimentation, and analysis, contributing to advancements in fields such as artificial intelligence, human-robot interaction, and educational psychology. We will delve into the research projects enabled by these robots and the broader impact on academic knowledge.

Student testimonials:To gain insights into the effectiveness of service robots in education, we will feature testimonials from students who have interacted with these robots. Their firsthand experiences will shed light on the benefits, challenges, and overall impact of robotic assistance on their educational journeys.

c. K-12 Education Initiatives

K-12 education initiatives involving service robots represent a forward-thinking approach to enhancing primary and secondary education. These initiatives introduce students to robotics at an early age, fostering both STEM (Science, Technology, Engineering, and Mathematics) skills and a sense of wonder about technology. In this section, we will explore K-12 programs that integrate service robots into the classroom, examine the improvements in academic performance, and gather feedback from teachers and parents on the outcomes of these initiatives.

Robotics in primary and secondary schools:Service robots are finding their way into classrooms, bringing interactive and experiential learning opportunities to students in primary and secondary schools. We will explore how robots are integrated into the curriculum, the types of robots used, and the educational benefits they offer to students.

 One of the key goals of K-12 education initiatives involving service robots is to enhance academic performance. We will examine data and studies that demonstrate how robot-assisted learning positively affects students' academic achievements, including improvements in test scores and subject comprehension.

Teacher and parent feedback: To provide a comprehensive view of the impact of service robots in K-12 education, we will gather feedback from both educators and parents. Teachers will share their experiences with integrating robots into their teaching methods, while parents will reflect on the changes they observe in their children's learning habits and outcomes.

d. Online Learning Platforms

Online learning platforms have revolutionized education, making it more accessible and flexible. In recent years, the integration of service robots into Massive Open Online Courses (MOOCs) and other online learning environments has opened up new possibilities for engaging and supporting learners in virtual spaces. In this section, we will explore how service robots are being integrated into online learning platforms, their impact on student engagement and course completion rates, and the scalability of robot-assisted online courses.

Online learning platforms

Robotics integration in MOOCs: MOOCs have become a popular way for individuals to access high-quality education online. Service robots are being employed to enhance these courses by providing real-time assistance, feedback, and interactivity. We will examine how MOOC providers are incorporating robots into their platforms to create more engaging and effective learning experiences.

Engagement and completion rates: The presence of service robots in online courses can significantly impact student engagement and completion rates. We will analyze data and studies that highlight how robots contribute to increased motivation, participation, and successful course completions among online learners.

Scalability of online robot-assisted courses: Scalability is a crucial factor in the effectiveness of robot-assisted online courses. We will explore the technologies and strategies used to ensure that robot-assisted learning remains accessible and manageable, even as the number of online learners continues to grow.

e. Policy and Regulation

Policy and regulation in the realm of educational service robotics are crucial aspects that ensure the safe, ethical, and effective integration of robots within educational environments. These policies guide the development, deployment, and operation of educational robots, promoting responsible usage and mitigating potential risks. This section delves into the overarching policies and regulations needed to harness the potential of service robotics in education responsibly.

f. Education Robotics Standards

Education robotics standards set the framework for the design, functionality, and behavior of robots used in educational settings. These standards encompass safety measures, ethical guidelines, and performance expectations. Establishing standardized parameters ensures that educational robots meet specific criteria, fostering a secure and conducive learning environment. In this section, we explore the foundational standards that govern educational robots' design and operation.

Safety Guidelines for Educational Robots: This subtopic delves into protocols and safety measures required to ensure the physical and emotional safety of students and educators interacting with robots in educational settings.

Ethical Guidelines for Robot Behavior: This subtopic explores ethical considerations governing how robots interact with individuals, emphasizing respect, empathy, and appropriate conduct within educational contexts.

Certification and Compliance: This subtopic discusses the process of certification and compliance that educational robots must undergo to meet established standards, ensuring adherence to safety and ethical guidelines.

g. Funding and Support

Funding and support are vital aspects that drive the development and implementation of educational service robotics. Financial backing and collaborative efforts facilitate research, innovation, and sustainable integration of robotics into educational practices. This section addresses the crucial role of funding and support mechanisms in advancing educational service robotics.

Government Grants for EdTech Robotics: This subtopic explores the various grants and funding opportunities provided by governments to support the research, development, and deployment of educational robotics within the edtech sector.

Investment in Research and Development: This subtopic discusses the necessity of private and public sector investments in research and development to propel advancements in educational robotics, promoting innovation and scalability.

Public-Private Partnerships: This subtopic emphasizes the collaborative efforts between public institutions and private entities to foster innovation and drive the successful integration of educational service robotics into the education sector.

These introductions provide a glimpse into the critical topics of policy and regulation, education robotics standards, and funding and support that play pivotal roles in shaping the responsible integration and growth of educational service robotics.

Certainly! Here are the introductions for the technical aspects of educational service robotics related to robot design and hardware, as well as software and AI:

h. Technical Aspects of Educational Service Robotics

The development and integration of service robotics in educational settings represent a promising frontier in modern pedagogy. As technological advancements continue to reshape education, incorporating robots into the learning environment opens up innovative opportunities to enhance engagement, interaction, and personalized learning experiences. This section delves into the intricate technical aspects that form the foundation of educational service robotics, focusing on both robot design and hardware, as well as software and AI technologies.

Robot Design and Hardware

The design and hardware of educational service robots play a pivotal role in shaping their functionality and effectiveness within learning environments. Tailoring the physical attributes of robots to suit educational needs is essential for seamless integration and engagement. This includes exploring various robot form factors specifically designed to enhance interaction and learning outcomes. Additionally, the integration of cutting-edge sensor and actuator technologies is fundamental, enabling robots to perceive and interact with their surroundings effectively. Equally important are power management solutions that optimize robot operation and durability, ensuring sustained performance over extended periods of use within educational contexts.

Robot Form Factors for Education: In the realm of educational service robotics, robot form factors are meticulously designed to be conducive to learning experiences. These forms often incorporate a balance of size, mobility, and appearance, allowing for easy maneuverability within classrooms and interaction with students. The design also considers ergonomic aspects, ensuring that the robot's physical structure promotes engagement and minimizes any potential intimidation, making it approachable and friendly for learners.

Sensor and Actuator Technologies: Sensors and actuators constitute the sensory and motor capabilities of educational service robots. Sensors, ranging from cameras and LiDAR to microphones and touch sensors, empower robots to perceive their environment and gather data critical for interaction and learning. Actuators, including motors and arms, enable robots to move, gesture, and interact with objects and learners. The integration of advanced sensor and actuator technologies is vital to the robot's ability to adapt and respond appropriately to dynamic educational scenarios.

Power Management and Durability: Efficient power management and durability are cornerstones of educational service robot design. These robots must operate over extended periods without frequent charging, making effective power management systems crucial. Moreover, durability ensures that robots withstand the demands of daily use in educational settings. Robustness in design and material selection contributes to prolonged robot lifecycles, minimizing maintenance efforts and associated costs.

Robot Design and Hardware

Software and AI

The software and artificial intelligence (AI) components of educational service robotics are at the heart of their functionality and adaptability. These components define how robots interact with users, process information, and make intelligent decisions to enhance the learning experience. Utilizing machine learning and natural language processing, educational robots can personalize teaching methodologies, provide real-time feedback, and adapt to individual learner needs, ultimately revolutionizing the way students engage with educational content.

Machine Learning for Educational Robots: Machine learning algorithms are pivotal in enabling educational service robots to analyze and learn from data, facilitating the customization of educational interactions. Through continuous analysis of student performance, preferences, and interactions, machine learning algorithms can tailor teaching strategies, content delivery, and assessment methods. This personalization enhances engagement and comprehension, fostering a more effective and personalized learning journey.

Natural Language Processing for Tutoring: Natural language processing (NLP) empowers robots to understand and generate human-like language, enabling seamless communication and tutoring interactions. Educational service robots equipped with NLP capabilities can engage in dialogue, answer questions, and provide explanations in a manner that resonates with students. This fosters an environment where learners feel comfortable seeking assistance and clarification, thereby promoting a deeper understanding of educational material.

Algorithmic Advancements: Ongoing advancements in algorithms significantly contribute to the sophistication and efficiency of educational service robots. These advancements encompass a wide array of areas, including data processing, pattern recognition, and decision-making algorithms. As algorithms become more refined and specialized, educational robots can offer increasingly sophisticated features, adaptability, and learning enhancements, ultimately enriching the educational experience for learners of all ages.

These technical aspects underscore the pivotal role of robot design, hardware, software, and AI in shaping the future of educational service robotics. By focusing on optimizing the design and leveraging advanced technologies, educational service robots can contribute to a transformative and personalized learning journey for students, educators, and institutions alike.

User Experience and Feedback

In the realm of educational service robotics, understanding and optimizing the user experience is paramount. This entails considering how students and teachers interact with robots, how these interactions influence learning and teaching, and how feedback from users can drive improvements. User experience and feedback play a pivotal role in tailoring robotic systems to meet the specific needs and preferences of the educational community.

i. Student Perspectives:

Student perspectives are crucial in evaluating the impact and effectiveness of educational service robotics.

Student Surveys and Feedback: Conducting surveys and collecting feedback directly from students allows for insights into their experiences with robotic-assisted learning. Questions may cover aspects like comfort levels, perceived benefits, and areas for improvement.

Usability and Robot Preferences: Understanding how easily students can use and interact with robots is key to optimizing usability. Additionally, discerning students' preferences regarding robot characteristics, interfaces, and functionalities helps in refining robot designs.

Learning Outcomes and Engagement: Evaluating how robotics influence learning outcomes, academic performance, and overall student engagement in educational settings is fundamental. This involves assessing if robotic integration enhances comprehension, participation, and enthusiasm for learning.

j. Teacher Perspectives

The perspectives of educators are pivotal in shaping the successful integration of robotics into the educational landscape.

Teacher Perspectives

Educator Feedback on Robot Efficacy: Collecting feedback from teachers about the effectiveness of robots in achieving educational objectives is essential. This feedback informs developers and researchers about the pedagogical value of the robotics systems.

Professional Development with Robots: Understanding how robots can support teachers in their professional development is critical. This involves exploring how robots can assist in lesson planning, classroom management, and the overall teaching-learning process.

Integrating Robots into Teaching Strategies: Gaining insights into how teachers integrate robots into their teaching methodologies and curriculum is key. This includes understanding the strategies, approaches, and creative ways educators use robots to enhance the learning experience.

In summary, this section focuses on understanding and analyzing the experiences and viewpoints of both students and teachers in the context of educational service robotics. By gaining valuable insights from these perspectives, it becomes possible to refine and enhance the role of robotics in education, ensuring it aligns with the needs and expectations of its users.

12.5. Entertainment Robotics

Real-Time Applications of Entertainment Robotics

Entertainment robotics revolutionize various sectors, captivating audiences and enhancing user experiences through interactive and engaging applications. These robots are designed to seamlessly integrate with real- time scenarios, adding a touch of innovation and excitement. Let's explore some of the prominent applications of entertainment robotics:

Interactive Theme Park Attractions:

Entertainment robots play a pivotal role in theme parks by creating interactive attractions. Imagine a robot guiding visitors through a theme park, providing information, entertaining with stories, or even participating in rides. These interactive experiences elevate the enjoyment and engagement levels for visitors, making their theme park adventure memorable and immersive.

Gesture-based Interaction: Enabling robots to respond to human gestures, enhancing engagement and interactivity during performances.

Speech Recognition and Response: Allowing robots to understand and respond to spoken commands or prompts, enhancing the natural interaction between robots and humans.

Emotion Recognition: Integrating systems that enable robots to recognize and respond to human emotions, leading to more empathetic and engaging interactions.

Robotics for Rides and Attractions:
Ride Automation and Control:

Description: Ride automation involves utilizing robotics and automated systems to control the functioning of amusement park rides. Advanced sensors, control systems, and robotics ensure precise and safe operation of the rides, including starting, stopping, speed control, and other movements.

Importance: Automation enhances safety, efficiency, and overall ride experience. It minimizes human error and allows for more complex and synchronized ride sequences.

Safety Measures and Monitoring:

Description: This involves using robotics and sensors to constantly monitor the ride and its surroundings for any safety hazards. Automated safety protocols can be triggered in real-time to halt or modify the ride's operation to prevent accidents.

Importance: Safety is paramount in amusement parks. Robotics help in real-time safety monitoring, reducing risks and ensuring a secure environment for visitors.

Theming and Storyline Integration:

Description: Robotics play a crucial role in bringing the theme and storyline of attractions to life. Animatronics and robotic elements are designed to fit within the theme, enhancing the immersive experience for visitors.

Importance: Theming and storyline integration create a captivating and cohesive experience, making attractions more engaging and memorable for visitors.

Character Design and Animation:

Description: Character design in robotics involves creating physical and animated representations of characters, often using robotics and animatronics technology. The design includes aspects like appearance, movement, facial expressions, and body language to mimic the character's traits and personality.
Importance: Well-designed characters enhance audience engagement and immersion, especially in theme parks and entertainment venues. Detailed animation and design captivate the audience and make the characters more relatable.

Costuming and Aesthetics:

Description: Costuming involves designing and creating the physical appearance and outfits for the robotic characters, ensuring they align with the character's image and theme. Aesthetics focus on the overall look, ensuring it's visually appealing and consistent with the intended character.
Importance: Aesthetically pleasing designs and costumes contribute to the overall entertainment value by attracting and engaging the audience. A well-designed costume makes the character more appealing and recognizable.

Performance and Interactivity:

Description: This aspect involves programming the robotic characters to perform specific actions, movements, and interactions in line with their designated characters. Interactivity entails enabling the characters to respond to audience actions or cues, creating a dynamic and engaging experience.
Importance: The performance and interactivity of robotic characters elevate the entertainment level, making the audience feel more connected to the characters. It adds a layer of excitement and unpredictability to the overall entertainment experience.

In summary, robotics in rides and attractions focus on automation, safety, and theming, ensuring a seamless and engaging experience for visitors. On the other hand, character and mascot robotics involve design, costuming, and programming to create compelling, interactive characters that enhance audience engagement and immersion in entertainment settings.

Live Event Performances and Shows:

Robots have stepped onto the stage, literally! In live events such as concerts, theatrical performances, or presentations, robots are becoming performers. With programmed movements and often a dash of AI, these robots can dance, play musical instruments, or engage in scripted acts, adding a futuristic and mesmerizing dimension to live entertainment.

Live Event Performances and Shows

Augmented Reality (AR) and Virtual Reality (VR) Integration:

Integrating entertainment robots with AR and VR technologies enhances the sensory experience. Imagine wearing AR glasses that allow you to see virtual characters interacting with real-world entertainment robots. This integration blurs the lines between the virtual and physical world, creating an immersive and captivating encounter for users.

Immersive Experiences: Incorporating AR to create captivating and immersive experiences for audiences, enhancing entertainment value.

Interactive Storytelling: Using AR to tell interactive stories, providing a dynamic and engaging narrative for users.

Gaming and Simulation: Introducing AR-based games and simulations, blending the virtual world with the real world for an entertaining experience.

360-Degree Content: Developing entertainment content that offers a 360-degree view, providing a more immersive and encompassing experience for users.

VR Game Development: Creating virtual reality games that engage users in a fully immersive gaming environment, enhancing entertainment possibilities.

Virtual Tours and Experiences: Offering virtual tours of various locations or scenarios, allowing users to explore and experience new environments through VR.

Interactive Museum Exhibits:

Museums are transforming their exhibits with the introduction of entertainment robots. These robots act as guides, providing interactive tours, historical insights, or playful interactions, enriching visitors' learning experiences. Visitors can engage with history and art in a more dynamic and interactive manner, making museum visits educational and entertaining.

Entertainment Robots in Shopping Malls and Retail Spaces:

Robots are making shopping experiences more enjoyable and efficient. In malls and retail spaces, robots can assist customers by providing directions, suggesting products, or even entertaining with dances and gestures. This not only enhances the shopping experience but also attracts more customers to the retail environment, creating a buzz and increasing foot traffic.

These real-time applications illustrate how entertainment robotics is transforming the way we engage with various forms of entertainment, ranging from theme parks and live events to museums and retail spaces. These applications showcase the potential of entertainment robotics to enhance engagement, learning, and overall enjoyment in different entertainment domains.

Revolutionary Outcomes of Entertainment Robotics

Entertainment robotics have revolutionized various industries by providing novel and engaging experiences to customers. These outcomes go beyond mere amusement, significantly impacting customer engagement, foot traffic, revenue generation, and marketing strategies for businesses.

Enhanced Customer Engagement and Experience:

Entertainment robots profoundly enrich customer engagement and experience. Through interactive performances, personalized interactions, and dynamic storytelling, these robots captivate and immerse customers in a unique and unforgettable journey. This heightened engagement fosters a positive perception of the brand or venue, leaving a lasting impression on the visitors.

Increased Foot Traffic and Revenue Generation:

The integration of entertainment robots often leads to a surge in foot traffic and subsequent revenue growth. These futuristic attractions draw crowds, enticing both regular patrons and curious first- time visitors. The buzz generated around these robots, along with positive word-of-mouth, spreads rapidly, attracting a broader audience and translating into increased revenue for the venue or business.

Personalized and Adaptive Interactions:

Entertainment robots excel in providing personalized and adaptive interactions with individuals. Using advanced algorithms and data analysis, these robots tailor their responses and actions to suit the preferences and behaviors of each user. This level of personalization fosters a deeper connection, making the interaction more meaningful and enjoyable for the individual, enhancing overall satisfaction.

Integration with Social Media and Online Platforms:

Entertainment robots are seamlessly integrated with social media and online platforms, extending the experience beyond the physical interaction. Users can share their encounters with these robots, accompanied by photos and videos, on various online platforms. This integration amplifies the reach, creating a viral effect and generating substantial online attention, which further boosts the venue's visibility and attracts more visitors.

Enhanced Branding and Marketing Strategies:

Leveraging entertainment robots in branding and marketing strategies is a game-changer. These robots serve as unique brand ambassadors, embodying the values and character of the brand they represent. Their appeal and innovative nature garner significant media coverage, driving a surge in brand recognition. Incorporating entertainment robots into marketing campaigns elevates the brand's image, positioning it as forward-thinking, tech-savvy, and customer-focused.

Entertainment robotics not only infuses excitement and innovation into the entertainment industry but also has a profound impact on business outcomes. They elevate customer experiences, draw larger crowds, generate revenue, and revolutionize branding efforts. As technology continues to evolve, the potential for entertainment robotics to reshape the landscape of customer engagement and business success remains both promising and exciting.

Human-Robot Interaction (HRI) in Entertainment Natural Language Processing   (NLP):

Conversational Agents: Conversational agents are robotic systems equipped with NLP capabilities to engage in meaningful dialogues with humans. They can hold conversations, provide information, answer questions, and assist users in various tasks. Through advancements in NLP, conversational agents become more natural, understanding context and nuances, leading to seamless interactions in entertainment settings like virtual tours or interactive performances.

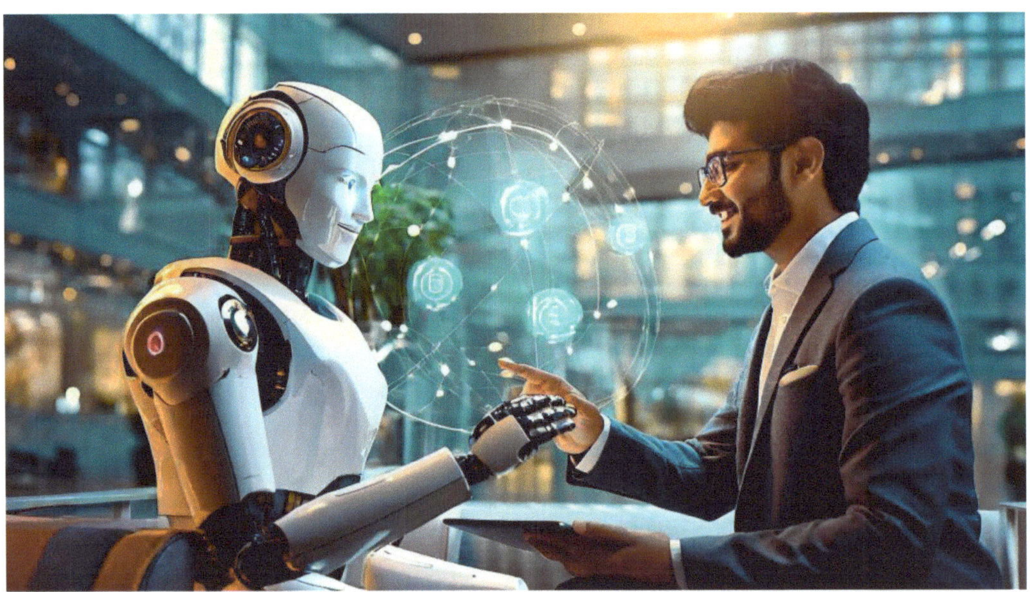

Human-Robot Interaction (HRI) in Entertainment

Language Understanding and Generation: NLP enables robots to not only understand human language but also generate appropriate responses.

Understanding the context, sentiment, and intent behind a user's input allows robots to tailor responses, making the interaction more human-like and effective. This aspect is crucial for creating engaging dialogues during interactive storytelling or educational applications.

Multilingual Support: Multilingual support allows robots to comprehend and respond in multiple languages. This inclusivity broadens the reach of entertainment robotics, catering to diverse audiences globally. In multicultural settings, robots can engage with visitors or audiences in their native languages, enhancing accessibility and user satisfaction.

Facial Expression Recognition:

Emotion Analysis: Emotion analysis through facial expression recognition involves using technology to detect and interpret human emotions based on their facial expressions. Entertainment robots can utilize this technology to gauge audience reactions during performances or engagements. Understanding emotions allows for real-time adjustments, ensuring a more personalized and engaging experience.

Expressive Avatar Creation: Entertainment robots can be designed as avatars capable of expressing a range of emotions through facial mimicry. Advanced facial expression recognition technology enables these avatars to display happiness, sadness, excitement, and more, enhancing the connection and relatability between the robot and the audience. This is particularly valuable in animated storytelling or interactive displays.

Personalized User Engagement: By recognizing individual facial expressions and emotions, robots can tailor their interactions to suit the emotional state of each user. For instance, if a user appears happy, the robot can respond in a joyful manner, creating a personalized and empathetic engagement. This fosters a stronger bond between the user and the robot, improving the overall entertainment experience.

Social Robotics:

Social Interaction and Engagement: Social robots are designed to interact with humans in a socially acceptable and engaging manner. They possess the ability to engage in conversations, express emotions, and respond to human behavior appropriately. In entertainment, these robots can be utilized to engage with audiences, making events more interactive and enjoyable.

Companion Robots: Companion robots are social robots designed to provide companionship and emotional support. In the context of entertainment, they can serve as interactive companions during events or performances. These robots offer engaging interactions, conversation, and even personalized recommendations, enhancing the overall experience for individuals.

Social Skills Learning and Development: Some robots are specifically designed to assist in the learning and development of social skills, especially for children or individuals with social interaction difficulties. Through interactive engagement and guided interactions, these robots help users learn how to express emotions, engage in conversations, and understand social cues, making the process enjoyable and educational.

HRI in entertainment encompasses a range of technologies and capabilities aimed at making robot- human interactions more natural, engaging, and emotionally resonant. These advancements enhance the quality of entertainment experiences, whether through conversational agents, facial expression recognition, or the development of social robots capable of fostering companionship and social skill development.

Dance and Choreography:

In the realm of entertainment robotics, dance and choreography involve programming robots to execute intricate and coordinated dance moves, steps, and routines. These robots are designed to mimic human-like movements and gestures, allowing them to perform alongside human dancers in a synchronized and visually appealing manner. The choreography is meticulously designed and programmed, ensuring that the robot seamlessly integrates into the dance performance, captivating the audience with a blend of technology and artistry.

Key Aspects:

Programming Complexity: Choreographing robot movements involves precise programming to achieve synchronized and complex dance routines, often requiring advanced algorithms and motion control techniques.
Synchronization: Ensuring seamless synchronization between human dancers and robots to maintain the fluidity and grace of the performance.
Aesthetic Design: Designing robots with an aesthetic appeal that complements the artistic aspects of the dance performance.

Music and Instrument Playing:

This facet of robot-assisted performances focuses on designing robots capable of playing musical instruments or contributing to live musical performances. These robots are programmed to replicate the actions of a musician, producing music and enhancing the auditory experience for the audience. They can perform solo pieces, accompany human musicians, or form a robotic ensemble, creating a unique blend of technology and musical artistry.

Key Aspects:

Instrument Mastery: Equipping robots with the ability to play a variety of musical instruments, including string, percussion, wind, or electronic instruments.
Musical Interpretation: Programming robots to interpret musical scores and play with appropriate timing, tempo, and expression, closely mimicking human musicians.
Integration with Humans: Facilitating seamless collaboration and synchronization between robotic and human musicians to create a harmonious musical performance.

Theatrical Performances:

Theatrical performances involving robots entail incorporating these machines into stage productions, plays, or performances to fulfill specific roles or portray characters. Robots are designed and programmed to act, emote, and interact with other characters or performers, enriching the storyline and engaging the audience in a novel and captivating way.

Key Aspects:

Character Representation: Creating robots that embody and bring to life various characters within the storyline, enhancing the immersive and theatrical experience.

Emotional Expression: Programming robots to display a range of emotions and expressions suitable for their assigned roles, contributing to the depth and authenticity of the performance.

Interaction with Human Actors: Ensuring seamless interaction and coordination between robotic and human actors to maintain the coherence and flow of the theatrical narrative.

Enhanced Audience Engagement with Robotics

Enhanced Audience Engagement with Robotics refers to the integration of robotic technologies into live events, performances, or entertainment settings to create a more interactive, personalized, and immersive experience for the audience. The goal is to captivate the audience by involving them actively in the event, allowing them to influence and shape their entertainment experience through interactions with robots. This level of engagement enhances the overall entertainment value and creates memorable experiences for the audience.

Enhanced Audience Engagement with Robotics

Audience Interaction Devices

Audience Interaction Devices involve the use of wearable technologies and IoT (Internet of Things) integration to facilitate seamless and interactive engagements between the audience and robots during events or performances. These devices empower the audience, enabling them to participate, provide feedback, and influence the course of the event, thereby enhancing their engagement and enjoyment.

Wearables and IoT Integration

Providing the audience with wearables or integrating IoT devices that interact with robots, allowing for a more personalized and interactive experience during events or performances.

Wearable Technologies:

- Wearables could include devices such as smart wristbands, headsets, or clothing embedded with sensors, LEDs, or haptic feedback mechanisms.
- These wearables can communicate with robots and other event systems, allowing for personalized experiences based on individual preferences and interactions.

IoT Integration:

- Integration with the Internet of Things (IoT) enables real-time data exchange between wearables and centralized event systems.
- IoT devices can collect and transmit data related to audience preferences, movements, or biometric information, enhancing the responsiveness of robots to audience actions.

Personalization and Interactivity:

- Wearables and IoT integration enable robots to tailor interactions and responses based on the data received from the audience's devices.
- The personalized experience enhances audience engagement by making them an active part of the event, where their actions influence the robots and the overall entertainment.

Real-Time Feedback Systems

Implementing systems that gather real-time feedback from the audience during events, enabling robots to adapt and customize their interactions accordingly.

Audience Sentiment Analysis:

- Real-time feedback systems can analyze audience emotions or sentiments through facial expressions, gestures, or wearable sensors.
- This analysis helps robots understand the audience's mood and adapt their interactions to match the overall sentiment of the crowd.

Instant Feedback Loop:

- By providing a mechanism for immediate feedback, robots can adjust their behavior, performance, or responses based on audience reactions in real time.
- This iterative feedback loop ensures that the audience's preferences are

accommodated, enhancing their engagement and satisfaction.

Audience-Driven Experiences

- Allowing the audience to actively influence and participate in the performance or event through interactions with robots, enhancing engagement and entertainment.

Interactive Elements:

- Robots can feature interactive elements that encourage audience participation, such as voting, choosing storylines, or controlling aspects of the performance.
- Audience members can use their wearables or IoT-integrated devices to actively influence the course of the event, making them feel more engaged and invested.

Collaborative Activities:

- Robots can facilitate collaborative games, challenges, or activities that require collective input or cooperation from the audience.
- This collaborative engagement fosters a sense of community and involvement, amplifying the overall entertainment experience.

Customized Narratives:

- Robots can adapt their storytelling or performance based on audience choices, creating a unique narrative for each performance.
- This customization allows the audience to feel a sense of agency and ownership over their entertainment experience.

Incorporating these audience interaction devices and strategies into entertainment events effectively blurs the line between the audience and the performance, resulting in a more interactive, engaging, and memorable entertainment experience for everyone involved.

Haptic Feedback and Sensation:

Haptic feedback involves the use of robots to provide tactile sensations or touch-based feedback to the audience. Here's an elaboration:

Tactile Sensations: Robots are designed to simulate touch or physical sensations that can be felt by users. For example, during a live performance, a robot might gently touch an audience member's hand to convey an emotion or provide a physical connection with a character or element in the show.

Enhanced Immersion: Haptic feedback enhances the immersion of the audience by allowing them to physically feel vibrations, textures, or movements that align with the performance. For instance, in a virtual reality-based theater, a robot might simulate the sensation of raindrops falling on the audience during a rainy scene.

Feedback for Accessibility: Haptic feedback can also be used to provide feedback to individuals with visual or hearing impairments, making performances more accessible and inclusive.

Multi-Sensory Robotics

Scent Dispersion:

Scent dispersion involves the use of robots to release specific scents relevant to the event or performance. Here's an elaboration:

Olfactory Dimension: Robots equipped with scent dispersion systems release fragrances or scents that are related to the storyline, setting, or theme of the performance. For example, during a theater production set in a forest, a robot might disperse the scent of pine trees or fresh earth to immerse the audience in the woodland atmosphere.

Enhancing Emotional Impact: Scent dispersion can be strategically used to evoke emotions and memories. For instance, in a nostalgic scene, a robot could release scents associated with childhood or a particular time period to trigger emotional responses in the audience.

Customizable Experiences: Scent dispersion systems can be customized based on the preferences of the audience or the requirements of the performance, allowing for a tailored and engaging sensory experience.

Audiovisual Enhancements:

Audiovisual enhancements involve robots that contribute to improving the auditory and visual aspects of an event or performance. Here's an elaboration:

Enhancing Sound: Robots can be equipped with advanced sound systems to deliver high- quality audio, creating a more immersive auditory experience for the audience. This can involve spatial audio, surround sound, or directional audio effects.

Visual Effects: Robots with visual enhancements may include projectors, LED displays, or other visual technologies that enhance the visual elements of the performance. For instance, robots may project dynamic visuals onto screens or surfaces, synchronized with the storyline or music.

Interactive Lighting: Some robots can control lighting effects, adjusting colors, intensity, and patterns to match the mood or theme of the performance. Interactive lighting can add drama and visual impact to the show.

Ethical and Social Implications of Entertainment Robotics

The rise of entertainment robotics brings ethical and social considerations to the forefront. Balancing technological innovation with respect for privacy, data security, and consent is paramount. Society grapples with the potential impact of automation on jobs, necessitating proactive reskilling efforts. Ethical considerations encompass fairness, bias prevention, and the responsible use of robotic technology. Additionally, ensuring accessibility allows diverse populations to engage and benefit from these advancements.

Ethical and Social Implications of Entertainment Robotics

Privacy and Data Security:

Privacy and data security in entertainment robotics address safeguarding user information during interactions with robots. It involves ensuring responsible data collection, explicit consent mechanisms, and robust encryption methods to protect sensitive data. This focus promotes trust and confidence in users regarding their personal information handling, fostering a secure and ethical engagement with entertainment robots.

User Data Collection and Handling:

Explanation: Entertainment robots often collect data from users during interactions, including personal information, preferences, and behaviors. It's crucial to handle this data responsibly, respecting user privacy and security.

Importance: Transparent and responsible handling of user data establishes trust and ensures that individuals are comfortable engaging with entertainment robots without concerns about misuse or unauthorized access to their information.

Consent and Permissions:

Explanation: Users should be informed about the data being collected and how it will be used. Obtaining explicit consent and providing clear information about data usage respects users' autonomy and allows them to make informed decisions.

Importance: This fosters a sense of control and trust, as users are aware of the extent to which their data is being used, empowering them to decide whether or not to engage with the robot.

Data Encryption and Anonymization:

Explanation: Encryption and anonymization techniques help safeguard user data by making it unreadable to unauthorized parties. Anonymization ensures that personally identifiable information is removed or altered to protect user identities.

Importance: These security measures are vital in preventing data breaches and unauthorized access, promoting a safer environment for user data and upholding privacy.

Ethical Considerations in Robotics:

Explanation: The field of entertainment robotics raises ethical concerns related to robot rights, bias in programming, and responsible technology use. Ensuring that robots are treated ethically, avoiding bias, and using technology for societal benefit is critical.

Importance: Addressing these ethical considerations is essential to create a balanced and fair interaction between robots and humans, promoting a responsible and inclusive integration of robotics in entertainment.

Challenges in Entertainment Robotics

Entertainment robotics has witnessed remarkable growth and innovation, offering captivating and interactive experiences to audiences worldwide. However, this burgeoning field also faces several significant challenges and holds promising future prospects.

Cost and Accessibility Challenges:

Elaboration: The development and deployment of advanced entertainment robots often involve substantial costs in research, design, manufacturing, and maintenance. This cost can pose accessibility challenges for smaller entertainment venues or startups looking to incorporate robots into their offerings.

Future Prospects: Advances in technology and increased demand may lead to economies of scale, potentially reducing costs over time. Innovations like modular robot designs could make entertainment robots more affordable and accessible to a broader range of businesses.

Safety and Liability Concerns:

Elaboration: Ensuring the safety of human-robot interactions is paramount. Malfunctions, unexpected behaviors, or accidents involving entertainment robots can lead to liability issues. Designing robots to operate safely in dynamic, crowded environments is a complex challenge.

Future Prospects: The development of robust safety protocols, improved sensor technologies, and real-time monitoring systems can mitigate safety concerns. Additionally, industry standards and regulations may evolve to address liability issues and establish clear responsibilities.

Future Trends and Innovations in Entertainment Robotics

The future of entertainment robotics is poised for transformative advancements. Embracing cutting- edge technologies like AI, quantum computing, and innovative materials, robots will achieve heightened intelligence, agility, and sustainability. This evolution promises dynamic, immersive interactions, enhancing audience engagement and enjoyment. Energy efficiency and eco-friendly designs underline a commitment to a greener, more responsible entertainment industry. The convergence of innovation and sustainability propels entertainment robotics into an exciting and environmentally conscious future.

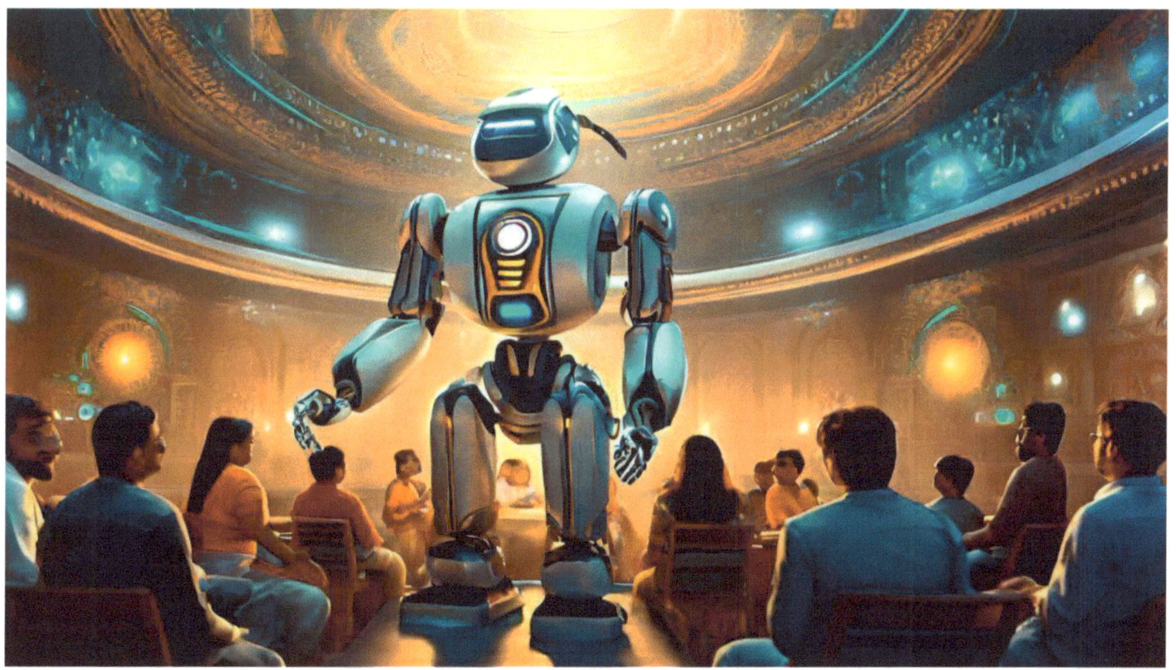

Future Trends and Innovations in Entertainment Robotics

Advanced Robotics Technologies:

AI and Machine Learning Integration:

Incorporating advanced AI and machine learning techniques: This involves integrating state-of-the-art artificial intelligence (AI) and machine learning algorithms into entertainment robots. By doing so, robots can analyze data, learn from experiences, and make intelligent decisions. This integration enhances the robot's ability to adapt to various situations and interact more effectively with humans.

Enabling intelligent decision-making: AI and machine learning allow robots to understand and respond to user preferences and behaviors, making their interactions more personalized and engaging. For example, a robot can adapt its conversation style or actions based on the user's mood or preferences.

Facilitating adaptive behavior: These technologies enable robots to adapt to changing environments and tasks. For instance, a robot at an amusement park could adjust its performance based on the crowd's energy level, creating a more dynamic and enjoyable experience.

Quantum Computing Applications:

Exploring the potential applications of quantum computing: Quantum computing is a cutting-edge technology that has the potential to revolutionize computing power. In the context of entertainment robotics, it can be used to process vast amounts of data in real-time, allowing robots to perform complex calculations and simulations rapidly.

Enhancing capabilities and performance: Quantum computing can significantly boost the capabilities of entertainment robots. For instance, it can enable robots to process and analyze large datasets from sensors, cameras, and user interactions in real-time, leading to more immersive and responsive experiences.

Pushing the boundaries of real-time interactions: Quantum computing's computational speed can push the boundaries of what's possible in real-time interactions. Robots can provide instantaneous responses, simulate complex scenarios, and generate highly detailed virtual environments, all contributing to more captivating entertainment experiences.

Advanced Materials and Components:

Researching and implementing cutting-edge materials and components: Advances in materials science and engineering can lead to the development of stronger, more flexible, and lighter materials for building robots. These materials can improve the overall durability, agility, and functionality of entertainment robots.

Improving durability: Sturdier materials can make robots more resilient to wear and tear, allowing them to operate reliably in various environments, including outdoor and theme park settings.

Enhancing agility: Lighter and more agile materials can enable robots to move with greater precision and agility, enhancing their ability to perform intricate tasks or interact with the audience in dynamic ways.

Functionality enhancements: Innovative components, such as high-resolution sensors, more efficient actuators, and advanced power sources, can enable robots to perform tasks with greater accuracy and energy efficiency.

Sustainable and Green Robotics:

Energy-Efficient Robotics:

Developing energy-efficient robots: Sustainable robotics involves designing robots that consume minimal energy during operation. This not only reduces operational costs but also minimizes their environmental impact.

Sustainable power sources: Exploring alternative and renewable energy sources, such as solar power, to supply energy to robots. This reduces the reliance on fossil fuels and decreases greenhouse gas emissions.

Minimizing the environmental footprint: Energy-efficient robots contribute to a greener future by reducing their carbon footprint, making them more environmentally friendly for long-term use in entertainment settings.

Eco-Friendly Materials and Design:

- Emphasizing eco-friendly materials: Using materials that are sustainable, biodegradable, or recyclable in robot design and manufacturing. This reduces the environmental impact associated with the production and disposal of robotic components.
- Promoting sustainability throughout the robot's lifecycle: Ensuring that every stage of a robot's lifecycle, from production to disposal, considers sustainability. This includes selecting materials that are responsibly sourced and promoting recycling and reuse of components.

Recycling and Reusability in Robotics Design:

Focusing on designing robots with recyclability and reusability: Designing robots with components that can be easily disassembled and recycled at the end of their life cycle. This reduces electronic waste and promotes a circular economy.

Reducing waste: By emphasizing recycling and reusability, the robotics industry can significantly reduce waste generation, contributing to a more sustainable and environmentally responsible approach to robotics manufacturing.

These trends and innovations in entertainment robotics encompass the integration of advanced technologies like AI, machine learning, and quantum computing to enhance robot intelligence and performance. Additionally, a focus on sustainability includes energy-efficient designs, eco-friendly materials, and recycling practices to minimize the environmental impact of entertainment robotics. These advancements pave the way for more engaging, eco-conscious, and technologically advanced entertainment experiences.

12.6. Domestic Robotics

Evolution and History of Domestic Robotics

The evolution of domestic robotics can be traced back to the mid-20th century when the concept of automated devices to aid in household chores started to gain traction. Early examples include automated vacuum cleaners and robotic lawnmowers. However, the real surge in domestic robotics came in the late 20th and early 21st centuries with advancements in technology, particularly in robotics, artificial intelligence, and miniaturization of components. Robots evolved from simple, single-function devices to more sophisticated machines capable of learning, adapting, and performing a multitude of tasks.

Domestic Robotics

a. Importance and Relevance in Modern Society

The importance and relevance of domestic robotics in modern society are profound and multi- faceted:

Efficiency and Time-Saving: Domestic robots significantly reduce the time and effort required to complete various household tasks. Automated devices can work tirelessly, allowing individuals to focus on other essential activities.

Improved Quality of Life: Domestic robots enhance the overall quality of life by minimizing mundane and repetitive chores. This allows people to spend more time engaging in meaningful and enjoyable activities, ultimately leading to a better lifestyle.

Assistance for the Elderly and Disabled: Domestic robots play a crucial role in assisting the elderly and those with disabilities in their daily routines. They provide assistance in mobility, medication management, and other activities, enabling a sense of independence and improved well-being.

Sustainability and Resource Efficiency: Many modern domestic robots are designed to be energy-efficient and eco-friendly. By automating and optimizing resource usage, these robots contribute to a more sustainable and environmentally conscious way of living.

Technological Advancements and Innovation: The field of domestic robotics drives technological advancements and fosters innovation. Researchers and developers continually push boundaries to create smarter, more capable, and safer robots, leading to a positive ripple effect in related industries.

Addressing Labor Shortages: In some regions facing labor shortages, domestic robots can help fill the gap by performing tasks that may be challenging to find human labor for. This can be particularly important for tasks that require specific skills or are physically demanding.

domestic robotics is an evolving field with immense potential to transform daily life by automating tasks, improving efficiency, and enhancing overall well-being. Its importance in modern society is increasingly recognized as it addresses fundamental aspects of convenience, health, sustainability, and technological progress.

b. Real-Time Applications of Domestic Robotics

Real-Time Applications of Domestic Robotics explores the evolving landscape of robotics designed for household use. These robots, equipped with advanced technology, cater to a range of daily tasks. From cleaning and cooking to security and healthcare, they revolutionize how we manage our homes. This exploration delves into the practicality and impact of these real-time applications on modern living.

1) Cleaning and Maintenance:

The field of domestic robotics has revolutionized cleaning and maintenance tasks within households. Through innovative technologies and automation, robots now play a pivotal role in keeping homes clean and organized. From vacuuming and mopping floors to cleaning windows, these robots offer convenience and efficiency. This introduction explores the transformative impact of domestic robotics on cleaning and maintenance.

Vacuuming robots: These robots are equipped with sensors and brushes to autonomously clean floors by navigating around a space, avoiding obstacles, and efficiently covering the area. They can be scheduled or activated on-demand for routine cleaning, helping maintain a clean living environment.

Mopping robots: These robots automate the mopping or floor-cleaning process. They are designed to dispense water or cleaning solution, scrub the floor, and effectively remove dirt and stains. They provide a hands-free way of maintaining floors in homes.

Window-cleaning robots: These robots are specialized for cleaning windows. They use suction or other mechanisms to adhere to windows and move across surfaces, cleaning them effectively without the need for human intervention, especially for hard-to-reach or high windows.

2) Cooking and Food Preparation:

The realm of domestic robotics has revolutionized cooking and food preparation. Automated cooking appliances and meal planning robots have become integral, simplifying meal preparation for modern households. These innovative solutions streamline cooking processes, optimize ingredient use, and cater to diverse culinary preferences. With their time- saving and convenience-enhancing features, they embody a significant advancement in kitchen technology.

Automated cooking appliances: These appliances include smart ovens, microwaves, and other cooking devices that can be controlled and programmed to prepare meals. They may adjust cooking settings based on the type of food and desired preferences, streamlining the cooking process.

Meal planning and preparation robots: These robots can assist in meal planning by suggesting recipes based on dietary preferences and available ingredients. They can also help in preparing ingredients, chopping, and performing routine cooking tasks, making meal preparation more efficient.

Kitchen cleaning robots: These robots can assist in keeping the kitchen clean and organized. They can pick up spills, wipe surfaces, and organize utensils, contributing to a tidy cooking space.

3) Home Security and Surveillance:

The integration of domestic robotics in home security and surveillance marks a significant advancement in safeguarding residences. These robotic systems employ cutting-edge technology to enhance security measures and ensure peace of mind for homeowners. From automated surveillance to intruder detection, they offer real-time monitoring and response capabilities. This evolution reflects a proactive approach towards creating safer living environments.

Automated security systems: These systems use sensors, cameras, and alarms to detect unauthorized entry, fire, or other potential threats. They can automatically alert homeowners or emergency services, enhancing home security.

Surveillance and monitoring robots: These robots can patrol and monitor a home, providing real-time video feeds and alerts to homeowners' devices. They contribute to a proactive approach to security by keeping an eye on the premises, especially when the homeowners are away.

Intruder detection and response robots: These robots are equipped to detect suspicious activity or intruders and respond accordingly. They may have capabilities to deter intruders or alert the appropriate authorities.

4) Elderly and Patient Care:

In the realm of domestic robotics, a significant and compassionate application emerges in elderly and patient care. These robots offer essential aid, companionship, and monitoring, catering to the unique needs of older adults and individuals with health concerns. Their role extends from medication management to providing vital alerts, elevating the quality of life and ensuring a safer, more independent living for those in need. This integration of technology signifies a promising paradigm shift in caregiving within the comfort of one's home.

Assistance and companionship robots: These robots can provide companionship and assistance to the elderly or patients. They can engage in conversation, remind individuals to take medications, and even provide emergency assistance.

Medication management robots: These robots assist in organizing and dispensing medications at the right time, ensuring that individuals adhere to their prescribed medication schedules.

Monitoring and alert systems: These systems can monitor vital signs, movement, and activity levels of the elderly or patients. In case of irregularities or emergencies, they can send alerts to caregivers or healthcare professionals, enabling timely intervention.

5) Enhanced Convenience and Time Savings:

- Domestic robotics, by automating mundane and time-consuming tasks such as cleaning, cooking, and organizing, significantly enhance convenience for individuals and families.
- Automation allows people to reclaim valuable time that can be redirected towards more meaningful or productive activities, ultimately improving work-life balance.

6) Improved Quality of Life:

- By handling repetitive chores, domestic robots free up individuals to engage in activities that contribute to their personal growth, well-being, and relationships.
- For elderly or disabled individuals, domestic robots can provide invaluable assistance, promoting a higher quality of life by enabling them to live independently and perform daily tasks.

7) Independence and Empowerment:

Domestic robots can offer a sense of independence to individuals who may have limitations due to age, disability, or other factors. They enable self-sufficiency in managing household tasks and routines.

This newfound independence fosters a sense of empowerment and boosts self-confidence, particularly among those who require assistance in their daily lives.

8) Sustainability and Energy Efficiency:

Many domestic robots are designed with energy-efficient features, promoting a greener and more sustainable lifestyle.For instance, automated home climate control systems optimize energy usage, reducing overall consumption and minimizing the household's environmental footprint.

9) Economic Implications and Market Growth:

The growth of the domestic robotics market stimulates economic activity, generating employment opportunities in manufacturing, research, development, maintenance, and related sectors. Additionally, the increased demand for domestic robots drives innovation and competition, potentially leading to cost reductions and making these technologies more accessible to a broader consumer base.

10)   Ethical Considerations and Privacy Concerns:

As domestic robots become more integrated into daily life, ethical considerations regarding their use and impact on society, privacy, and human relationships are crucial.

Privacy concerns arise regarding the data collected by these robots, prompting discussions on data security, consent, and responsible handling of personal information. Striking a balance between convenience and privacy becomes paramount.

c. Technological Advancements Driving Domestic Robotics

Technological advancements are the driving force behind the rapid evolution of domestic robotics. These innovations are reshaping our homes and enhancing the capabilities of robotic devices. Several key areas of technological development are propelling the field of domestic robotics forward.

Artificial Intelligence and Machine Learning:

AI and machine learning algorithms empower domestic robots to learn and adapt to their environments. They enable robots to recognize patterns, make informed decisions, and improve their performance over time.

These technologies are fundamental for tasks such as navigation, object recognition, and natural language processing, making robots more capable and user-friendly.

Sensors and Perception Technologies:

- Sensors, including cameras, LiDAR, ultrasonic sensors, and touch sensors, provide robots with a wealth of data about their surroundings.

- Perception technologies enable robots to interpret this data, allowing them to navigate safely, avoid obstacles, and interact with objects and people in a more sophisticated manner.

Connectivity and IoT Integration:

- Domestic robots are increasingly integrated into the Internet of Things (IoT) ecosystem. They can connect to other smart devices and systems in the home, allowing for seamless coordination and control.
- IoT integration enhances the overall smart home experience, enabling users to manage various aspects of their home through a single interface.

Human-Robot Interaction and User Experience:

Advances in human-robot interaction focus on making robots more intuitive to use and interact with.

Natural language processing, gesture recognition, and emotional intelligence algorithms enable robots to understand and respond to human communication and emotions, making them more user-friendly and engaging.

Power Management and Energy Efficiency:

- Power management systems in domestic robots are designed for energy efficiency, optimizing battery usage and reducing environmental impact.
- Energy-efficient components and software algorithms extend the operating time of robots and reduce the need for frequent recharging.

Technological advancements in artificial intelligence, sensors, connectivity, human-robot interaction, and energy efficiency are driving domestic robotics to new heights. These innovations are not only expanding the capabilities of robots but also improving the user experience and making these robots an integral part of modern homes.

d. Challenges and Future Prospects

The field of domestic robotics is undoubtedly promising, offering innovative solutions to various everyday challenges. However, it also faces a range of challenges and uncertainties that must be addressed to unlock its full potential. In this section, we will delve into these challenges and explore the future prospects of domestic robotics. These challenges include:

Cost and Affordability:

Domestic robots often come with a significant price tag, which can limit their accessibility to a wider audience.

Finding cost-effective ways to manufacture and distribute these robots is essential to make them more affordable and widely adopted.

Innovative business models, such as robot-as-a-service (RaaS), may help mitigate the upfront cost burden on consumers.

Technical Limitations and Performance Challenges:

Domestic robots must perform reliably and effectively in dynamic and unstructured environments, which presents technical challenges.

Issues like navigation in cluttered spaces, object recognition, and adapting to diverse user preferences require ongoing research and development.

Advances in sensor technology, machine learning algorithms, and robotic hardware are critical to overcoming these limitations.

Social Acceptance and Psychological Factors:

Acceptance of robots as part of the household ecosystem is influenced by cultural, psychological, and sociological factors.

Some users may experience discomfort or unease when interacting with robots, while others may readily embrace them.

Designing robots with user-friendly interfaces, natural interactions, and effective communication can help foster social acceptance.

Potential Integrations with Other Technologies (e.g., AR/VR, 5G):

The synergy between domestic robotics and other emerging technologies can lead to transformative outcomes.

Integrating augmented reality (AR), virtual reality (VR), and 5G connectivity can enable more immersive and efficient interactions with domestic robots.These integrations may open up new possibilities for remote control, enhanced user experiences, and real-time data sharing.

In conclusion, addressing these challenges and capitalizing on future prospects is crucial for realizing the full potential of domestic robotics. As technology advances and societal attitudes evolve, the role of domestic robots in our homes is likely to become increasingly significant, impacting our daily lives in ways we are only beginning to imagine.

e. Case Studies and Success Stories

In the realm of domestic robotics, several remarkable case studies and success stories have emerged, showcasing the transformative power of robotic technology in everyday life. These examples not only demonstrate the practical applications of robotics but also highlight the innovative solutions they provide to address various needs and challenges in the home environment. Here are five notable case studies and success stories:

Roomba: The Robotic Vacuum Cleaner

Overview: Roomba, developed by iRobot, is a pioneering robotic vacuum cleaner designed to autonomously clean floors in homes and offices.

Technology: Roomba utilizes advanced sensors, navigation algorithms, and AI to map and clean spaces efficiently.

Impact: Roomba has revolutionized home cleaning by saving time and effort, ensuring cleaner living spaces, and reducing the burden of manual vacuuming.

Success: Roomba's success has led to the development of various models, making it a household name and a symbol of domestic robotics.

Nest: Smart Home Devices for Automation

Overview: Nest, acquired by Google, offers a range of smart home devices that enhance automation, energy efficiency, and security.

Technology: Nest products include smart thermostats, cameras, smoke detectors, and doorbells, all connected through a central hub.

Impact: Nest devices enable homeowners to remotely control and monitor their homes, reducing energy consumption and increasing security.

Success: Nest's integration with voice assistants and its ecosystem of compatible devices has made it a leader in the smart home market.

Nest: Smart Home Devices for Automation

Pepper: The Emotional Assistance Robot

Overview: Pepper, developed by SoftBank Robotics, is a humanoid robot designed to interact with humans emotionally and provide companionship.

Technology: Pepper is equipped with cameras, microphones, and sensors to recognize and respond to human emotions and gestures.

Impact: Pepper has found applications in healthcare, retail, and education, offering emotional support and improving customer engagement.

Success: Pepper's unique ability to understand and respond to human emotions has made it a breakthrough in human-robot interaction.

Anki Vector: A Robot Companion

Overview: Anki Vector is a small, AI-powered robot designed to serve as a personal companion and assistant.
Technology: Vector features voice recognition, computer vision, and machine learning capabilities to engage with users.
Impact: Anki Vector has gained popularity as a cute and interactive robot that can provide information, play games, and assist with tasks.
Success: Although Anki, the company behind Vector, faced challenges, the robot's popularity among enthusiasts highlighted the potential for personal robots in the home.

iRobot Braava: Robotic Mopping Solution

Overview: iRobot Braava is a series of robotic mopping devices that automate the floor cleaning process.
Technology: Braava uses navigation and mapping technology to efficiently mop floors, whether wet or dry.
Impact: iRobot Braava simplifies floor maintenance, making it easier for homeowners to keep their floors clean and polished.
Success: Braava complements the Roomba vacuum cleaners and has contributed to iRobot's reputation as a leader in domestic robotics.

These case studies and success stories exemplify how domestic robotics have transcended the realm of novelty and have become integral to improving the quality of life and convenience in modern households. They showcase the diverse applications and the potential for further innovation in this exciting field.

f. Pioneering the Future: The Impact and Prospects of Domestic Robotics
Recap of the Impact of Domestic Robotics:

Domestic robotics has had a significant impact on modern living. These machines have revolutionized the way we approach household chores, manage our time, enhance security, and provide care for the elderly or ill. Robotic vacuum cleaners, for instance, have automated and simplified the cleaning process, saving valuable time and effort for individuals. Similarly, robots aiding in cooking and food preparation have not only made meals more convenient but have also contributed to healthier eating habits. The integration of AI and machine learning has elevated these robots to be more adaptive and efficient, further improving their impact.

Future Outlook and Trends in Domestic Robotics:

The future of domestic robotics is immensely promising. Advancements in technology will drive increased autonomy, intelligence, and versatility of robots. We can anticipate robots that not only perform tasks but also understand and adapt to our preferences, habits, and surroundings. Integration with smart home systems and IoT will make them a seamless part of our daily lives. Furthermore, there will likely be a surge in robots designed for companionship, mental health support, and specialized care for the aging population.

Encouragement for Further Research and Development:

To propel the field of domestic robotics forward, continued research and development are crucial. Researchers and engineers should focus on refining the user experience, making these robots more intuitive and user-friendly. Additionally, addressing privacy and security concerns is imperative to ensure widespread acceptance and adoption. Collaborations between academia, industry, and government bodies can foster innovation and streamline regulatory frameworks.

Encouraging interdisciplinary research that integrates fields like psychology, sociology, and engineering will help in creating robots that are not only technically advanced but also socially and emotionally intelligent, enhancing their acceptance and impact on society.

Domestic robotics has the potential to shape the future of how we live and interact with our surroundings. With ongoing research, innovative designs, and a thoughtful approach, we can harness the full potential of these robots to improve our lives in countless ways.

12.7. Lessons Learned And Best Practices

Introducing service robotics requires understanding user needs, designing for reliability and safety, and prioritizing sustainability. Lessons learned emphasize user-centric design, robustness, and data security. Best practices include thorough user research, rigorous testing, and adherence to industry standards for interoperability. Success hinges on balancing functionality with energy efficiency, prioritizing user safety, and fostering a culture of continuous improvement to meet evolving demands and enhance human-robot interaction.

Service robotics requires understanding user needs

a. Designing for Success: Customization and Adaptability

The field of service robotics has seen rapid advancements in recent years, transforming the way we interact with technology in various sectors such as healthcare, logistics, hospitality, and more. To ensure the success of service robotics, designing with customization and adaptability in mind is paramount. This article delves into key aspects related to the design principles that revolve around customization and adaptability in service robotics, explaining how these principles contribute to the overall success of the technology.

b. Understanding the Landscape of Service Robotics:

- Provide an overview of the current state of service robotics, emphasizing its growing importance in diverse industries.
- Highlight the critical role of customization and adaptability in enhancing the usability, effectiveness, and acceptance of service robots.

c. Tailoring Solutions to Specific Applications:

- Discuss the need to design robots that are tailored to specific applications, such as healthcare assistance, cleaning, or logistics, to maximize their efficiency and effectiveness.
- Explain how customization at the design stage helps in addressing the unique requirements of different industries.

d. Modular Design: Enabling Flexibility and Scalability:

- Describe the benefits of modular design, allowing components or features to be easily interchanged or upgraded to adapt to evolving needs or new applications.
- Showcase examples of modular robots and their versatility in adapting to different tasks and environments.

e. Human-Centered Design for Enhanced User Experience:

- Emphasize the importance of considering the end-users during the design phase, tailoring the robot's interactions, interfaces, and functionalities to meet their needs and preferences.
- Discuss the role of user feedback and iterative design processes in refining the robot's design for optimal user experience.

f. Machine Learning and AI Integration:

- Discuss the integration of machine learning and artificial intelligence in service robots to enhance adaptability and customization.
- Explain how AI enables robots to learn and adapt to new tasks, environments, and user preferences, ultimately improving performance and efficiency.

g. Interoperability and Standardization:

- Highlight the significance of interoperability and standardization in service robotics, allowing robots to seamlessly work with other systems and technologies.
- Discuss how adhering to industry standards facilitates customization and adaptability by ensuring compatibility and ease of integration.

h. Continuous Improvement and Upgrades:

- Stress the importance of ongoing research, development, and updates in

the field of service robotics to keep up with technological advancements and changing needs.
- Discuss the role of data analytics and user feedback in driving improvements and updates to the robots' functionalities and features.

i. Ethical Considerations in Customization and Adaptability:

- Address ethical concerns related to customization and adaptability, such as privacy, data security, and ensuring the responsible use of robots in various settings.
- Discuss the importance of designing robots with ethical guidelines and societal implications in mind.

Summarize the critical role of customization and adaptability in the success of service robotics, emphasizing that a thoughtful and user-centric approach to design is essential for creating robots that effectively meet the needs of specific applications and industries. Encourage further research and collaboration to advance the field and ensure that service robots continue to evolve and benefit society.

Chapter 13: Service Robotics: The Ultimate Career Guide

13.1. Introduction To Careers In Robotics

Robotics is indeed a rapidly growing field with a vast spectrum of career opportunities. The integration of robots into various industries, such as manufacturing, healthcare, logistics, and agriculture, has revolutionized the way tasks are performed. As robotics technology advances, it opens up new and innovative applications, expanding the field's potential. This dynamic landscape creates a surging demand for skilled professionals in robotics.

In manufacturing, robots enhance efficiency and precision, reducing production costs and improving product quality. In healthcare, they assist in surgeries, deliver medications, and provide companionship for the elderly. In logistics, autonomous robots streamline warehouse operations and delivery processes. In agriculture, they cultivate crops and monitor livestock, optimizing food production.

The diversity of applications underscores the need for robotics experts, from mechanical and electrical engineers to software developers and data scientists. A career in robotics offers the opportunity to be at the forefront of technological advancement, solving complex challenges and contributing to automation and innovation across various sectors. As the field continues to evolve, those with the skills to design, build, and program robots will find themselves in high demand, making robotics a promising and rewarding career choice.

Why Choose a Career in Robotics?

There are many reasons to choose a career in robotics.
Here are just a few:

- Robotics is both an intellectually stimulating and gratifying domain. Robotics engineers and technicians engage with state-of-the-art technology, grappling with intricate challenges to devise innovative solutions for practical dilemmas.

- The field of robotics is experiencing substantial growth, assuring a plethora of job prospects. According to the Bureau of Labor Statistics, employment opportunities for robotics engineers are projected to burgeon by 22% from 2020 to 2030, significantly surpassing the average growth rate across all occupations.

- One of the most enticing facets of a career in robotics is the compensation. The median annual salary for robotics engineers in 2021 was $111,160, offering financial stability and recognition for their expertise and contributions.

Moreover, robotics is an exceptionally versatile field, presenting a wide array of career options. From designing autonomous vehicles to developing robotic prosthetics or exploring the depths of artificial intelligence, individuals can select roles that align with their personal interests and proficiencies. In this rapidly evolving landscape, there's a niche for everyone, making robotics a compelling choice for those with a passion for innovation and technology.

Here Are Some Additional Reasons Why Careers In Robotics Are So Appealing:

- Robotics is a global field, transcending borders and cultures. The ubiquity of robots worldwide ensures that job opportunities for robotics engineers and technicians exist in every corner of the globe. Whether it's manufacturing robots in Asia, healthcare robots in Europe, or agricultural robots in South America, the demand for skilled professionals in robotics is universal, making it a versatile and geographically diverse career path.

- Moreover, robotics is inherently collaborative. Robotics engineers and technicians often find themselves working on interdisciplinary teams alongside scientists, designers, and other engineers. This environment fosters a culture of collaboration, enabling professionals to learn from one another's expertise and collectively address complex challenges. It's a field where innovation thrives through teamwork.

- Additionally, robotics is a creative endeavor. Engineers and technicians have the unique opportunity to design and construct robots with diverse functionalities. This creative freedom allows them to leverage their problem-solving skills and imaginative thinking. Whether developing robots for space exploration, underwater exploration, or assisting people with disabilities, the scope for innovation is boundless.

For those aspiring to join the world of robotics, there are abundant resources available. Information on robotics programs at universities and colleges can be easily accessed, while job listings and networking opportunities can be found online or through participation in professional conferences. These resources help aspiring robotics professionals embark on an exciting journey within this dynamic and ever-evolving field.

Why Careers In Robotics Are Contemporary And Growth-Oriented

Robotics is a dynamic and forward-looking field that remains in a constant state of evolution and expansion. This continual progression is fueled by the development of novel robotic technologies and their integration into an ever-widening array of applications across diverse industries. As a result, there is an escalating demand for skilled professionals within the robotics domain.

Moreover, the role of robotics in addressing global challenges is becoming increasingly prominent. It plays a pivotal part in mitigating urgent issues like climate change and healthcare. For instance, robots contribute to the advancement of renewable energy sources, enhancing sustainability efforts and combating climate change. In healthcare, they're employed in complex surgeries, facilitating precision and minimally invasive procedures, ultimately improving patient outcomes.

Careers in robotics are not only contemporary and promising, but they also carry a profound societal impact. The individuals working in this field are instrumental in ushering in technological innovations that directly address some of the world's most pressing problems, making it a field of immense relevance and importance in our modern era.

Major Career Options In Robotics Explore Key Career Paths In Robotics: Service Robotics Roles:

- ★ **Robotics engineer:** Robotics engineers design, build, and test robots. They work in a variety of industries, including manufacturing, healthcare, and logistics.
- ★ **Robotics technician:** Robotics technicians assist robotics engineers with the design, construction, and testing of robots. They also help to install and maintain robots in industrial settings.
- ★ **Robotics programmer:** Robotics programmers develop the software that controls robots.
  They work with robotics engineers to design and implement the robot's behavior.
- ★ **Robotics researcher:** Robotics researchers develop new robotic technologies and applications. They work in universities, research labs, and private companies.
- ★ **Robotics Ethicist:** Robotics ethicists address ethical concerns related to the use of robots, AI, and automation, ensuring that they adhere to moral and societal norms.
- ★ **Robotics System Integrator:** System integrators focus on assembling and configuring various components to create functional robotic systems, making them ready for specific applications.
- ★ **Robotics Project Manager:** Project managers oversee the planning, execution, and delivery of robotics projects, ensuring they meet objectives, deadlines, and budget constraints.
- ★ **Robotics Simulation Engineer:** Simulation engineers develop virtual environments to test and validate robotic systems and algorithms before they are deployed in the real world.
- ★ **Space Robotics Engineer:** Space robotics engineers design robots and autonomous systems for use in space exploration, including missions to other planets and space stations.
- ★ **Underwater Robotics Engineer:** These engineers develop robots for underwater exploration and tasks like deep-sea research, offshore drilling, and marine resource management.
- ★ **Robotic Vision Specialist:** Robotic vision specialists focus on enhancing a robot's

ability to perceive and interpret visual data through advanced computer vision technologies.

★ **Humanoid Robot Designer:** Humanoid robot designers create robots that closely resemble humans in appearance and movement, often used in research, entertainment, and human- robot interaction studies.

★ **Robotics Product Manager:** Product managers in the robotics industry oversee the development and marketing of robotic products, ensuring they meet market demands and performance expectations.

Ai Role Related To Service Robotics:

★ **Machine Learning Engineer:** These professionals focus on developing machine learning algorithms and models that enable robots to perceive and make decisions based on their environment.

★ **Computer Vision Engineer:** Computer vision engineers specialize in creating systems that allow robots to interpret and understand visual data, which is crucial for tasks like object recognition and navigation.

★ **Control Systems Engineer:** Control systems engineers design algorithms and systems that govern the behavior of robots, ensuring they move and interact with precision and accuracy.

★ **Automation Engineer:** Automation engineers work on integrating robots and other automated systems into industrial processes to enhance productivity and efficiency.

★ **Artificial Intelligence (AI) Researcher:** AI researchers focus on advancing the intelligence and decision-making capabilities of robots through cutting-edge AI techniques and algorithms.

★ **Human-Robot Interaction Specialist:** These experts design and implement interfaces and interactions that enable humans to work with and alongside robots safely and efficiently, such as in collaborative environments.

★ **Biomedical Roboticist:** Specializing in healthcare, biomedical roboticists develop robots and robotic systems for medical applications, such as surgical assistance and rehabilitation.

★ **Agricultural Robotics Engineer:** Agricultural robotics engineers work on developing robots and automation solutions for precision agriculture, optimizing crop management and harvesting processes.

★ **Drone Engineer:** While not always considered traditional robots, drones are a form of autonomous machines. Drone engineers design and develop unmanned aerial vehicles for various applications, including surveillance, delivery, and photography.

★ **Drone Delivery Specialist:** These professionals work on the development and implementation of autonomous drone delivery systems for logistics and e-commerce companies.

- ★ **Autonomous Vehicle Engineer:** While not traditional robots, autonomous vehicle engineers design self-driving cars, trucks, and drones, which involve many robotic and AI technologies for navigation and control.

Career options showcase the wide range of roles available within the robotics field, allowing individuals to explore areas that align with their interests and career goals.

Robotic Pioneers: Profiling Prominent Roles

1. Robotics Engineering Career: Pioneering the Future

Robotics engineering is a dynamic and forward-looking field that plays a pivotal role in shaping the technological landscape of the present and future. As the demand for automation and artificial intelligence continues to surge across industries, the role of a robotics technician has become increasingly crucial.

Role Of A Robotics Engineering:

Robotics Engineering is an interdisciplinary field that combines elements of mechanical engineering, electrical engineering, computer science, and artificial intelligence to design, build, and program intelligent machines, commonly known as robots. These robots are designed to move, interact with their environment, and perform a wide range of tasks autonomously or under human guidance. Robotics engineers work to create machines that can replicate, and in some cases surpass, the capabilities of robots often portrayed in science fiction movies.

Unlocking The Future: The Promise And Excitement Of Robotics Engineering

Increasing Automation: As technology continues to advance, there's a growing demand for automation in various industries. Robots are increasingly being used in manufacturing, logistics, agriculture, healthcare, and even everyday households to perform repetitive or dangerous tasks. Robotics engineers play a pivotal role in developing these automation solutions.

Artificial Intelligence Integration: With the rapid progress of artificial intelligence and machine learning, robots are becoming more intelligent and capable of adapting to new situations. Robotics engineers are at the forefront of integrating AI algorithms into robots, enabling them to learn, make decisions, and improve their performance over time.

Aging Population: Many developed countries are experiencing an aging population, which has led to a greater need for assistance in healthcare and daily living. Service robots are being developed to provide care and support for the elderly and disabled. Robotics engineers are essential in creating these robots to address the needs of an aging society.

Space Exploration: Space agencies and private companies are increasingly relying on robotics for missions to explore other planets, asteroids, and celestial bodies. The development of advanced robotic systems is vital for the success of these missions.

Disaster Response: In disaster-stricken areas, such as during natural disasters or industrial accidents, robots are used to perform tasks that are too dangerous for humans. Robotics engineers are instrumental in creating robots that can navigate and operate in these challenging environments.

Collaborative Robotics: Collaborative robots, or cobots, are designed to work alongside humans. They are used in manufacturing and other industries to assist workers in various tasks. Robotics engineers are essential in ensuring the safety and effectiveness of these human-robot collaborations.

Job Market: The job market for robotics engineers is robust and is expected to grow in the coming years. This field offers numerous opportunities for innovation and problem-solving, making it attractive to those with a passion for technology.

Interdisciplinary Skills: Robotics engineering requires a mix of mechanical, electrical, and software skills, making it an excellent choice for individuals interested in a diverse skill set and a dynamic work environment.

Robotics Engineering is a promising field due to its ability to address contemporary and future challenges across various sectors, from manufacturing to healthcare, space exploration, disaster response, and more. The integration of advanced technologies like AI, combined with the demand for automation and assistance in an aging society, positions robotics engineers as instrumental players in shaping the world of tomorrow. Those entering this field have the opportunity to contribute to groundbreaking advancements and innovations that will continue to redefine how we live and work.

2. Robotics Researcher:

Robotics researchers develop new robotic technologies and applications. They work in universities, research labs, and private companies.

Robotics Researcher: Pioneering The Future Of Automation

A career as a robotics researcher is a gateway to the forefront of technological innovation. These professionals are the driving force behind the development of cutting-edge robotic technologies and applications, shaping the way we interact with automation. Operating in diverse environments like universities, research labs, and private companies, robotics researchers are instrumental in revolutionizing industries across the globe.

Key Contributions Of A Robotics Researcher:

Innovative Solutions: Robotics researchers are the creative minds behind the conception and design of novel robotic systems. They devise solutions that address real-world challenges, spanning from advanced manufacturing processes to medical interventions and beyond.

Experimental Validation: These experts conduct rigorous experiments to test and validate the functionality, efficiency, and safety of robotic prototypes. This critical phase ensures that the technology can perform reliably in a variety of conditions.

Academic Insight: In academic settings, robotics researchers not only contribute to technological advancements but also play a vital role in educating the next generation of engineers and innovators. They inspire students with the latest research findings and mentor them in their pursuits.

Cross-Disciplinary Collaboration: Robotics is an interdisciplinary field, intersecting with computer science, mechanical engineering, artificial intelligence, and more. Researchers often collaborate across various disciplines to harness collective expertise and develop holistic solutions.

3. **Artificial Intelligence (AI) Researcher:** AI researchers focus on advancing the intelligence and decision-making capabilities of robots through cutting-edge AI techniques and algorithms.

Role Of A Artificial Intelligence (Ai) Researcher:

The role of an Artificial Intelligence (AI) Researcher involves delving into the development and enhancement of artificial intelligence capabilities in robots. This entails leveraging state-of-the-art AI techniques and algorithms to significantly improve the intelligence and decision-making capacities of these machines.

In the current technological landscape and looking into the near future, the field of AI Researcher holds immense promise. Here are several reasons why it is considered a promising career:

Technological Advancements: AI technologies are advancing at an exponential pace, driven by breakthroughs in machine learning, neural networks, natural language processing, computer vision, and more. AI researchers play a crucial role in harnessing these advancements and applying them to improve robot intelligence.

Problem-Solving and Innovation: AI researchers are at the forefront of solving complex problems related to robotics, automation, and decision-making. Their innovative solutions contribute to making robots more efficient, adaptable, and capable of performing a diverse range of tasks across various industries.

Industry Applications: AI is being integrated into a wide array of industries, including healthcare, finance, manufacturing, transportation, and entertainment, among others. AI researchers are instrumental in tailoring AI solutions to meet the unique needs and challenges of each industry, expanding the possibilities for robot deployment.

Societal Impact: AI has the potential to positively impact society by addressing critical issues such as healthcare, climate change, accessibility, and more. AI researchers are at the forefront of developing AI-driven solutions that can revolutionize these sectors, making the field not only promising but also socially impactful.

Job Demand and Growth: The demand for AI researchers is steadily increasing as more industries recognize the value of incorporating AI into their operations. This demand is expected to grow substantially in the near future, providing ample job opportunities and career growth for professionals in this field.

Collaborative Interdisciplinary Work: AI researchers often collaborate with professionals from diverse disciplines, including computer science, neuroscience, mathematics, and more. This interdisciplinary collaboration fosters a rich and dynamic work environment, allowing for cross-pollination of ideas and accelerated innovation.

Continuous Learning and Development: The field of AI is ever-evolving, requiring AI researchers to stay updated with the latest advancements and continuously enhance their skills. This constant learning and development contribute to personal and professional growth, keeping the field intellectually stimulating.

An AI Researcher in the realm of robotics offers an exciting and promising career path, driven by rapid technological advancements, its critical role in various industries, and the potential to create a positive impact on society.

4. **Human-Robot Interaction Specialist:** These experts design and implement interfaces and interactions that enable humans to work with and alongside robots safely and efficiently, such as in collaborative environments.

Robot engineering is a promising field in both the present and the near future due to several key factors that are driving its growth and relevance. One specific subfield within this domain is "Human-Robot Interaction" (HRI) specialist, which focuses on designing and implementing interfaces and interactions to facilitate safe and efficient cooperation between humans and robots. Here's an elaboration on the robot engineering career and why it's promising:

Automation And Industry 4.0: The increasing demand for automation in industries, often referred to as Industry 4.0, is one of the primary drivers of the robot engineering field. As companies seek to enhance productivity and efficiency, they are turning to robots to handle various tasks, from manufacturing to logistics. Robot engineers are essential in developing and maintaining these robotic systems.

Collaborative Robots (Cobots): The concept of collaborative robots, or cobots, is becoming increasingly popular. These robots are designed to work alongside humans, and Human-Robot Interaction Specialists play a crucial role in ensuring that these interactions are safe, intuitive, and productive. This demand for cobots is expected to rise significantly, creating numerous opportunities in HRI.

Healthcare And Rehabilitation: The healthcare sector is embracing robotics for tasks such as surgery, patient care, and physical therapy. Robot engineers working in this field develop robots to assist healthcare professionals and improve patient outcomes. With an aging population, the demand for such technologies is set to grow.

Consumer Robotics: In addition to industrial and healthcare applications, there is a growing market for consumer robotics. From robotic vacuum cleaners to personal assistants like smart speakers, these consumer products require engineers to design the hardware and software that drive them.

Environmental Applications: Robots are being used for environmental monitoring, disaster response, and other applications that are hazardous for humans. Robot engineers play a role in designing and developing robots for these critical tasks, making the field relevant for addressing environmental and safety concerns.

Interdisciplinary Nature: Robot engineering is an interdisciplinary field, encompassing mechanical engineering, electrical engineering, computer science, and more. As technology continues to evolve, professionals from diverse backgrounds can find a place within this field, making it accessible to a broader range of individuals.

Global Demand: The need for automation and robotic solutions is not limited to any specific region. As industries and economies worldwide seek to remain competitive, the demand for robot engineers is global, offering career opportunities on a global scale.

Research And Development: The field of robot engineering is heavily research-driven. Organizations and institutions are investing in R&D to push the boundaries of what robots can do. This offers career opportunities for individuals interested in pushing the technological envelope.

Robot engineering, particularly the subfield of Human-Robot Interaction, is a promising and multifaceted career choice due to the increasing need for robots in various sectors, the rise of collaborative robots, advancements in AI, and its interdisciplinary nature. As automation and robotics continue to shape the future, professionals in this field can expect a wide range of opportunities and the chance to contribute to transformative technological advancements.

5. **Robotics Product Manager:** Product managers in the robotics industry oversee the development and marketing of robotic products, ensuring they meet market demands and performance expectations.

Robotics engineering is a dynamic and rapidly evolving field that encompasses the design, development, testing, and implementation of robotic systems. It combines elements of mechanical engineering, electrical engineering, computer science, and even artificial intelligence. Here's why a career in robotics engineering is promising both now and in the near future:

High Demand: As automation continues to play an increasingly vital role in various industries, the demand for skilled robotics engineers is on the rise. This demand spans across manufacturing, healthcare, logistics, agriculture, and even consumer electronics.

Versatility: Robotics engineers have the opportunity to work in a wide range of industries and applications. They can develop robots for tasks as varied as medical surgeries, autonomous vehicles, industrial automation, and even household chores.

Innovation And Technological Advancements: The field of robotics is in a constant state of innovation. Engineers are continually pushing boundaries, creating robots with enhanced capabilities, improved sensors, and more sophisticated AI algorithms. This constant innovation keeps the field exciting and full of opportunities for growth.

Addressing Real-World Challenges: Robotics is at the forefront of solving real-world problems. For instance, in healthcare, robots are assisting in surgeries and providing care for the elderly. In agriculture, they're revolutionizing precision farming. In disaster scenarios, they're being used for search and rescue missions.

Interdisciplinary Nature: Robotics engineering involves knowledge and skills from various disciplines. This interdisciplinary nature allows engineers to work on cutting-edge projects that require expertise in mechanical design, electronics, software development, and sometimes even bioengineering.

Global Impact: Robots have the potential to make a positive impact on society by increasing efficiency, safety, and productivity across various industries. They can also address labor shortages in sectors where demand is high but the workforce is limited.

Competitive Salaries: Due to the specialized skills and expertise required, robotics engineers often command competitive salaries. This trend is expected to continue as the demand for skilled professionals in this field grows.

Ethical Considerations And Human-Robot Interaction: As robots become more integrated into our daily lives, there are critical ethical considerations surrounding their use. This opens up new avenues for research and development in areas like human-robot interaction, ethics in AI, and robot safety.

Continual Learning And Growth: Given the rapid pace of technological advancement, professionals in robotics engineering are encouraged to engage in lifelong learning. This could involve staying updated with the latest technologies, attending conferences, and participating in ongoing professional development.

Entrepreneurial Opportunities: The field of robotics is ripe with opportunities for entrepreneurial ventures. Engineers who have innovative ideas can start their own companies or collaborate with others to develop and commercialize new robotic technologies.

A career in robotics engineering offers a blend of innovation, problem-solving, and the potential for meaningful societal impact. It's a field that promises not only job security and competitive salaries but also the chance to be at the forefront of technological advancements that will shape the future.

6. **Biomedical Roboticist:** Specializing in healthcare, biomedical roboticists develop robots and robotic systems for medical applications, such as surgical assistance and rehabilitation.

A career in robot engineering, particularly in the field of biomedical robotics, is indeed promising for the present and the near future. Here's an elaborate explanation of this field and why it holds significant potential:

Healthcare Revolution: The healthcare industry is undergoing a significant transformation, with an increasing focus on efficiency, precision, and accessibility. Robotic technology is playing a crucial role in this revolution by automating and enhancing various aspects of healthcare.
Biomedical roboticists contribute to this transformation by developing robots that can assist in surgeries, rehabilitate patients, and even perform diagnostic tasks.

Surgical Assistance: One of the most critical applications of biomedical robotics is in surgical assistance. Robotic surgical systems, such as the da Vinci Surgical System, are now widely used to assist surgeons in minimally invasive procedures. These robots provide greater precision, dexterity, and the ability to perform complex tasks that can be challenging for human hands. The field is constantly evolving with new robots and applications, making it a hotbed for innovation.

Rehabilitation: Biomedical roboticists also work on devices for patient rehabilitation. These robots help individuals recovering from injuries, surgeries, or strokes to regain their mobility and strength. Exoskeletons and robotic prostheses are examples of such devices. As the aging population grows, the demand for rehabilitation robotics is expected to increase, providing ample opportunities for engineers in this field.

Personalized Medicine: The integration of robotics with data analytics and artificial intelligence is opening new avenues for personalized medicine. Robots can assist in collecting and analyzing patient data, leading to better treatment plans tailored to individual needs. This personalized approach to healthcare is likely to become more prevalent in the future, offering exciting prospects for those in the field.

Telemedicine: With the rise of telemedicine, robotics is becoming an integral part of remote healthcare delivery. Robots can be used for remote monitoring of patients, drug dispensing, and even remote surgery. The COVID-19 pandemic has accelerated the adoption of telemedicine, highlighting the relevance of robotic systems in healthcare.

Addressing Labor Shortages: The healthcare industry is facing a shortage of skilled professionals, and robotics can help bridge this gap. Robots can perform routine tasks, allowing healthcare professionals to focus on more complex and critical aspects of patient care. This not only improves the efficiency of healthcare delivery but also enhances patient safety.

Research and Development: Biomedical roboticists are actively involved in research and development, which ensures a constant demand for new and improved robotic systems. This aspect of the career offers opportunities for innovation and the chance to contribute to groundbreaking advancements in healthcare technology.

Interdisciplinary Nature: This field is inherently interdisciplinary, requiring expertise in engineering, computer science, biology, and medicine. This broad range of skills makes it highly versatile and allows professionals to collaborate across various domains, opening doors to diverse career options.

A career in biomedical robotics is promising for several reasons. It aligns with the ongoing transformation of the healthcare industry, the need for advanced surgical and rehabilitation solutions, the rise of personalized medicine, and the increasing importance of telemedicine. As technology continues to advance, the demand for skilled biomedical roboticists is expected to grow, making it an exciting and promising field for both the present and the near future.

7. Role Of Agricultural Robotics Engineer

Agricultural Robotics Engineer is in the field of agricultural robotics engineering, professionals focus on designing, developing, and implementing robotic solutions and automation technologies tailored specifically for the agricultural sector. These engineers work on creating innovative robotic systems that can optimize various aspects of farming, ranging from crop management to harvesting processes.

One of the key objectives of agricultural robotics engineers is to enhance efficiency and productivity in agriculture. They achieve this by leveraging robotics and automation to streamline repetitive and labor-intensive tasks. For instance, robots can be designed to autonomously plant seeds, monitor crop growth, apply fertilizers or pesticides, and even harvest crops when they're ready, all with precision and accuracy.

Explore The Bright Prospects Of A Career As An Agricultural Robotics Engineer In The Present And Near Future:

Efficiency And Productivity: Agricultural robotics can significantly improve efficiency by automating tasks that would traditionally require extensive human labor. Robots can work around the clock, increasing the speed and precision of various farming operations.

Sustainability: With a growing global population, the demand for food is increasing. Agricultural robotics engineers are contributing to sustainability by developing technologies that optimize resource usage. This includes precise application of water, fertilizers, and pesticides, reducing waste and environmental impact.

Precision Agriculture: Precision agriculture is a rapidly evolving field where agricultural robotics play a pivotal role. These engineers develop robots equipped with advanced sensors and data analytics capabilities, enabling farmers to make data-driven decisions for better crop management, ultimately leading to higher yields.

Labor Shortages And Cost Reduction: Many regions are facing a shortage of farm labor due to factors such as urbanization and changing demographics. Agricultural robots can fill this gap, reducing the dependency on human labor and mitigating associated costs.

Technological Advancements: The advancements in robotics, artificial intelligence, machine learning, and sensor technologies are creating exciting opportunities for innovation in the agricultural sector. Agricultural robotics engineers are at the forefront of integrating these technologies to enhance farming practices.

Global Demand: The need for efficient and sustainable agriculture is a global concern. Agricultural robotics engineers have the opportunity to work on projects that impact food production worldwide, making this field both impactful and fulfilling.

Interdisciplinary Collaboration: Agricultural robotics engineering involves collaboration with various fields such as mechanical engineering, electronics, computer science, and agronomy. This interdisciplinary approach fosters a dynamic work environment and encourages a holistic understanding of agricultural challenges.

The role of an Agricultural Robotics Engineer is crucial in shaping the future of agriculture, ensuring that the industry meets the growing demand for food in a sustainable, efficient, and technologically advanced manner.

8. Role Of Drone Engineer:

Drone Engineer is a drone engineer is a professional responsible for designing, building, and maintaining unmanned aerial vehicles (UAVs), also known as drones. They work on various aspects of drone technology, including hardware, software, communications systems, and autonomous capabilities. The role involves designing drones for specific applications, optimizing flight performance, enhancing battery efficiency, integrating sensors, and ensuring compliance with safety and regulatory standards.

In the present and near future, the field of drone engineering holds significant promise due to several compelling factors:

Technological Advancements: Rapid advancements in drone technology, including lightweight materials, advanced sensors, longer-lasting batteries, and sophisticated software, have expanded the capabilities and applications of drones. Engineers play a crucial role in harnessing these advancements to design drones with improved performance and functionalities.

Diverse Applications: Drones find applications in various industries such as agriculture, construction, environmental monitoring, search and rescue, logistics, surveillance, and entertainment. Drone engineers can specialize in any of these sectors, providing tailored solutions to address specific industry needs and challenges.

Cost-effectiveness And Efficiency: Drones offer cost-effective and efficient solutions for tasks that were traditionally time-consuming, expensive, or dangerous. For instance, in agriculture, drones can monitor crop health and optimize resource usage. This cost- effectiveness attracts businesses and industries to integrate drone technology into their operations, creating a demand for skilled drone engineers.

Environmental And Social Impact: Drones are increasingly used for environmental monitoring, disaster response, and conservation efforts. Drone engineers contribute to developing solutions that aid in assessing and mitigating environmental impact, responding to disasters more effectively, and conserving natural resources.

Job Opportunities And Growth: As the demand for drones continues to rise, so does the demand for skilled drone engineers. This field offers diverse job opportunities in research and development, manufacturing, operations, maintenance, and consulting. Moreover, with the ongoing advancements in drone technology, the potential for career growth and specialization within the field is significant.

Regulatory Adaptations: Governments and international bodies are continually adapting regulations to ensure safe and responsible drone operations. Drone engineers are essential in ensuring compliance with these regulations and developing technology that meets the evolving legal and ethical frameworks.

The field of drone engineering is promising due to technological advancements, diverse applications, cost-effectiveness, positive environmental and social impact, ample job opportunities, and evolving regulations. Drone engineers play a vital role in driving innovation and shaping the future of this rapidly growing industry.

9. Role Of Robotics System Integrator:

A Robotics System Integrator is a professional who specializes in the assembly and configuration of various components to create fully functional robotic systems tailored for specific applications. This career is promising for both the present and the near future for several reasons:

Increased Automation Demand: As industries across the spectrum seek to automate tasks and processes, there's a growing demand for customized robotic solutions. These solutions often require the expertise of system integrators to bring them to life. This demand is particularly pronounced in fields like manufacturing, logistics, healthcare, and agriculture.

Customization And Flexibility: Robotics System Integrators play a pivotal role in tailoring robotic systems to the unique needs of businesses. They ensure that robots are well-suited to perform specific tasks efficiently and effectively. As businesses strive to optimize their operations, customization is key.

Technological Advancements: The field of robotics is constantly evolving. New technologies, materials, and components are being developed at a rapid pace. System integrators need to stay updated with the latest advancements to remain competitive. This keeps the field intellectually stimulating and offers opportunities for growth.

Cost Efficiency: Investing in a custom robotic system can lead to significant cost savings in the long run. These systems can streamline processes, reduce errors, and enhance productivity. Companies are increasingly recognizing the potential for return on investment (ROI) when it comes to integrating robotics into their operations.

Diverse Applications: The applications of robotics are vast and expanding. From traditional industrial robots to service robots, such as those used in healthcare, retail, and hospitality, the range of possibilities is extensive. This diversity means that Robotics System Integrators can work in various industries and roles.

Collaborative Robots (Cobots): The rise of collaborative robots, or cobots, is a notable trend. These robots can work alongside humans and require fine-tuned integration into existing workflows. As the adoption of cobots grows, the need for skilled integrators to ensure safe and efficient collaboration becomes increasingly important.

Global Market Growth: The global market for robotics and automation is on the rise. As more countries and industries embrace automation, the demand for professionals who can integrate and maintain robotic systems is expected to continue to increase.

Skills Shortage: With the rapid growth of the robotics industry, there's a shortage of skilled professionals who can integrate and maintain robotic systems. This shortage creates a favorable job market for those with the necessary skills and knowledge.

Robotics System Integration is a promising career for the present and near future due to the increasing demand for customized robotic solutions, ongoing technological advancements, the cost- efficiency of automation, and the expanding applications of robotics. It's a dynamic field that offers a wide range of opportunities for individuals with the right skill set and a keen interest in robotics.

10. Role Of Robotics Project Manager:

A Robotics Project Manager is a professional responsible for coordinating and overseeing all aspects of robotics projects, from inception to completion. Their role involves meticulous planning, efficient execution, and successful delivery of projects within predefined objectives, timelines, and budgetary constraints.

In recent years, the field of robotics has witnessed a significant surge in advancements and applications across various industries. This rapid growth is largely driven by technological innovations, such as artificial intelligence, machine learning, and advanced sensors. Robotics is increasingly being integrated into numerous sectors, including healthcare, manufacturing, logistics, agriculture, and even consumer electronics.

Discover The Bright Prospects Of A Robotics Project Manager Career, Now And Beyond

Technological Advancements: The ongoing advancements in robotics technologies are expanding the potential and capabilities of robots. This translates to a broader range of projects and applications that require effective management to ensure successful outcomes.

Increased Demand For Automation: Industries are increasingly adopting automation to enhance efficiency, accuracy, and productivity. Robotics plays a crucial role in automation, and skilled project managers are essential to drive and oversee these automation initiatives.

Cross-Industry Applications: Robotics is not limited to a specific sector; its applications are versatile and span various industries. Robotics Project Managers have the opportunity to work on diverse projects, gaining valuable experience and exposure to different domains.

Cost-Effectiveness And ROI: Implementing robotics can lead to cost savings and improved return on investment (ROI) for organizations. Robotics Project Managers are vital in ensuring projects are delivered on time and within budget, maximizing cost-effectiveness and ROI.

Addressing Complex Challenges: Robotics projects often involve tackling complex challenges related to hardware, software, integration, and system optimization. A Robotics Project Manager's expertise in project management methodologies and technical understanding is crucial in effectively addressing these challenges.

Sustainable And Green Technologies: With a growing focus on sustainability and environmental responsibility, robotics is being used to develop eco-friendly solutions in various sectors. Robotics Project Managers can contribute to and lead projects that align with sustainable development goals.

Global Opportunities: The demand for robotics project management expertise is not limited to a specific geographic location. The field offers opportunities for professionals to work on projects globally, collaborating with diverse teams and gaining a global perspective on robotics applications.

In conclusion, a career as a Robotics Project Manager is promising due to the rapid advancements in robotics technology, increased adoption of automation, cross-industry applications, cost- effectiveness, addressing complex challenges, focus on sustainability, and the availability of global opportunities. As industries continue to embrace robotics, the need for skilled project managers to guide and lead these projects will only grow, making this a promising field for the present and foreseeable future.

11.Role Of Robotics Simulation Engineer:

A Robotics Simulation Engineer is a professional responsible for designing and creating virtual environments to simulate and evaluate the performance of robotic systems and algorithms. These simulations mimic real-world scenarios and conditions, allowing engineers to test various aspects of the robot's functionality, behavior, and interactions without the need for physical prototypes.

Why Robotics Simulation Engineering Holds Great Promise Now And In The Near Future:

Cost-Effective Prototyping: Developing physical prototypes for robots can be expensive and time-consuming. Simulation engineers help in reducing costs by allowing extensive testing and validation in a virtual environment, minimizing the need for physical prototypes.

Rapid Iteration And Optimization: Simulations enable rapid iteration and optimization of robot designs and algorithms. Engineers can tweak parameters and variables quickly to improve efficiency, performance, and safety, accelerating the development process.

Risk Mitigation: By simulating various scenarios, engineers can identify potential issues and risks in the design and functionality of robotic systems. Addressing these concerns before deployment mitigates risks and enhances the overall safety and reliability of the robots.

Algorithm Validation And Testing: Complex algorithms that govern the behavior and decision-making of robots can be thoroughly tested and refined in a simulation environment. Engineers can observe how different algorithms perform under diverse conditions, aiding in fine-tuning and improving the software.

Machine Learning And AI Integration: As machine learning and AI play a critical role in modern robotics, simulation engineers are essential for training and testing AI algorithms. Simulated environments provide a diverse range of scenarios for training AI models, making them more adaptable and capable in real-world situations.

Scalability And Flexibility: Simulations offer scalability, allowing engineers to simulate multiple robots or entire fleets in a variety of environments. This flexibility is vital as robotics applications continue to expand in various industries, from healthcare and agriculture to manufacturing and logistics.

Optimizing Human-Robot Interaction (HRI): Simulation engineers can simulate human- robot interactions to enhance the design and usability of robots. This optimization is crucial for service robots, ensuring they can effectively assist and collaborate with humans in different settings.

Job Market Demand: With the growing demand for robotics in various industries, the need for skilled Robotics Simulation Engineers is on the rise. Companies are actively seeking professionals who can efficiently develop and utilize simulation environments to improve robotic system performance.

In conclusion, Robotics Simulation Engineering is a promising field that significantly contributes to the advancement and successful deployment of robotics technology. It offers a cost-effective, efficient, and safe way to design, test, and optimize robotic systems, making it a crucial profession for the continued growth and integration of robotics into our daily lives.

12. Role Of Space Robotics Engineer:

A Space Robotics Engineer is a professional responsible for designing, developing, and maintaining robotic systems and autonomous technologies specifically tailored for space exploration. This career involves creating robots that can operate in the challenging and unique conditions of outer space, such as in missions to other planets (like Mars) or within space stations.

Exploring The Promise And Significance Of A Career As A Space Robotics Engineer In The Present And Near Future:

Advancement In Space Exploration: With an increasing focus on space exploration, there's a growing need for advanced robotic systems to assist and augment human capabilities in space. Robots can perform tasks that are dangerous, monotonous, or in environments inhospitable to humans. Space agencies like NASA, ESA, Roscosmos, and private companies like SpaceX and Blue Origin are investing heavily in robotic missions.

Cost-Effectiveness And Efficiency: Utilizing robots in space missions can significantly reduce costs compared to sending humans for every task. Robots don't require life support systems, food, or rest, making them cost-effective and efficient for prolonged operations in space.

Exploration Beyond Earth: As humanity looks beyond Earth for exploration and potential colonization, robots will play a vital role in scouting, mapping, and preparing environments for human presence. Robots can conduct preliminary studies and set up infrastructure in advance, ensuring the safety and preparedness of human missions.

Remote Sensing And Data Collection: Space robots equipped with various sensors and instruments can collect invaluable data from celestial bodies, helping scientists and researchers to understand the composition, atmosphere, and potential resources of other planets and moons. This data is crucial for planning future missions and informing scientific research.

Human-Robot Collaboration: The future of space exploration involves seamless collaboration between humans and robots. Space Robotics Engineers design robots that can work alongside astronauts, enhancing productivity, safety, and the success of missions. These robots can be remotely operated or programmed for autonomous decision-making.

Technological Innovation: The field of robotics is continually evolving, with advancements in artificial intelligence, machine learning, and materials science. Space Robotics Engineers are at the forefront of integrating these technologies into robotic systems, resulting in more capable, versatile, and intelligent robots for space exploration.

Private Sector Involvement: The rise of private companies in space exploration, with ambitious goals of reaching other planets and even mining asteroids, presents numerous opportunities for Space Robotics Engineers. These companies require expertise in developing and maintaining robotic systems for their missions.

In conclusion, the role of a Space Robotics Engineer is promising due to the increasing interest in space exploration, the need for cost-effective solutions, the potential for human colonization beyond Earth, the importance of data collection from celestial bodies, the synergy between humans and robots, technological advancements, and the active involvement of private entities in the space sector. This career path offers exciting opportunities to contribute to humanity's understanding of the cosmos and the potential for our species to become an interplanetary civilization.

13. Role Of Underwater Robotics Engineer:

The role of an Underwater Robotics Engineer involves the design, development, and maintenance of robots specifically tailored for underwater use. These engineers are responsible for creating advanced robotic systems capable of operating in challenging underwater environments for various purposes, such as deep-sea exploration, offshore drilling operations, marine resource management, environmental monitoring, and underwater data collection.

In Recent Years, The Field Of Underwater Robotics Engineering Has Gained Significant Momentum Due To Several Factors:

Technological Advancements: Rapid advancements in technology have led to the development of sophisticated underwater robotic systems with enhanced capabilities. These robots can now navigate through complex underwater terrain, gather high-resolution data, and perform intricate tasks with precision.

Growing Demand For Exploration And Research: The need for understanding and exploring the vast and largely uncharted underwater world is on the rise. Underwater robots play a crucial role in conducting research, exploring marine life, studying geological formations, and investigating historical shipwrecks.

Offshore Industries: The offshore industry, including oil and gas exploration, renewable energy installations, and undersea mining, heavily relies on underwater robotics for inspection, maintenance, and repair operations. These robots can reach extreme depths and perform critical tasks, ensuring efficient and safe operations.

Environmental Monitoring And Conservation: With increasing concerns about the health of marine ecosystems and climate change, underwater robots are used for monitoring and assessing the state of underwater environments. They help in collecting data related to water quality, biodiversity, and climate patterns, contributing to conservation efforts.

Cost-Effectiveness And Efficiency: Advancements in materials and robotics technology have made underwater robots more cost-effective to produce and operate. They offer a viable alternative to human divers, reducing risks and costs associated with underwater tasks.

Global Demand For Resources: As the global demand for marine resources such as minerals, fish, and energy continues to rise, underwater robots play a pivotal role in exploring and extracting these resources sustainably and efficiently.

Considering the increasing need for underwater exploration, resource management, and environmental conservation, the field of underwater robotics engineering presents promising opportunities. Professionals in this field can contribute to addressing critical challenges related to climate change, resource scarcity, and sustainable development. As technology continues to evolve, underwater robotics engineers will be at the forefront of innovation, making significant contributions to our understanding of the underwater world and its potential for future advancements and discoveries.

14. Role Of Robotic Vision Specialist:

A Robotic Vision Specialist is a professional dedicated to advancing the capabilities of robots in perceiving and understanding visual data by utilizing cutting-edge computer vision technologies. This field involves developing and implementing algorithms, systems, and hardware that enable robots to effectively process, analyze, and respond to visual information from their surroundings. Here's an elaboration on the career and why it holds promise for the present and the near future.

Rapid Technological Advancements: Computer vision and related technologies have been advancing rapidly, driven by breakthroughs in machine learning, deep learning, and artificial intelligence. Robotic Vision Specialists play a critical role in adapting and integrating these advancements into robotic systems, enabling robots to perceive and interpret visual data with greater accuracy and efficiency.

Automation And Industry 4.0: With the emergence of Industry 4.0, automation is becoming increasingly prevalent across various industries. Robots are being deployed for tasks ranging from manufacturing and logistics to healthcare and agriculture. Enhanced robotic vision is essential for these robots to navigate complex environments, collaborate with humans, and perform intricate tasks with precision. Robotic Vision Specialists are instrumental in making this happen.

Enhanced Robotics Applications: As robots become more sophisticated and versatile, their applications broaden. From autonomous vehicles and drones to surgical robots and household assistance robots, there is a growing need for advanced visual perception capabilities. Robotic Vision Specialists are essential in developing these capabilities, expanding the scope and potential impact of robotic applications.

Improved Human-Robot Interaction: Human-robot interaction is a crucial aspect of robotics, particularly in domains like healthcare, elderly care, and education. Robots need to understand human gestures, emotions, and intents for effective communication and collaboration. Robotic Vision Specialists work on algorithms that enable robots to interpret human expressions and behavior accurately, facilitating seamless and intuitive interaction.

Safety And Security: Robotic vision is vital for ensuring the safety and security of both robots and the environment they operate in. Vision systems can detect hazards, obstacles, and anomalies, enabling robots to navigate and operate in a secure and controlled manner. In applications like surveillance and disaster response, robust visual perception is critical, and Robotic Vision Specialists contribute to developing solutions to address these needs.

Research And Innovation: The field of robotic vision is still evolving, offering ample room for research and innovation. Ongoing research leads to the development of new techniques, algorithms, and hardware that continually enhance a robot's vision capabilities. Robotic Vision Specialists are at the forefront of such innovation, contributing to advancements that will shape the future of robotics.

In conclusion, the role of a Robotic Vision Specialist is pivotal in shaping the future of robotics by leveraging advancements in computer vision technologies. The integration of enhanced visual perception capabilities in robots opens up vast possibilities across various industries and applications, making this a promising and dynamic field for professionals in the present and the near future.

15. Role Of Humanoid Robot Designer:

The field of Humanoid Robot Design is a promising and evolving career path that involves creating robots that closely resemble humans in appearance and movement. These humanoid robots are designed to mimic human behaviors, motions, and even emotions to a certain extent. This line of work is becoming increasingly relevant and attractive due to several factors in the present and foreseeable future.

Advancements In Technology: The rapid advancements in robotics, artificial intelligence (AI), materials science, and mechatronics have significantly enhanced the capabilities and potential of humanoid robot design. Cutting-edge technologies enable designers to create robots with more human-like features, agility, and intelligence.

Applications In Various Industries: Humanoid robots find applications in a variety of industries, including healthcare, entertainment, education, customer service, space exploration, disaster response, and more. For instance, in healthcare, humanoid robots can assist in patient care, physical therapy, and medical research. In entertainment, they can perform tasks like acting, dancing, or engaging with an audience.

Human-Robot Interaction Studies: Understanding how humans interact with robots is a crucial aspect of designing effective humanoid robots. Researchers need to study and analyze human-robot interactions to improve the design and functionality of these robots. Humanoid robot designers play a vital role in creating robots that can seamlessly integrate into society and interact with humans in a natural and intuitive manner.

Research And Innovation: The field offers immense opportunities for research and innovation. Designers work on improving the mechanics, sensors, AI algorithms, and overall performance of humanoid robots. Research breakthroughs can lead to robots that are more capable, versatile, and adaptable to various tasks and environments.

Collaboration With AI And Machine Learning: Integration with AI and machine learning technologies enables humanoid robots to learn and adapt, making them more intelligent and responsive over time. This collaboration ensures that robots can understand human behavior, anticipate actions, and provide more tailored and meaningful interactions.

Addressing Societal Challenges: Humanoid robots have the potential to address societal challenges, such as aging populations and labor shortages in certain industries. They can assist the elderly, perform dangerous tasks in hazardous environments, and support individuals with disabilities, thereby contributing to societal well-being.

Demand And Career Opportunities: As the demand for humanoid robots increases across various sectors, there is a growing need for skilled professionals in this field. Humanoid robot designers, engineers, programmers, and researchers are in high demand, offering ample career opportunities and potential for growth.

The field of Humanoid Robot Design is promising due to rapid technological advancements, diverse applications, the need for in-depth human-robot interaction studies, the scope for research and innovation, collaboration with AI, addressing societal challenges, and the increasing demand for skilled professionals. It is an exciting and evolving field with the potential to shape the future of robotics and its integration into our daily lives.

A Drone Delivery Specialist is a professional who specializes in the development and implementation of autonomous drone delivery systems. These systems are utilized by logistics and e-commerce companies to enhance their delivery processes by employing drones to transport packages and goods.

Roles And Responsibilities:

System Development: Drone Delivery Specialists are responsible for designing and developing autonomous drone delivery systems. This includes creating the hardware and software necessary for safe and efficient drone operation.

Navigation And Control: They work on algorithms and systems that enable drones to navigate through various terrains, avoid obstacles, and ensure accurate and timely delivery.

Regulatory Compliance: Drone Delivery Specialists stay updated with the evolving regulations related to drone operations and ensure that their systems comply with legal and safety requirements.

Operational Optimization: They analyze data and optimize delivery routes to improve efficiency, reduce delivery times, and enhance the overall logistics process.

Integration And Testing: Specialists integrate various components of the drone delivery system and conduct thorough testing to ensure reliability, safety, and performance.

Maintenance And Upgrades: They oversee the maintenance of drones and their systems, making necessary upgrades to keep up with technological advancements and operational demands.

Why Drone Delivery Is Promising:

Efficiency and Speed: Drones offer a rapid and efficient mode of delivery, significantly reducing transit times. This is particularly advantageous in delivering time-sensitive goods like medical supplies or perishable items.

Cost-Effectiveness: Drone deliveries can be cost-effective for short to medium distances, providing an economical alternative to traditional delivery methods, especially in remote or challenging-to-reach areas.

Reduced Environmental Impact: Drones are generally electrically powered, reducing the carbon footprint associated with traditional delivery vehicles. This aligns with the increasing focus on sustainability and environmental responsibility.

Last-Mile Delivery Solutions: Drone delivery is an excellent solution for the "last mile" problem, where delivering to individual homes or businesses can be expensive and time- consuming using traditional methods.

Advancements In Technology: Ongoing advancements in drone technology, including longer flight times, improved battery life, obstacle detection, and artificial intelligence, are further propelling the potential and feasibility of drone deliveries.

Pandemic Response And Remote Areas: The COVID-19 pandemic highlighted the need for contactless delivery options, making drone deliveries even more relevant. Drones can also reach remote or underserved areas, ensuring accessibility to essential goods.

Market Demand And Investment: The increasing demand for faster and more convenient deliveries, coupled with substantial investments and interest from major e-commerce and logistics companies, indicate a strong growth trajectory for drone delivery services.

The role of a Drone Delivery Specialist is promising due to the potential of drones to revolutionize the logistics and e-commerce sectors by offering efficient, cost-effective, and environmentally friendly delivery solutions. As technology continues to advance and regulations become more accommodating, the drone delivery field is expected to see substantial growth and opportunities in the near future.

17. Role Of Autonomous Vehicle Engineer:

Autonomous Vehicle Engineering is a rapidly evolving and promising field that combines cutting- edge technologies in robotics, artificial intelligence (AI), machine learning, computer vision, and advanced sensors to design and develop self-driving vehicles like cars, trucks, and drones. The goal is to create vehicles that can operate without human intervention, making transportation safer, more efficient, and convenient.

Exploring The Bright Future Of A Career As An Autonomous Vehicle Engineer:

Safety Advancements: Autonomous vehicles have the potential to significantly reduce accidents caused by human error, which is a leading cause of road accidents. Engineers in this field work on developing sophisticated AI algorithms and sensor technologies to enhance safety on the roads.

Efficiency And Sustainability: Self-driving vehicles can optimize traffic flow, reduce congestion, and minimize fuel consumption by implementing efficient route planning and driving strategies. This can lead to a more sustainable and environmentally friendly transportation system.

Technological Advancements: Advances in AI, deep learning, machine learning, and sensor technologies have paved the way for more sophisticated and reliable autonomous systems. Engineers continuously work on improving these technologies to enhance the capabilities and performance of autonomous vehicles.

Innovative Job Opportunities: The rapid growth of autonomous vehicle technology has created a high demand for skilled professionals in various domains, including software development, machine learning, hardware design, data analysis, and system integration. This demand is expected to rise as the industry continues to expand.

Government And Industry Support: Governments and industries worldwide are investing heavily in autonomous vehicle research, development, and infrastructure. Regulations and policies are evolving to accommodate and encourage the adoption of self-driving vehicles, providing a supportive environment for professionals in this field.

Improving Mobility And Accessibility: Autonomous vehicles have the potential to transform mobility for people with disabilities, the elderly, and those who are unable to drive. This technology can provide them with newfound independence and accessibility to essential services.

Integration With Smart Cities: Autonomous vehicles can seamlessly integrate with smart city initiatives by communicating with traffic systems, pedestrians, and other vehicles. This integration can further enhance safety and efficiency, making cities more livable and sustainable.

Research And Innovation: Autonomous vehicle engineering offers ample opportunities for research and innovation. Engineers constantly work on solving complex problems related to perception, decision-making, human-vehicle interaction, and adapting to diverse driving conditions.

Global Impact: As the adoption of autonomous vehicles spreads globally, engineers in this field have the opportunity to work on projects that can have a transformative impact on societies and economies worldwide.

A career as an Autonomous Vehicle Engineer is promising due to its potential to revolutionize the transportation industry, improve safety, enhance efficiency, contribute to sustainability, and create numerous innovative job opportunities. With the continuous advancements in technology and widespread adoption of autonomous vehicles on the horizon, professionals in this field are poised to play a crucial role in shaping the future of transportation.

13.2. Educational And Skill Requirements For Careers In Robotics

Education: Most of these roles require a bachelor's degree in robotics engineering, mechanical engineering, electrical engineering, or computer science. Some roles, such as robotics researcher and robotics ethicist, may require a master's degree or PhD.

Experience: Most of these roles also require some experience in the field. This experience can be gained through internships, co-ops, or entry-level jobs.

Skills: In addition to education and experience, you will also need to have the necessary skills for the role you are interested in. For example, robotics engineers need to have strong knowledge of robotics principles, such as mechanics, electronics, and programming. Robotics programmers need to have strong knowledge of programming languages, such as C++, Python, and Java.

How To Advance Your Career In Robotics

Education: Most robotics roles require a bachelor's degree in robotics engineering, computer science, electrical engineering, or a related field. Some employers may prefer candidates with a master's degree or Ph.D. in robotics engineering or a related field.

Internships: Internships are a great way to gain experience in robotics and to learn from professionals in the field. Many companies offer robotics internships to undergraduate and graduate students.

Entry-Level Jobs: After completing their education and internships, robotics professionals can start their careers in entry-level positions such as robotics engineer trainee, robotics technician, or robotics software engineer. In these roles, they will learn about the different aspects of robotics and gain experience working on real-world projects.

Mid-Level Jobs: With experience, robotics professionals can move up to mid-level positions such as robotics engineer, robotics software engineer, or robotics project manager. In these roles, they will take on more responsibility and lead or contribute to complex robotics projects.

Senior-Level Jobs: Experienced robotics professionals can move up to senior-level positions such as lead robotics engineer, senior robotics engineer, or robotics director. In these roles, they will lead teams of engineers and oversee complex robotics projects.

Some robotics professionals may also choose to pursue careers in academia or research. To do this, they will typically need to complete a Ph.D. in robotics engineering or a related field.

Tips For Advancing Your Career In Robotics:

- Stay up-to-date on the latest trends and technologies in robotics. The field of robotics is constantly evolving, so it is important to stay up-to-date on the latest trends and technologies. You can do this by attending industry events, reading trade publications, and taking online courses.
- Network with other robotics professionals. Networking is a great way to learn about new job opportunities and to meet people who can help you advance your career. Attend industry events, connect with people on LinkedIn, and reach out to people you admire.
- Take on leadership roles. Taking on leadership roles is a great way to demonstrate your skills and abilities to potential employers. Volunteer to lead projects at work or join a professional organization.
- Publish your work. Publishing your work in research papers and presenting at conferences is a great way to build your reputation as an expert in robotics.

With hard work and dedication, you can have a successful and rewarding career in robotics, regardless of the specific role you choose.

Tips For Meeting The Eligibility Requirements For

Robotics Roles Education In Robotics:

Robotics Engineer: Take courses in robotics principles, such as mechanics, electronics, and programming. Get experience with CAD software and robotics simulation software.

Robotics Technician: Take courses in robotics principles and programming. Get experience with CAD software and robotics simulation software.

Robotics Programmer: Take courses in programming languages, such as C++, Python, and Java. Get experience with robotics frameworks, such as ROS or Gazebo.

Robotics Researcher: Conduct research on robotics technologies and applications. Publish your research in academic journals and present your work at conferences.

Robotics Ethicist: Conduct research on the ethical implications of robotics, AI, and automation. Publish your research in academic journals and present your work at conferences.

Robotics System Integrator: Get experience in designing and integrating robotic systems. Get experience with CAD software and system integration tools.

Robotics Project Manager: Get experience in managing robotics projects. Get experience with project management principles and practices.

Robotics Simulation Engineer: Get experience in developing and using robotics simulation software. Get experience with CAD software and simulation tools.

Space Robotics Engineer: Take courses in aerospace engineering, robotics engineering, and space systems engineering. Get experience with CAD software and space robotics simulation tools.

Underwater Robotics Engineer: Take courses in ocean engineering, robotics engineering, and marine engineering. Get experience with CAD software and underwater robotics simulation tools.

Robotic Vision Specialist: Take courses in computer vision, machine learning, and artificial intelligence. Get experience with programming languages, such as Python and MATLAB.

Humanoid Robot Designer: Take courses in mechanical engineering, robotics engineering, and human-robot interaction. Get experience with CAD software and human-robot interaction.

Robotics Product Manager: Get experience in product management or a related field. Get experience with product development processes and marketing strategies.

Becoming a robotics professional requires hard work and dedication, but it is a rewarding career. By following the tips above, you can increase your chances of success. Get a good education. This is the most important step. Make sure to choose a program that will give you the skills and knowledge you need for the role you are interested in.

Academic Programs In Robotics

Bachelor's Programs

- Bachelor of Science in Robotics Engineering (BSRE)
- Bachelor of Science in Mechanical Engineering with a robotics concentration
- Bachelor of Science in Electrical Engineering with a robotics focus
- Bachelor of Science in Computer Science with robotics-related coursework

Master's Programs

- Master of Science in Robotics Engineering (MSRE)
- Master of Science in Computer Science with a specialization in robotics
- Master of Science in Electrical Engineering with a focus on robotics
- Master of Science in Artificial Intelligence with a robotics track

Ph.D. Programs

- Doctor of Philosophy (Ph.D.) in Robotics or related fields like mechatronics or automation
- Specialized Robotics Institutes
- Carnegie Mellon University's Robotics Institute
- Georgia Institute of Technology's Institute for Robotics and Intelligent Machines
- Massachusetts Institute of Technology's Computer Science and Artificial Intelligence Laboratory
- University of California, Berkeley's Robotics and Intelligent Machines Institute
- University of Pennsylvania's GRASP Laboratory

Interdisciplinary Programs:

Master's Programs

- Master of Science in Robotics and Artificial Intelligence at Carnegie Mellon University
- Master of Science in Robotics and Computer Science at Stanford University
- Master of Science in Robotics and Mechanical Engineering at University of California, Berkeley
- Research-Centric Programs
- Master of Science in Robotics and Master of Business Administration at

Carnegie Mellon University
  - Master of Science in Robotics and Master of Science in Computer Science at Stanford University
  - Master of Science in Robotics and Master of Science in Mechanical Engineering at University of California, Berkeley

Ph.D. Programs:

  - Ph.D. in Robotics at Carnegie Mellon University
  - Ph.D. in Robotics and Intelligent Machines at Georgia Institute of Technology
  - Ph.D. in Computer Science and Artificial Intelligence with a focus on Robotics at Massachusetts Institute of Technology
  - Ph.D. in Electrical Engineering and Computer Sciences with a focus on Robotics at University of California, Berkeley
  - Ph.D. in Computer and Information Science with a focus on Robotics at University of Pennsylvania
  - Dual-Degree Programs

Summer Schools And Workshops

  - Carnegie Mellon University's Robotics Summer School
  - Georgia Institute of Technology's Robotics Summer Camp
  - Massachusetts Institute of Technology's Robotics Summer School
  - University of California, Berkeley's Robotics Summer School
  - University of Pennsylvania's GRASP Robotics Summer School
  - Continuing Education and Certifications
  - Robotics Engineering Technology Certificate at Purdue University
  - Robotics and Automation Certificate at University of California, Berkeley Extension
  - Robotics Certificate at University of Pennsylvania

Online Programs

  - edX's Master of Science in Robotics
  - Coursera's Master of Science in Computer Science with a specialization in Artificial Intelligence and Robotics

Get Experience In Robotics:

Get experience. Try to get as much experience as possible in the field. This can be done through internships, co-ops, or entry-level jobs.

Importance Of Robotics Research Internships

Robotics research internships are a valuable way to gain experience in the field, develop your skills, and network with leading experts. They also offer the opportunity to work on cutting-edge research projects and make a real impact on the development of robotics technology.

Key Takeaways For The Applicant

If you are fortunate enough to be selected for one of the below listed internships, you can expect to have a challenging and rewarding experience. You will learn from some of the best minds in robotics, and you will have the opportunity to work on projects that could have a real impact on the world.

Anticipated Internship Insights: A Preview Of Key Takeaways:

Deep Understanding Of Robotics Research: You will gain a deep understanding of the current state of robotics research and the challenges that researchers are working to address.

Hands-On Experience With Robotics Technologies: You will have the opportunity to work with a variety of robotics technologies, including hardware, software, and sensors.

Collaboration With Leading Experts: You will have the opportunity to collaborate with leading experts in the field of robotics.

Publication And Presentation Opportunities: You may have the opportunity to publish your work in research papers and present it at conferences.

Networking Opportunities: You will have the opportunity to network with other interns and professionals in the field of robotics.

The internships listed above are some of the best opportunities available for students who are interested in a career in robotics research. If you are selected for one of these internships, you can expect to have a challenging and rewarding experience that will prepare you for a successful career in the field.

Internship Programs By Top Organizations :

- Google AI Robotics Internship
- Microsoft AI Robotics Internship
- Amazon Robotics Internship
- Tesla AI Robotics Internship

- Nvidia Robotics Internship
- Intel AI Robotics Internship
- IBM AI Robotics Internship
- Boston Dynamics Robotics Internship
- Alphabet X Robotics Internship
- OpenAI Robotics Internship
- Carnegie Mellon University Robotics Institute Internship
- Georgia Institute of Technology Institute for Robotics and Intelligent Machines Internship
- Massachusetts Institute of Technology Computer Science and Artificial Intelligence Laboratory Robotics Internship
- University of California, Berkeley Robotics and Intelligent Machines Institute Internship
- University of Pennsylvania GRASP Laboratory Robotics Internship

13.3. Networking With Other Interns And Professional :

Networking with other interns and professionals in the field of robotics is important for a number of reasons:

- **Learn From Others:** Networking is a great way to learn from other people's experiences and expertise. You can learn about different career paths, research projects, and industry trends.
- **Get Feedback:** Networking can help you get feedback on your work and career goals. This feedback can be invaluable in helping you improve your skills and make informed decisions about your future.
- **Find Job Opportunities:** Networking can help you find new job opportunities. Many companies hire through their networks, so it's important to get to know people in the field.
- **Build Relationships:** Networking can help you build relationships with other people in the field. These relationships can be helpful for getting support, mentorship, and collaboration opportunities.

The Advantages Of Building Connections With Fellow Interns:

Maximizing Your Internship Through Networking And Collaboration

- **Learn About Different Internship Experiences:** Networking with other interns can help you learn about different internship experiences and what to expect. You can also get advice on how to make the most of your internship.
- **Find Mentors:** Networking with other interns can help you find mentors who can provide guidance and support.

- **Collaborate On Projects:** Networking with other interns can help you find collaborators for your research projects or other creative endeavors.

Unlocking The Advantages Of Professional Networking:

- **Get Career Advice:** Networking with professionals can help you get career advice on topics such as resume writing, interviewing, and salary negotiation.
- **Learn About Different Career Paths:** Networking with professionals can help you learn about different career paths in robotics and how to get started in those fields.
- **Find Job Opportunities:** Networking with professionals can help you find new job opportunities. Many companies hire through their networks, so it's important to get to know people in the field.
- **Build Relationships With Potential Employers:** Networking with professionals can help you build relationships with potential employers. These relationships can be helpful in getting your foot in the door for job opportunities.

Networking is an essential skill for anyone interested in a career in robotics. By networking with other interns and professionals, you can learn from others, get feedback, find job opportunities, and build relationships.

Effective Networking Tips For Professionals In The Robotics Industry:

- **Attend Robotics Events And Conferences:** This is a great way to meet other people in the field and learn about new opportunities.
- **Connect With People On Linkedin:** LinkedIn is a great platform for connecting with professionals in robotics. You can search for people by company, industry, and other criteria.
- **Reach Out To People Directly:** Don't be afraid to reach out to people directly via email or LinkedIn. Most people are happy to help others, especially interns and students.

Networking can be intimidating at first, but it's important to remember that everyone is there to connect and learn from each other. Just be yourself and be genuine, and you'll be sure to make some valuable connections.

Network with people in the field. Attend industry events and meetups. Get to know people who work in the roles you are interested in.

Key Robotics Networking Events And Industry Associations

- ICRA - International Conference on Robotics and Automation(https://www.ieee-ras.org/)
- RSS - Robotics: Science and Systems (https://roboticsconference.org/)
- AuRoNS - Australasian Conference on Robotics and Automation
- IEEE/RSJ   International Conference on Intelligent Robots and Systems(https://iros2022.org/)
- RoboCup (https://www.robocup.org/)
- RoboBusiness Conference(https://www.robobusiness.com/)
- Field Robotics Conference(https://www.conferenceseries.com/Field-Robotics.php)
- Human-Robot Interaction Conference(https://humanrobotinteraction.org/2023/)
- Cognition, Learning, and Adaptive Behavior Conference
- International Conference on Mobile Robotics and Autonomous Systems
- International Conference on Artificial Intelligence (http://icoai.org/)

- Robotics Industries Association (RIA) (https://www.automate.org/companies/robotic- industries-association)
- Association for Advancing Automation (A3) (https://www.automate.org/)
- International Federation of Robotics (IFR) (https://ifr.org/)
- European Robotics Association (EURobotics) (https://eu-robotics.net/)
- Japan Robot Association (JARA) (https://www.jara.jp/e/)
- China Robot Industry Alliance (CRIA) (https://www.therobotreport.com/suppliers/cria- china-robot-industry-alliance-cn/)
- Korean Association of Robot Industry (KARI) (https://www.k-robot.co.kr/)
- German Robotics Association (VDMA Robotics + Automation)(https://www.vdma.org/en/board-robotics-automation)
- French Robotics Association (IFR) (https://ifr.org/)
- Italian Robotics Association (AIR) (https://airlab.deib.polimi.it/)
- Spanish Robotics Association (AER) (https://www.aer-automation.com/automation_review/congratulations-spanish-robotics-automation- association/)

Networking Success: Key Strategies For Effective Professional Interaction

Prepare Your Elevator Pitch: Be ready to introduce yourself and your work in a clear and concise way.

Research The People You Want To Meet: Do some research on the people you are interested in meeting, so that you can start a conversation with them about something specific.

Bring Business Cards: Business cards are a great way to exchange contact information with people you meet.

Follow Up After The Event: Be sure to follow up with the people you meet after the event to thank them for their time and to continue the conversation.

Top Online Learning Platforms For Robotics Enthusiasts:

Empower Your Robotics Journey: Top Online Learning Platforms And Resources

Develop your skills. Keep your skills up-to-date by taking online courses, reading articles, and attending workshops.

Coursera: Coursera offers a variety of robotics courses from top universities around the world, including Stanford, MIT, and Carnegie Mellon. Some popular robotics courses on Coursera include:
Robotics: Fundamentals and Applications
Modern Robotics: Mechanics, Planning, and Control Introduction to Robotics for Intelligent Machines
edX: edX also offers a variety of robotics courses from top universities, including Harvard, Berkeley, and Georgia Tech. Some popular robotics courses on edX include:
Robotics: Science and Systems Robotics: From Perception to Action Robot Motion Planning
Udacity: Udacity offers a variety of robotics courses, including self-driving cars, robotics engineering, and artificial intelligence.
Some Popular Robotics Courses On Udacity Include:
Self-Driving Car Engineer Nanodegree Robotics Software Engineer Nanodegree
Artificial Intelligence For Robotics Nanodegree

Udemy: Udemy is a platform where anyone can create and sell online courses. There are a wide variety of robotics courses available on Udemy, covering topics such as programming, hardware, and machine learning. Some popular robotics courses on Udemy include:

The Complete Robotics Course: From Beginner to
Advanced
Robotics for Beginners: Learn to Build and
Program Robots Robotics with Python
YouTube: There are many great YouTube channels that offer robotics tutorials and video lectures.

In addition to the above resources, there are many other online learning sources related to robotics. A quick Google search will reveal a wealth of information on the topic.

Certifications: Industry-Recognized Certifications Can Boost Your Credentials

Certifications and Training

Certified Robotics System Architect (CRSA)

The Certified Robotics System Architect (CRSA) certification is designed to validate the skills and knowledge necessary to design, develop, and deploy robotic automation solutions. The CRSA exam covers a wide range of topics, including:

- Robotics fundamentals
- Robotic architecture
- Robotic process automation
- Robotic systems engineering
- Robotic project management

To be eligible for the CRSA certification, candidates must have a bachelor's degree in engineering, computer science, or a related field, and two years of experience in robotic automation. Alternatively, candidates may qualify with a master's degree in engineering, computer science, or a related field, and one year of experience in robotic automation.

Certified Robotics Software Developer (CRSD)

The Certified Robotics Software Developer (CRSD) certification is designed to validate the skills and knowledge necessary to develop and maintain robotic automation solutions. The CRSD exam covers a wide range of topics, including:

- Robotic programming
- Robotic debugging
- Robotic testing
- Robotic performance tuning
- Robotic security

To be eligible for the CRSD certification, candidates must have a bachelor's degree in engineering, computer science, or a related field, and one year of experience in robotic software development. Alternatively, candidates may qualify with a master's degree in engineering, computer science, or a related field, and six months of experience in robotic software development.

Certified ROS Developer (CRD)

The Certified ROS Developer (CRD) certification is designed to validate the skills and knowledge necessary to develop and maintain ROS-based robotic applications. The CRD exam covers a wide range of topics, including:

- ROS fundamentals
- ROS programming
- ROS debugging
- ROS testing
- ROS performance tuning

To be eligible for the CRD certification, candidates must have a bachelor's degree in engineering, computer science, or a related field, and one year of experience in ROS development. Alternatively, candidates may qualify with a master's degree in engineering, computer science, or a related field, and six months of experience in ROS development.

Project Management Professional (PMP)

The Project Management Professional (PMP) certification is designed to validate the skills and knowledge necessary to manage projects effectively. The PMP exam covers a wide range of topics, including:

- Project initiation
- Project planning
- Project execution
- Project monitoring and control
- Project closure

To be eligible for the PMP certification, candidates must have a high school diploma or equivalent, three years of project management experience, and 35 hours of project management education. Alternatively, candidates may qualify with a bachelor's degree or equivalent, three years of project management experience, and 35 hours of project management education.

Certified Robotics Engineer (CRE)

The CRE certification is the most advanced and comprehensive robotics certification. It is designed to validate the skills and knowledge necessary to design, develop, deploy, and maintain robotic systems. The CRE exam covers a wide range of topics, including:

- Robotics fundamentals
- Robot kinematics and dynamics
- Robot programming
- Robot sensors and actuators
- Robot safety
- Robot systems engineering

To be eligible for the CRE certification, candidates must have a bachelor's degree in engineering or computer science, and three years of experience in robotics engineering. Alternatively, candidates may qualify with a master's degree in engineering or computer science, and one year of experience in robotics engineering.

Certified Robotics Technician (CRT)

Certified Robotics Technician (CRT)

The CRT certification is designed to validate the skills and knowledge necessary to operate and maintain robotic systems. It is a good option for individuals who are new to robotics or who want to focus on the practical aspects of robotics. The CRT exam covers a wide range of topics, including:

- Robot safety
- Robot programming
- Robot troubleshooting
- Robot maintenance

To be eligible for the CRT certification, candidates must have a high school diploma or equivalent, and one year of experience in robotics or a related field. Alternatively, candidates may qualify with a two-year associate's degree in robotics or a related field.

Certified Robotics Safety Professional (CRSP)

The CRSP certification is designed to validate the skills and knowledge necessary to assess and mitigate the hazards associated with robotic systems. It is a good option for individuals who want to focus on the safety of robotic systems. The CRSP exam covers a wide range of topics, including:

- Robotics fundamentals
- Robot hazards and risk assessment
- Robot safety standards
- Robot safety design and engineering
- Robot safety programs

To be eligible for the CRSP certification, candidates must have a bachelor's degree in engineering, safety, or a related field, and three years of experience in robotics safety. Alternatively, candidates may qualify with a master's degree in engineering, safety, or a related field, and one year of experience in robotics safety.